Iron-Sulfur Proteins

VOLUME I

Biological Properties

MOLECULAR BIOLOGY

An International Series of Monographs and Textbooks

Editors: BERNARD HORECKER, NATHAN O. KAPLAN, JULIUS MARMUR, AND HAROLD A. SCHERAGA

A complete list of titles in this series appears at the end of this volume.

Iron-Sulfur Proteins

VOLUME I

Biological Properties

Edited by Walter Lovenberg

SECTION ON BIOCHEMICAL PHARMACOLOGY
EXPERIMENTAL THERAPEUTICS BRANCH
NATIONAL HEART AND LUNG INSTITUTE
NATIONAL INSTITUTES OF HEALTH
BETHESDA, MARYLAND

ACADEMIC PRESS New York and London 1973

A Subsidiary of Harcourt Brace Jovanovich, Publishers

ACADEMIC PRESS, INC.
111 Fifth Avenue, New York, New York 10003

United Kingdom Edition published by
ACADEMIC PRESS, INC. (LONDON) LTD.
24/28 Oval Road, London NW1

Library of Congress Cataloging in Publication Data

Lovenberg, Walter.
Iron-sulfur proteins.

(Molecular biology)
Includes bibliographies.
1. Iron-sulfur proteins. I. Title. II. Series: Molecular biology; an international series of monographs and textbooks.
QP552.I7L69 574.1′9245 72-13613
ISBN 0–12–456001–6 (v. 1)

PRINTED IN THE UNITED STATES OF AMERICA

Contents

List of Contributors

Numbers in parentheses indicate the pages on which the authors' contributions begin.

JEFFREY BARON (193), *Department of Biochemistry, The University of Texas, Southwestern Medical School, Dallas, Texas*

HELMUT BEINERT (1), *Institute for Enzyme Research, Madison, Wisconsin*

BOB B. BUCHANAN (129), *Department of Cell Physiology, University of California, Berkeley, California*

R. C. BURNS (65), *Central Research Department, E. I. du Pont de Nemours and Company, Wilmington, Delaware*

MINOR J. COON (173), *Department of Biological Chemistry, Medical School, The University of Michigan, Ann Arbor, Michigan*

RONALD W. ESTABROOK (193), *Department of Biochemistry, The University of Texas, Southwestern Medical School, Dallas, Texas*

I. C. GUNSALUS (151), *Biochemistry Department, School of Chemical Sciences, University of Illinois, Urbana, Illinois*

M. GUTMAN (225), *Molecular Biology Division, Veterans Administration Hospital, San Francisco, California*

R. W. F. HARDY (65), *Central Research Department, E. I. du Pont de Nemours and Company, Wilmington, Delaware*

J. D. LIPSCOMB (151), *Biochemistry Department, School of Chemical Sciences, University of Illinois, Urbana, Illinois*

EGLIS T. LODE* (173), *Department of Biological Chemistry, Medical School, The University of Michigan, Ann Arbor, Michigan*

* Present address: Biochemistry Department, University of California, Berkeley, California.

JOHN McCARTHY* (193), *Department of Biochemistry, The University of Texas, Southwestern Medical School, Dallas, Texas*

J. IAN MASON (193), *Department of Biochemistry, The University of Texas, Southwestern Medical School, Dallas, Texas*

LEONARD E. MORTENSON (37), *Department of Biological Sciences, Purdue University, Lafayette, Indiana*

VINCENT MASSEY (225, 301), *Department of Biological Chemistry, The University of Michigan, Ann Arbor, Michigan*

GEORGE NAKOS (37), *Department of Biological Sciences, Purdue University, Lafayette, Indiana*

JOHN PURVIS† (193), *Department of Biochemistry, The University of Texas, Southwestern Medical School, Dallas, Texas*

ANTHONY SAN PIETRO (111), *Department of Plant Science, Indiana University, Bloomington, Indiana*

JAMES N. SIEDOW‡ (111), *Botany Department, Indiana University, Bloomington, Indiana*

EVAN R. SIMPSON§ (193), *Department of Biochemistry, The University of Texas, Southwestern Medical School, Dallas, Texas*

THOMAS P. SINGER (225), *Department of Biochemistry and Biophysics, University of California, San Francisco, and Molecular Biology Division, Veterans Administration Hospital, San Francisco, California*

KOJI SUZUKI|| (193), *Department of Biochemistry, The University of Texas, Southwestern Medical School, Dallas, Texas*

WAYNE E. TAYLOR (193), *Department of Biochemistry, University of Texas, Southwestern Medical School, Dallas, Texas*

CHARLES F. YOCUM** (111), *Botany Department, Indiana University, Bloomington, Indiana*

* Present address: Department of Biology, Southern Methodist University, Dallas, Texas.

† Present address: Department of Biochemistry, University of Rhode Island, Kingston, Rhode Island.

‡ Present address: Biophysics Research Division, Institute of Science and Technology, The University of Michigan, Ann Arbor, Michigan.

§ Present address: Department of Physics Applied to Medicine, Middlesex Hospital Medical School, London, England.

|| Present address: Department of Physics and Medicine, Jutendo Medical University, Tokyo, Japan.

** Present address: Section of Biochemistry and Molecular Biology, Cornell University, Ithaca, New York.

Preface

Our understanding of the biological role and molecular properties of iron–sulfur proteins has grown rapidly during the past decade. In fact, from the time that this treatise was first conceived (about three years ago), our knowledge on detailed structure has proceeded from a meager level to a reasonably sophisticated one; but even this level may seem meager within several years.

The objective of this two-volume treatise is to present a detailed account by outstanding scientists of the biological importance and the physical and chemical properties of this group of proteins. While it is meaningless to compare the relative importance of any group of molecules in complex biological systems, it should be noted that iron–sulfur proteins provide the cornerstone for such fundamentally important processes as nitrogen fixation and photosynthesis; at the same time they participate in innumerable key reactions in nature, ranging from the most primitive bacterium to man. It is for this reason that students and/or investigators in all biological disciplines will find valuable information in this work.

Volume I of "Iron–Sulfur Proteins" deals largely with the biological properties of these proteins; Volume II deals with the molecular properties. This division is somewhat arbitrary since many of the authors are deeply involved in investigating both aspects of these molecules, and it was the intent that all chapters represent the interests and philosophies of the individual laboratories.

In Chapter 1 of Volume I, Dr. Beinert presents a historical review of the subject and some details on the extremely important role played by electron paramagnetic resonance in developing our understanding of iron–sulfur proteins. Other chapters are devoted to the role of ferredoxin in nitrogen fixation and to the general aspects of bacterial metabolism. Two contributions deal with the role of iron–sulfur proteins in the photosynthetic process. Three chapters are concerned with the mechanism of hydroxylation reactions that require iron–sulfur proteins, specifically rubredoxin and alkane hydroxylation, putidaredoxin and camphor hydroxylation, and adrenodoxin and steroid hydroxylation. This first volume also includes very complete review chapters on complex iron–sulfur proteins.

Volume II provides an in-depth analysis of the chemical and physical properties of many of the iron–sulfur proteins. Particular emphasis is placed on the theory and use of physicochemical techniques in the study of metalloproteins. The chemical properties of the ferredoxins and a very complete structural and genetic analysis of iron–sulfur electron carriers are presented in individual chapters. Subsequent chapters are then devoted to the theory and use in iron–sulfur protein research of the following techniques: optical spectroscopy, electron paramagnetic resonance and electron nuclear double resonance spectroscopy, X-ray crystallography, proton magnetic resonance spectroscopy, and Mössbauer spectroscopy. The volume is concluded by a detailed discussion of our current knowledge on the active centers of each type of iron–sulfur protein.

The presentations have not been changed to conform to a particular style. It is hoped that the individualistic nature of each chapter will give the reader an idea of the approaches being pursued in some of the major laboratories.

WALTER LOVENBERG

Contents of Volume II

MOLECULAR PROPERTIES

CHAPTER 1

Development of the Field and Nomenclature

HELMUT BEINERT

I. INTRODUCTION

The accelerated pace of research in our days, sustained by laboratory and communications technology, ensures that in the time span from the early beginnings of a field of scientific endeavor to its rise, maturation, and final establishment—the time when "everybody knows" that it exists and is interesting—is continually shrinking. Thus all these events may now indeed fall into the active life span of a single generation of investigators. This is an experience which everyone who makes science his career finds agreeable and fulfilling.

Although we cannot safely anticipate the developments of the future, it seems likely that the area of endeavor and knowledge, with which these volumes are concerned, is of the very kind just alluded to. In view of the brief history of this research field, we can then hope and expect that in our attempt to cover the background and present knowledge in the field of iron–sulfur proteins, a number of those who have seen or helped this field to rise from its very beginnings will contribute here their thoughts and experience. While the development of the field of hemes and hemeproteins must probably be counted in terms of centuries, the field of iron–sulfur proteins barely dates back two decades—if one considers the finding of iron as a constituent as the crucial event, and not merely acquaintance with some protein which finally turned out to belong to this category. Nevertheless, the ubiquitous occurrence of iron–sulfur proteins and their function—mainly as electron carriers of unusually low potential—in the principal reactions sustaining life, such as photosynthesis, nitrogen fixation, or electron transport in subcellular organelles, would seem amply to justify consideration of their study as a research field in its own right, comparable to that of heme- or flavoproteins.

Although it will be set out in more detail later in Section V on the nomenclature of these compounds, it may be anticipated here that the name iron–sulfur proteins applies only to such iron proteins in which *sulfur is a ligand of the iron* and the iron is not simultaneously held by a much more powerful ligand, such as, e.g., porphyrin, which would dominate the iron environment.

II. OCCURRENCE OF IRON IN LIVING MATTER

When introducing what the authors of this volume consider to be a major class of iron proteins in living organisms, it seems appropriate to set a frame of reference. Where, in what form, and to what extent does iron occur in living things? With an ubiquitous element such as iron, it is, of course, difficult to distinguish between intended and accidental presence. In fact, it is likely that there are a number of species of iron, more or less tightly bound to proteins and to molecules of low molecular weight, which have thus far escaped our attention; some of these may have a significant function, the existence of others may be fortuitous. Our knowledge of iron that is very tightly and specifically bound to protein is probably more complete, but even here surprises are lingering on the horizon (Loach *et al.*, 1970; Amoore, 1963; Brown *et al.*, 1968). Table I shows a gross survey of the major families or forms of iron compounds and their functions. The sideramines (Prelog, 1963; Neilands, 1966) and enterobactins (Pollock and Neilands, 1970)

TABLE I
MOST COMMON FORMS OF IRON IN BIOLOGICAL SYSTEMS

Fe storage and transport forms	O_2 carriers	Enzymes and electron carriers
Ferritin[a]	Hemoglobin	Heme enzymes
Hemosiderin[b]	Myoglobin	Iron–sulfur proteins
Transferrins (siderophilins)[c]	Hemerythrin[d,e]	Oxygenases[f,g]
Sideramines[h,i]		
Enterobactins (enterochelins)[j,k]		

[a] Hofmann and Harrison (1963).
[b] Ludewig and Franz (1970).
[c] Feeney and Komatsu (1966).
[d] Groskopf *et al.* (1966).
[e] Garbett *et al.* (1969).
[f] Hayaishi (1962).
[g] Hayaishi and Nozaki (1969).
[h] Prelog (1963).
[i] Neilands (1966).
[j] Pollock and Neilands (1970).
[k] O'Brien and Gibson (1970).

[or enterochelins (O'Brien and Gibson, 1970)], which thus far have only been found in microorganisms, are included here, although they are small peptides. Nevertheless, they are well characterized and bind iron very tightly.

Table II, from Drabkin (1951), gives the distribution of iron in man. Since this work dates back some 20 years, we must consider that a number of compounds are not included. Needless to say, iron–sulfur proteins are not included; but even the category of "cytochrome," for instance, comprises only cytochrome c, which was readily assayed for. From analytical data (Green and Wharton, 1963), which admittedly are only available for some organs or organelles, we may assume that all cytochromes, including the more recently discovered cytochrome P-450, reach a total quantity of ~1%, instead of the 0.08% given for the c component only. According to iron and heme analyses (Green and Wharton, 1963; Brückmann and Zondek, 1940; McFarlane, 1934), there is more nonheme than heme iron, but the difference does not quite amount to an order of magnitude, except in organs with specific iron storage functions. Therefore, the iron–sulfur proteins, which are an unknown fraction of the total nonheme iron protein pool, may also account for roughly 1% of the total body iron in mammals. Obviously (Table II), the oxygen-carrying heme proteins as well as the storage

TABLE II
ALLOCATION OF IRON IN BODY OF 70-kg MAN[a]
(Total Body Iron = 4.2 gm)

Compound	Total in body (gm)	Total Fe in compound (gm)	Total body Fe (%)
Hemoglobin	900.0	3.12	74.3
Myoglobin	40.0	0.14	3.3
Haptoglobin—hemoglobin	10.6	0.0085	0.2
Cytochrome c	0.8	0.0034	0.08
Catalase	5.0	0.0045	0.11
Ferritin	3.0	0.69	16.4
Transferrin (β_1-globulin)	7.5	0.003	0.07
Total organic iron		3.97	94.5
Remaining iron[b]		0.23	5.5

[a] The data of this table have been copied from Drabkin (1951) (with permission) and amended according to additional information kindly supplied by Dr. Drabkin. For more details the original should be consulted.

[b] Includes (a) unevaluated organic iron in cytochromes other than c, cytochrome oxidase and possibly some iron-containing catabolic products of the chromoproteins, whose exact character is unestablished, and (b) so-called "inorganic" iron, hemosiderin, etc.

forms, ferritin and hemosiderin, are in such overwhelming abundance that none of the catalytic iron proteins is present beyond the error of most analytical methods aiming at the iron only. It is not surprising, therefore, that they did not attract attention in analytical or isotope work which attempted to establish occurrence and distribution of iron compounds.

The advantage of the existing balance of iron compounds to the organisms could be that under conditions of iron shortage, the vital catalytic proteins can easily be maintained at the expense of only minor economy in the more abundant iron proteins. Yet, this does not appear to be so under all conditions (Beutler, 1957; Beutler and Blaisdell, 1958; Gubler *et al.*, 1957). It is interesting in this context that in *Micrococcus denitrificans* and *Candida utilis* the cytochromes appear to be more tenaciously preserved during iron starvation than certain iron–sulfur proteins (Imai *et al.*, 1968; Ohnishi *et al.*, 1969). Again it should not be concluded that this is a general phenomenon, since it has been found that in iron-deficient *Rhodopseudomonas spheroides* heme iron was much more decreased than nonheme iron (Reiss-Husson *et al.*, 1971), although iron–sulfur proteins were not considered separately.

III. DEVELOPMENT OF THE FIELD OF IRON–SULFUR PROTEINS

A. Observations Leading to the Discovery of Individual Iron–Sulfur Proteins

It was a matter of surprise to those active in the old and well-trodden field of biological oxidations, that a major principle should have remained so completely unknown and without prior inklings of its existence. In fact, this situation repeated itself within the past two decades with the discovery of the ubiquinones as well as with the iron–sulfur proteins. Part of the initial reluctance of a number of colleagues to accept these newcomers to the scene probably stems from the shock of their unannounced appearance—in addition, of course, to the healthy attitude of caution which scientists traditionally show toward the unexplored.

As constituents of anaerobic organisms, the iron–sulfur proteins have probably been elaborated by nature very early, and one may thus ask, why it took up to the middle of this century, until even signs of the existence of the iron–sulfur proteins were gleaned. There are a number of reasons for this. First, one of the principal tools of biological chemistry is spectrophotometry, and the brownish iron–sulfur proteins are by no means optically as conspicuous as the bright red heme proteins; the absorptivities per iron atom of their prominent bands in the visible region of the spectrum fall into the range of 2–5 mM^{-1} cm^{-1} as compared to the range of 10–100 mM^{-1} cm^{-1} for the heme proteins. In addition, the bands are broad and readily disguised by features of flavin spectra and the tails of heme absorptions. Second, the great successes of "bioorganic chemistry" in the past century have been built on the approaches of classic organic chemistry: chemical degradation and analysis. These approaches were readily applicable to relatively stable molecules such as the vitamins, heme compounds, and hormones. (In many cases the first two classes of compounds mentioned are the active prosthetic groups of natural macromolecules.) These same approaches were bound to fail—at least as far as determining their composition goes—with the primitive type of catalyst that iron–sulfur proteins contain, namely, simple iron sulfide wrapped up delicately in a protein shell. At best these approaches could have led to the detection of iron by chemical analysis and of escaping hydrogen sulfide. Coming to grips with the structure of iron–sulfur proteins had to await the advent of nondestructive but nevertheless specific methods of analysis, such as electron paramagnetic resonance (EPR), NMR, and Mössbauer spectroscopies and their ramifications. A third reason for the long latency of this class of proteins can probably be seen in their quantitative relationships to other iron-containing compounds, at least in mammalian species. This was discussed above.

This is not to say that iron compounds, which were recognized not to be heme compounds—no doubt including iron–sulfur proteins—had escaped the attention of physiological chemists for some decades (Brückmann and Zondek, 1940; McFarlane, 1932, 1934; Schapira and Dreyfus, 1948). The expression "nonheme iron" dates back into the early days of this branch of chemistry. In many instances, however, the then unknown storage forms, particularly ferritin, are likely to have made up the bulk of this nonheme iron. With regard to the occurrence of iron–sulfur proteins, most meaningful among the earlier work are results on tissues where the content of hemeproteins and storage proteins is low. However, it remained for the period starting twenty years ago, when subcellular fractions were first prepared and analyzed, that the occurrence of nonheme iron—in this case including significant amounts of iron–sulfur proteins—was specifically associated with certain structures and enzymatic activities (Green, 1956; Crane *et al.,* 1956). The finding of nonheme iron in enzyme components solubilized from these subcellular fractions added weight to the significance of these observations (Mahler and Elowe, 1954; Kearney and Singer, 1955).

Scientific discoveries are rarely made in a single break through, planned or unplanned. More often there are encounters with the unknown at different points; different and apparently unrelated sides of the problem are seen by different people, and often in relatively unrelated fields of science. From the historical viewpoint it is then interesting to see where these beginnings were, and particularly how the synthesis came about, namely, the recognition that the different aspects that had been uncovered were in fact related to a single more general phenomenon or, as in the case at hand, a class of substances.

I shall try to picture this development and I may be forgiven if, despite my striving toward objectivity, a bias to the area of my own involvement shines through. Essentially, there were three independent paths which led to the field of iron–sulfur proteins. They originated from entirely unrelated endeavors and dealt with very different aspects of the field. One of these was research on photoreactions of chloroplasts which led, independently in different laboratories, to the recognition of biological activities of "factors" which were able to catalyze photoreductions of a number of compounds, notably, heme proteins and pyridine nucleotides. With the knowledge of the "Hill reaction" (the light-driven evolution of oxygen by chloroplasts with stoichiometric reduction of an electron acceptor) and the fact that illuminated chloroplasts are able to reduce NADP, the recognition of a catalyst active in these processes was a significant advance toward understanding the light reaction of photosynthesis. These developments have been described in detail in previous reviews by those who have directly participated in this work (Arnon, 1965a,b; Buchanan and Arnon, 1970; San Pietro and Black, 1965; Davenport, 1965), and I shall not attempt to repeat them here. Suffice it to say, that these factors were largely, and at the outset exclusively, known by virtue of their biological activities, from which names such as "methemoglobin reducing factor" (Davenport *et al.*, 1952), "TPN reducing factor" (Arnon *et al.*, 1957), and "photosynthetic pyridine nucleotide reductase" (PPNR) (San Pietro and Lang, 1958) originated. Although their protein nature was recognized early, their iron–protein nature was not appreciated until the early sixties. While these approaches characterized what we now know to be iron–sulfur proteins as biological activities, the second entranceway to the iron–sulfur protein field, as anticipated above, was the characterization of representatives of this class in terms of unaccounted for protein-bound iron. Analysis of whole tissues, and later on fractionation of tissue homogenates and analysis of subcellular fractions, showed that there was bound iron several times in excess of that accountable for as heme iron (Green, 1956; Crane *et al.*, 1956). Contamination of this magnitude seemed unlikely and could be ruled out experimentally (Beinert, 1965). At the same time, when nonheme iron was found in subcellular particles, enzymes such as succinate and NADH dehydrogenases isolated from these particles were found to contain "nonheme" iron in addition to their previously recognized flavin components (Kearney and Singer, 1955; Mahler and Elowe, 1953). Independently, nutritional studies showed that yet another well-known flavoprotein, namely, xanthine oxidase, contained "nonheme" iron (Richert and Westerfeld, 1954) in addition to molybdenum (Richert and Westerfeld, 1953; DeRenzo *et al.*, 1953). This last line of research then reaches to the establishment of liver aldehyde oxidase as a molybdenum–iron–flavoprotein (Rajagopalan *et al.*, 1962a) and dihydroorotate dehydrogenase as an iron–flavoprotein (Friedmann and Vennesland, 1960). Despite some early attempts to prove a redox function for the iron in some of the mentioned iron–flavoproteins by chemical analysis (Mahler and Elowe, 1954; Massey, 1957, 1958; cf. Singer and Massey, 1957)—as we now know, a futile undertaking—this whole line of research only established the presence of nonheme iron as a native constituent of these proteins, and not as a catalytic component. In the work on chloroplast photoreactions, the activity of the protein was the clue, and the presence of iron was initially unknown.

However, a functional role for the iron, or at least for part of the analytically determined iron, was suggested by EPR spectroscopic observations on the mitochondrial bound and solubilized iron–flavoproteins.

An intense, but unusual signal—now known colloquially as the $g = 1.94$ signal—was discovered with the reduced forms of all of these proteins (Beinert, 1965; Beinert and Sands, 1960; Beinert *et al.*, 1962; Beinert and Lee, 1961; Rajagopalan *et al.*, 1962b; Bray *et al.*, 1961). This opened the possibility of searching for similar signals, viz., related structures, in other materials and of studying oxidation–reduction stoichiometries and kinetics (Bray *et al.*, 1964; Beinert and Palmer, 1965a,b; Orme-Johnson and Beinert, 1969). Indeed, EPR signals of the $g = 1.94$ type were seen not only in materials of mammalian origin but also in a number of microorganisms (Beinert *et al.*, 1962; Nicholas *et al.*, 1962), indicating that the underlying structure was present in many forms of life.

The third entranceway into the field of iron–sulfur proteins was the one opened up most recently and also that leading most rapidly to progress. A protein of small molecular weight was discovered as a component of nitrogen fixing fractions from *Clostridium pasteurianum*. The function initially found for this protein was that of linking the phosphoroclastic split of pyruvate to the evolution of hydrogen via hydrogenase (Mortenson *et al.*, 1962). It will be recalled here that in the early work on cell-free nitrogen fixation the reducing power necessary for conversion of N_2 to NH_3 was derived from the so-called phosphoroclastic reaction; hence, the new protein was required for nitrogen fixation because of its vital role in the phosphoroclastic reaction (cf. Mortenson, 1963). The catalytic protein was found to have a brown color and a high iron content and was given the name ferredoxin, i.e., a redox catalyst containing iron.

B. Recognition of the Relationship of Iron–Sulfur Proteins Found in Different Organisms and Systems

Up to this point in the gross developments in each of the three main avenues leading to the field of iron–sulfur proteins, there was no indication that the materials involved were related beyond their property of containing nonheme iron. How were the links between these areas established? This again occurred in steps and from different kinds of observations and reasoning, The first two areas which merged were those of the clostridial ferredoxin, the factor active in nitrogen fixation, and that active in photoreactions of chloroplasts. It had been recognized by this time that the various factors from chloroplasts or algae which had been obtained in different laboratories

(Davenport *et al.*, 1952; Arnon *et al.*, 1957; San Pietro and Lang, 1958; Gewitz and Völker, 1962) and which were known under different names were in fact closely related and possibly differed only in minor features connected with their origin from different species. They were, however, thought to be specific constituents of photosynthetic plant tissues. A number of observations, however, led to a broader view of the activities and functional relationships of the plant and bacterial electron transfer factors: It had been known for some time that photosynthetic bacteria and algae are able to produce hydrogen gas in the light under special conditions. The analogous production of hydrogen by chloroplasts in the light was initially found to require one of the viologens as an electron carrier in addition to a hydrogenase of bacterial origin. When, however, a crude hydrogenase of *Clostridium pasteurianum* was used, the artificial carrier was no longer required. It soon became apparent that clostridial ferredoxin, which had since been discovered, was present in the hydrogenase and took over the carrier function (Tagawa and Arnon, 1962). It was then found that the pyridine nucleotide reducing factor of chloroplasts was also able to catalyze hydrogen production as well as the reduction of pyridine nucleotides by hydrogen gas (Tagawa and Arnon, 1962) under certain conditions. Furthermore, a pyridine nucleotide reducing factor was isolated from *Chromatium*, a photosynthetic bacterium, and this factor was capable of effectively replacing the native chloroplast factor in mediating photoreduction of NADP (Losada *et al.*, 1961). (Chloroplasts for such tests had been depleted of their endogenous iron protein.) It should be recalled that photosynthetic bacteria are unable to evolve oxygen in light. With this background, it was a logical step to try substituting the clostridial ferredoxin for the chloroplast iron protein in photoreactions catalyzed by chloroplasts. This replacement succeeded (Tagawa and Arnon, 1962), and its significance lies in the fact that ferredoxin from *Clostridium pasteurianum*, an organism which does not depend on light, was able to undergo the very reaction thought to be specific for factors from photosynthetic tissues. This left little doubt that the two types of proteins, those from nonphotosynthetic anaerobic bacteria, and those from plants, were closely related, at least functionally, and that the general catalytic principle of the chloroplast factors was not a specific one of photoreactive systems. Following this finding, it was proposed (Tagawa and Arnon, 1962) to extend the term ferredoxin to encompass also the proteins previously found active in chloroplast photoreactions. Although both kinds of proteins contained iron, the iron content of the clostridial protein ($\sim 7\%$) was almost an order of magnitude higher than that of the plant proteins, and the light absorption spectrum was distinctly different. On the other hand, both proteins were redox catalysts and had, among known catalysts, an unusually low oxidation–reduction potential, a fact which also invited the substitution experiment just referred to. A comparison of chemical composition and structural features was not convincing that the two types of proteins were closely related. However at this time, an additional, common structural feature was recognized. There was in the literature of the middle 1950's an isolated observation on the enzyme succinate dehydrogenase (Massey, 1957). This protein gave off hydrogen sulfide rather readily on acidification, a reaction not expected from the behavior of known sulfur-containing protein constituents. This same type of "labile sulfur" was rediscovered with the ferredoxins of either plant or bacterial origin [by that time ferredoxins had been isolated from or recognized in a number of anaerobic microorganisms (Lovenberg *et al.*, 1963; Buchanan and Rabinowitz, 1964; Whiteley and Woolfolk, 1962; Valentine *et al.*, 1962)]. The peculiar features then of the catalytic capability in electron transfers at low potential levels and the simultaneous occurrence of iron and labile sulfur eliminated all doubt that with these proteins a whole new class of hitherto unknown catalysts had been discovered.

While this work proceeded, the almost ubiquitous occurrence of the EPR signal at $g = 1.94$, which was mentioned above in the discussion of the mammalian nonheme iron proteins, its constant association with nonheme iron, and its ready response to natural reductants and oxidants, appeared to establish just this same notion for the nonheme iron proteins recognized or isolated originally from mammalian and later also from bacterial sources (Beinert, 1965; Beinert *et al.*, 1962; Beinert and Palmer, 1965a). By this time Shethna *et al.* (1964) had prepared a ^{57}Fe nonheme iron protein by growing *Azotobacter vinelandii* on this iron isotope. Using this protein they found that the EPR signal at $g = 1.94$ showed hyperfine interaction of the paramagnetic ^{57}Fe nucleus ($I = \frac{1}{2}$). This experiment established that iron was indeed represented in the signal at $g = 1.94$ which had thus far been concluded from indirect evidence only. Nevertheless, in 1965, when the first large symposium was called on nonheme iron proteins (San Pietro, 1965), there was no all-around convincing evidence that the ferredoxins were more than distantly related to the mammalian and bacterial nonheme iron proteins, which had the EPR signal at $g = 1.94$. This type of EPR signal had not been found with any ferredoxin. There were, however, two important observations which strongly pointed to a closer relationship between the ferredoxins and the EPR detectable nonheme iron proteins. Procedures were found for removing flavin from molybdenum–iron–flavoproteins, so that a soluble molybdenum–iron protein remained (Rajagopalan and Handler, 1964).

As molybdenum does not make a major contribution to the light absorption spectrum, these proteins showed the light absorption of the remaining iron chromophore. The spectra of these flavin-free proteins were strikingly similar, in band position and absorptivity per iron atom, to those of the plant ferredoxins. Furthermore, all of the nonheme iron proteins, which on reduction showed an EPR signal at $g = 1.94$, were found to contain labile sulfur (Rajagopalan and Handler, 1964; Rieske *et al.*, 1964; Hardy *et al.*, 1965; King, 1964; Zeylemaker *et al.*, 1965; Lusty *et al.*, 1965; Suzuki and Kimura, 1965; Brumby *et al.*, 1965). It remained then to explain why one group of proteins did show the distinct and characteristic EPR signal while the other one, the ferredoxins, did not. The answer to this question (Hall *et al.*, 1966; Palmer and Sands, 1966; Palmer *et al.*, 1966), which provided the missing link and finally unified the whole field of iron–sulfur proteins, was forthcoming, with some advances in biological quarters concerning EPR technique—advances commonplace to physicists practicing the art. It had been known from the early studies on the mammalian nonheme iron proteins that in most cases the EPR signals at $g = 1.94$ were unusually sensitive to temperature. This could be ascribed to broadening by rapid electron spin relaxation. It had also been observed that the temperature sensitivity varied considerably from protein to protein. In retrospect it is easy to see in these features the suggestion that the signals from the proteins in the upper range of temperature sensitivity may never have been seen in the attempts at liquid-nitrogen temperature. Nevertheless, it remained for those who had available in their laboratories, together with EPR equipment, either unusually high concentrations of ferredoxins (Hall *et al.*, 1966) or temperatures below those of liquid nitrogen (Palmer and Sands, 1966; Palmer *et al.*, 1966) to show that the ferredoxins of bacterial or plant origin did indeed have EPR signals of the general type seen in the mammalian nonheme iron proteins.

This discovery concluded what one might call the adolescent phase of the field of iron–sulfur proteins. The unification of the field stimulated research on the structure of the whole class of iron–sulfur proteins, as it now became obvious that results obtained on the ferredoxins, for instance, were in large measure applicable to the mammalian nonheme iron proteins as well, and vice versa. This is not to say that prior to this the small molecular weight and ready availability of the ferredoxins from a number of sources had not attracted the attention of many who probe into the structure of proteins by chemical and physical means. Thus, the amino acid sequences of a number of ferredoxins of plant and bacterial origin soon became available (Tanaka *et al.*, 1966; Benson *et al.*, 1967; Tsunoda *et al.*, 1968; Sugeno and Matsubara, 1968;

Matsubara *et al.*, 1967; Keresztes-Nagy *et al.*, 1969), stimulating genetic considerations (Matsubara *et al.*, 1967, 1968; Keresztes-Nagy *et al.*, 1969; Weinstein, 1969). The genetic tree of the ferredoxins has grown to respectable size and is probably the tallest and among the most dense we know at the present time. The coordination chemistry of the iron site had received attention among biochemists, spectroscopists, inorganic chemists, and chemical physicists alike, and attempts at determination of the structure by X-ray crystallography have been underway for some time (Sieker and Jensen, 1965; Guillard *et al.*, 1965). Yet, the finding that the typical and, until rather recently, mysterious, EPR signal at $g = 1.94$ was also characteristic of the ferredoxins, thus yielding an additional inroad to the iron site, has further concentrated and intensified these efforts.

In addition, then, to having become an interesting example in the history of scientific discovery, the development of this field of science also represents an excellent example of the need and usefulness of interdisciplinary collaboration and the application of physical tools to the exploration of complex and delicate biological structures. At the same time, the developments in this field underscore the value of comparative biochemistry, and demonstrate the maturity of biochemical techniques and know-how in supplying well-defined materials; last but not least, they constitute an example of the challenge to the more "exact," nonbiological disciplines that rests in the unexplored inventions of living nature.

The developments which remain yet to be considered all fall into the most recent period, and I shall try to single out those that have had or promise to have the greatest determining influence on the future course of events in this field.

Before doing so, two developments, which have remained somewhat out of the mainstream of events and were simultaneous with those directly related to ferredoxins, must yet be mentioned, namely, the discoveries of the rubredoxins (Lovenberg and Sobel, 1965; Bachmayer *et al.*, 1968) and the "high-potential" iron proteins (Bartsch, 1963; Dus *et al.*, 1967). These proteins are legitimate members of the class of iron–sulfur proteins, as defined above (see also Section V), but they have thus far not been found to be nearly as widely distributed in nature as ferredoxins, and, with one exception (Boyer *et al.*, 1971; Peterson *et al.*, 1966), their function is unknown. Genetically, however, these proteins appear to have some relationship to the ferredoxins (Weinstein, 1969; Dus *et al.*, 1971), and some of their structural features are therefore of interest to those exploring the structure of the ferredoxins (Volume II, Chapter 8). This assumes importance as the rubredoxins and high-potential iron proteins are more stable, and determination of their structures has progressed faster (Strahs and Kraut, 1968; Herriott *et al.*, 1970) than that of ferredoxins.

C. Models for the Iron Center of Iron–Sulfur Proteins

There is no doubt that the exploration of the structures of those iron–sulfur proteins which contain labile sulfide, i.e., the majority, suffered from the lack of models for the type of compound involved. Although the idea that a pair of iron atoms may be present at the metal site of these proteins was mentioned early (Beinert *et al.*, 1962), it was not until evidence had accumulated from chemical analytical and molecular weight data, showing the presence of at least two iron atoms per molecule of all of these proteins, that the iron-pair or iron-cluster concept was considered seriously and useful models were proposed. Particularly, the EPR resonance with $g = 1.94$, $g = 2.01$ was not understood to be that of an iron compound, as long as the idea of a complex with a single iron center dominated. The one successful attempt to design, by theoretical reasoning, a one-iron model—bis(hexamethylbenzene) iron (I) indeed showed resonances of the type exhibited by ferredoxins (Brintzinger *et al.*, 1966a)—thus deserves all the more credit. A more general model, with a center of two interacting iron atoms was postulated (Brintzinger *et al.*, 1966b) almost simultaneously with a more specific model (Gibson *et al.*, 1966) for which antiferromagnetic coupling between iron atoms was suggested, with both iron atoms present as high spin ferric in the oxidized forms, and as high spin ferric and ferrous, respectively, in the reduced forms. The latter theoretical model for the two iron ferredoxins has been the most successful to date, and this volume will show much evidence confirming its validity. A very attractive model substance for the type of structure present in the high-potential iron–sulfur proteins—which contain four iron and four labile sulfur atoms—is available in a polynuclear iron dithiolene complex (Balch, 1969). Related structures have also been proposed for the ferredoxins containing eight iron atoms (Poe *et al.*, 1970).

The advances that have been made very recently in probing into the nature of sulfur-bridged iron dimers and related iron-cluster compounds involving sulfur ligands (Connelly and Dahl, 1970; Balch *et al.*, 1968; Wei and Dahl, 1965; Coucouvanis *et al.*, 1969; Wei *et al.*, 1966; Toan *et al.*, 1971) will certainly contribute to our understanding of the specific properties of the iron–sulfur proteins, although the analogy to the structures present in the proteins is mostly rather limited. The experimental difficulties in providing model compounds in which the number and kind of ligands, the state of oxidation, and the oxidation–reduction and magnetic behavior of the iron centers are all similar to those of the natural compounds, are considerable. Although this seemed to indicate that the protein plays an essential part in determining the specific interactions between the partners of the naturally occurring metal-sulfur clusters, this notion is now questioned because of the recent demonstration that relatively simple model compounds may indeed have most of the salient properties of a four-iron center as it occurs in high potential iron proteins or—in duplicate—in the clostridial type of ferredoxin (Heskovitz *et al.,* 1972). The essential part of the model substances is the $[Fe_4S_4(SR)_4]^{2-}$ anion. The compound, in which R = benzyl, has been well characterized by a number of physical techniques.

Consideration of the properties of these model structures supported or even led to the conclusion that iron–sulfur proteins actually occur in 3, not 2, oxidation states with oxidized high-potential iron proteins in the highest oxidation state (paramagnetic), reduced high-potential iron protein and oxidized ferredoxins equivalent and in an intermediate oxidation state (diamagnetic), and reduced ferredoxins in the lowest oxidation state (paramagnetic) (Carter *et al.* 1972).

D. Primary Structure and Genetic Relationships

Studies of the primary structure of the ferredoxins and rubredoxins have rapidly advanced to an exciting state. Since these molecules can be found in anaerobes, it will be possible to trace back evolutionary trends much farther than a study of hemeproteins would allow us to do. It appears that the smallest molecules, namely, the ferredoxins of the clostridial type, are composed of two analogous halves, each containing between 26 and 29 amino acid residues (Tanaka *et al.,* 1966; Benson *et al.,* 1967; Matsubara *et al.,* 1968). The sequences of the plant and algal ferredoxins, whose molecular weight is roughly twice that of the clostridial type, show unquestionable similarities to the sequences of the clostridial proteins (Matsubara *et al.,* 1967, 1968; Keresztes-Nagy *et al.,* 1969), and it seems that a doubling mechanism is again involved. In turn, the sequences of rubredoxins appear to be related to those of the ferredoxins (Weinstein 1969; Dus *et al.,* 1971). It thus appears possible that all these molecules may have descended from a common ancestral type. Several suggestions for routes of descendance and relationships among organisms inferred from this have been made (Matsubara *et al.,* 1967, 1968; Keresztes-Nagy *et al.,* 1969; Weinstein, 1969; Lee *et al.,* 1970; Benson *et al.,* 1971; Tanaka *et al.,* 1971a, 1973). The more it came as a surprise when it was reported (Tanaka *et al.,* 1970) that the iron–sulfur protein from beef adrenals, which without doubt has the same type of metal center as the plant ferredoxins, had an amino acid sequence apparently entirely unrelated to those of the clostridial- and plant-type ferredoxins. Short of concluding that nature has invented the same working principle, viz., the iron–sulfur cluster, twice independently, one must assume that all perceptible similarity has disappeared with the years and that no single residue is indispensable in its position for maintenance of the iron–sulfur structure. This may not be so, when one considers function. It should be recalled that neither the adrenal iron–sulfur protein nor other related proteins, such as the iron–sulfur protein from *Azotobacter*, are able to catalyze photoreduction of NADP by chloroplasts (Kimura, 1968; Shethna *et al.*, 1968), nor are spinach or *Euglena* ferredoxins or putidaredoxin (Cushman *et al.*, 1967) capable of catalyzing steroid hydroxylation in the systems from adrenals, in which the adrenal iron–sulfur protein functions (Kimura, 1968; Gunsalus, 1968). In turn, the adrenal iron–sulfur protein cannot substitute for putidaredoxin in the camphor hydroxylase system (Tsai *et al.*, 1971). It will, therefore, be of interest to compare sequences, as they become available (Chapter 2, Volume I; Tanaka *et al.*, 1973; Tsai *et al.*, 1970) within the group of proteins, akin to the adrenal iron–sulfur protein.

E. Complex Iron–Sulfur Proteins

The field of complex—or as one might say "multiheaded"—iron–sulfur proteins, namely, of those that have a number of active centers or groups in a single molecule, does not by any means represent a new development. The iron–sulfur flavoproteins and those which in addition contain molybdenum were among the first to be recognized in that line of research which relied mainly on EPR spectroscopy. In a way, the eight iron ferredoxins of the clostridial type may be considered complex iron–sulfur proteins. There is no question that, even when the structure of the simpler ferredoxins is known in a definitive way, the field of those complex proteins which contain a number of active groups will still be a challenge for many years to come. The recent growth of the list of such materials indicates that we are only at the beginning in this area. We may mention here the iron–sulfur and molybdenum–iron–sulfur proteins of nitrogen fixing systems ["azoferredoxin" and "molybdoferredoxin" (Nakos and Mortenson, 1971a; Dalton *et al.*, 1971) or "fractions I and II" (Bulen and LeComte, 1966)]. The molybdenum–iron–sulfur protein contains one molybdenum atom for ten to twenty iron–sulfur groups. One of the most complex molecules known in this class is bacterial sulfite reductase (Siegel and Kamin, 1968) which contains some twelve to fourteen iron–sulfur groups, four FMN and four FAD molecules, and at least two heme groups per molecule. The mitochondrial iron–sulfur (–flavo–) proteins belong in this class, although with these membrane bound structures the definition of molecular size becomes uncertain. The hydrogenases of *Clostridium pasteurianum* (Nakos and Mortenson, 1971b) as well as of *Desulfovibrio desulfuricans* (LeGall *et al.*, 1971) have recently been found to be iron–sulfur proteins with four iron and four labile sulfur groups per molecule. These proteins may, however, have to be counted with the simpler ferredoxins.

The study of the complex iron–sulfur proteins will probably remain a domain of biochemists for some time, as most physical approaches, such as X-ray diffraction, NMR, ENDOR, and Mössbauer spectroscopies and even optical methods, must capitulate in view of the sheer size and consequent dilution of any single site of these proteins. One must admire the courage in a recent attempt to disentangle, by Mössbauer spectroscopy, events which occur during nitrogen fixation (Kelly and Lang, 1970).

F. Present and Expected Contributions of Physical, Particularly Resonance, Techniques to Our Picture of the Structure of the Iron Center

As pointed out above, the discovery of the EPR signal at $g = 1.94$ initially opened the way to the exploration of the mitochondrial type of iron–sulfur proteins. The signal not only lent itself well to the elaboration of reaction rates and stoichiometries, but, by the proper choice of materials, it was able to furnish a number of unique clues to the structure of the iron site of iron–sulfur proteins. While the story of the exploration of the structure of these proteins on the whole appears to emphasize the power and contributions of physical techniques of analysis, it is worth pointing out at this juncture the unique contributions in this work of the sometimes disparaged comparative biochemistry. Had it not been for the curiosity and versatility of the biologically oriented investigators, those unique materials which exhibited a specific feature more prominently than others would never have been available. It was a misfortune, from the standpoint of structural exploration of the iron site, that the complicated clostridial ferredoxin and spinach ferredoxin, with its broad EPR signal, were discovered first and that for a number of years so much effort by physically oriented investigators was devoted almost exclusively to these materials.

Up to now, the iron–sulfur protein elaborated by the pseudomonad,

P. putida (Cushman *et al.*, 1967), in its adaptation to growth on camphor as substrate, has probably been the most useful material for the exploration of the iron site of iron–sulfur proteins. (This would be more apparent if *P. putida* could be obtained as easily as spinach!) This protein is colloquially known as putidaredoxin—an expression discouraged by the forthcoming rules on nomenclature (see below)—and belongs to the general type of the plant ferredoxins. As a microbial protein, it allows isotopic substitutions to be made quite readily, and in addition, it has the least complicated EPR spectrum, that is, one of practically complete axial symmetry ($g_x = g_y$). Isotope substitutions with ^{57}Fe, ^{33}S, ^{80}Se, and ^{77}Se using this protein gave clear-cut information on the following points: (1) The observed hyperfine structure (hfs) clearly indicates interaction of two iron nuclei with a magnetic moment equivalent to a spin of $\frac{1}{2}$. The single electron which the protein is able to take up on reduction therefore resides in a cluster involving both iron atoms. Hence, for this type of iron–sulfur protein, viz., the plant-type ferredoxins, a structure with less than two iron atoms is not feasible (Tsibris *et al.*, 1968; Beinert and Orme-Johnson, 1969). From EPR spectroscopy alone, however, it could not be decided whether an electron is shared approximately equally by the two iron atoms or whether the observed hfs arises from a magnetic moment resulting from antiferromagnetic coupling of a ferric with a ferrous ion (Gibson *et al.*, 1966). (2) The unpaired spin simultaneously interacts with sulfur atoms in the molecule (DerVartanian *et al.*, 1967). (3) Among these sulfur atoms are the two labile sulfur atoms (Orme-Johnson *et al.*, 1968) and not more than six stable sulfur atoms from the protein backbone. Since there are iron–sulfur proteins that do not contain methionine and not more than four cysteine residues (Aggarwal *et al.*, 1971; Newman *et al.*, 1969), and since one is inclined to visualize a rather symmetrical structure, the most likely number of stable sulfur groups, which are part of cysteine residues, is four. The interaction of the labile and stable sulfurs with the unpaired electron is of the same order of magnitude (Tsai *et al.*, 1970).

It thus became apparent from this and similar work on related proteins that the iron site has the following properties: There is no paramagnetism in the oxidized state at low temperature ($<100°K$). On reduction, one electron can be accommodated; this electron is accounted for in the EPR spectrum as well as by magnetic susceptibility measurements (spin $\frac{1}{2}$). Both iron nuclei interact with an unpaired spin of $\frac{1}{2}$ resulting from the electronic structure of the reduced iron–sulfur cluster and so do the two labile sulfur atoms and a number of stable sulfur residues.

It is unlikely that EPR "powder spectra" (i.e., of randomly oriented molecules) could lead to much further information on ferredoxins of the plant type, except perhaps to the exact number of interacting stable sulfur atoms. EPR investigations of single crystals might add new information. This does not necessarily apply to the more complicated ferredoxins (see Section V), namely, those containing more than two iron atoms per molecule or in addition other metals or functional groups. Here "powder spectra" may yield new insights, as the possible heterogeneity of the iron sites (Orme-Johnson and Beinert, 1969b; Gibson and Bray, 1968; Palmer and Massey, 1969; Orme-Johnson *et al.*, 1971; Gutman *et al.*, 1971) and their interactions have barely been explored.

As shown in these volumes, other physical methods are able to complement and extend the information gained from "powder" EPR. For the purposes of this introduction, the capabilities of the various methods in the exploration of structures such as those of the ferredoxins will be stressed, rather than details which are found in the individual chapters of this treatise.

Whereas EPR permits us to look at the unpaired spin and to indirectly obtain information on neighboring nuclei via the hyperfine interaction with this spin, Mössbauer spectroscopy enables us to look at the iron nuclei directly and to obtain indirect information on the surrounding electrons through the isomer shift, quadrupole moment, and hyperfine interaction. Nuclear magnetic resonance, in turn, will potentially inform us of all nuclei of which paramagnetic species are present or can be incorporated. At the present time this technique informs us primarily about protons and indirectly about the surrounding unpaired spins via contact interaction. Mössbauer spectroscopy has the advantage of detecting any iron, irrespective of paramagnetism, and furnishing additional information about magnetic properties, when spectroscopy is carried out in an externally applied magnetic field. On the other hand, NMR has the advantage of looking at atoms more remote from the metal site and particularly of giving information on the state of the complex at ambient temperature. This is important since EPR is useless at this temperature, because of the rapid electron spin relaxation. Since Mössbauer spectroscopy can only be done on solids, applying this technique at room temperature implies the use of dry proteins. Spectroscopy on such samples is, of course, not ideal in every respect and may leave uncertainties. Therefore, NMR spectroscopy is able to provide information on the proteins for the temperature range at which they normally function, although this feature renders the problem of protein stability more serious for NMR than for other resonance methods. This, however, is not to say that the low-temperature behavior is of secondary importance; it is a specific characteristic of these iron complexes and may bring to light some sharpened detail which may not be apparent at higher temperatures. Therefore, the resonance methods are to some extent mutually complementary. Yet to be mentioned is the ENDOR technique (Fritz *et al.*, 1971; Volume II, Chapter 5 of the treatise) which is uniquely useful in establishing the types of magnetic nuclei that interact with an unpaired electron and their exact interaction parameters. While these resonance techniques are thus able to provide quite specific answers on particular points, they all have their limitations in applicability, sensitivity, temperature, and state of the sample, as briefly discussed above. A related method, though not a resonance technique, which lacks much of the pointed specificity of the resonance methods, but is able to provide a very valuable overall picture of the magnetic properties, is that of measurement of the bulk magnetic susceptibility. On the other hand, since this technique will measure magnetism at any temperature—which is not true for EPR—it is able to provide information on the temperature dependence of the magnetic susceptibility over a wide range. This is particularly valuable in attempts to understand complex systems which involve spin–spin coupling phenomena (Volume II, Chapter 8 of this treatise).

Although in general the optical absorption spectra of even the most simple iron–sulfur proteins could not be interpreted unambiguously, most recently a search in the near infrared for ligand–field transitions of iron atoms of iron–sulfur proteins has been successful. Since the single iron atom in reduced rubredoxin is in the high-spin state, according to its magnetic susceptibility and Mössbauer spectrum, and is, according to X-ray diffraction, in a tetrahedral ligand environment, the possibility for finding a *d–d* transition of this ferrous iron in the near infrared appeared particularly favorable. This was first verified with rubredoxin (Eaton and Lovenberg, 1970), where the spin state of the reduced iron is known; the same approach was then successfully extended to plant-type ferredoxins (Eaton *et al.*, 1971), indicating that indeed a high-spin ferrous iron atom is present in the reduced forms of these ferredoxins—in agreement with the results of Mössbauer spectroscopy (Johnson *et al.*, 1971; Dunham *et al.*, 1971a).

Details of the applications of these techniques will be found in Volume II, Chapters 3 and 6 of this treatise. It may suffice to summarize here the essential findings available to date: The picture of the iron site of plant-type ferredoxins (two Fe, two labile S) drawn above on the basis of the EPR results was confirmed and significantly extended by the other techniques (cf. Dunham *et al.*, 1971b).

The measurements of the magnetic susceptibility at low temperature agreed with the results of EPR spectroscopy concerning the magnetic states of the oxidized and reduced forms of the ferredoxins (Moss *et al.*, 1969; Kimura *et al.*, 1970; Palmer *et al.*, 1971; Moleski *et al.*, 1970; Ehrenberg, 1966). Deviations were, however, observed at or close to room temperature, in the sense that paramagnetism increased with both the oxidized and reduced forms (Palmer *et al.*, 1971). This is a clear-cut sign of increased spin exchange coupling at the lower temperatures. From the temperature dependence of the paramagnetism, an estimate of the spin exchange energy could be obtained (Kimura *et al.*, 1970; Palmer *et al.*, 1971). This energy is of such magnitude that with all the proteins investigated to date at room temperature, spin coupling is not entirely overcome by thermal energy. Mössbauer spectroscopy confirmed equivalence (or near equivalence) of the iron atoms and diamagnetism in the oxidized state (Johnson *et al.*, 1968, 1969, 1971; Dunham *et al.*, 1971a; Münck *et al.*, 1972) but indicated inequivalence of the iron atoms in the reduced state (Johnson *et al.*, 1971; Dunham *et al.*, 1971a). In addition, the results of Mössbauer spectroscopy lend support to the idea that the iron atoms in both forms are in the high-spin state (Johnson *et al.*, 1971; Dunham *et al.*, 1971a). In spin-coupled systems—and the ferredoxins belong to these—such information is not available from other measurements.

These results are also compatible, in general, with those of NMR spectroscopy on the ferredoxins from spinach and parsley, alfalfa and soybean (Poe *et al.*, 1971a; Glickson *et al.*, 1971). However, NMR added the important information that, at ambient temperature, contact-shifted proton resonances (attributed to the β protons of cysteine residues) are observed even in the oxidized forms. In agreement with the measurements of magnetic susceptibility, this indicated that, in this temperature range, the iron atoms in the metal center are no longer completely spin coupled. In addition, the temperature dependence of the observed contact shifts is typical for spin-coupled systems in both oxidation states. The number of contact-shifted proton resonances, viz., eight under optimal conditions of resolution, supports the notion, put forth on the basis of composition data and ^{33}S hyperfine interactions observed by EPR, that four cysteine residues are bound to the iron atoms (cf. Dunham *et al.*, 1971b). The inequivalence of these resonances also indicates that the unpaired spin, which is detected in the reduced forms, is not equally distributed over the two iron atoms, the conclusion which was also reached from the results of ENDOR (Fritz *et al.*, 1971) and Mössbauer spectroscopy in accordance with the model of Gibson *et al.* (1966). It is useful at this juncture to recall that this model is based on the use of crystal field theory in explaining the g values of reduced spinach ferredoxin. The results of Mössbauer and ENDOR spectroscopy show, however, that the iron bonding at all sites has considerable covalent character (Fritz *et al.*, 1971). Therefore, the fact that the crystal field model can only represent an approximation—albeit a surprisingly successful one—must always be kept in mind. Detailed molecular orbital calculations on models for the structure of plant-type ferredoxin clearly bear out this point (Loew and Steinberg, 1971). Nevertheless, Dunham *et al.* (1971b), in their excellent summary, "On the Structure of the Iron–Sulfur Complex in the Two-Iron Ferredoxins," formulate their conclusions regarding this point as follows: "Although the assignment of definite valences to the iron atoms in the active site is presumptuous in view of the covalent bonding present, any assignment of electron configuration other than two high-spin d^5 ions in the oxidized proteins and one each of high-spin d^5 and d^6 in the reduced proteins is much more misleading, since the electric field gradient tensor (from Mössbauer spectroscopy) at the reduced protein ferrous iron is characteristic of the high-spin ferrous iron."

Nuclear magnetic resonance spectroscopy has also made inroads with the more complex ferredoxins of the clostridial type (Poe *et al.*, 1970, 1971a). Contact-shifted proton resonances were observed, the behavior of which has been interpreted in terms of nonequivalent iron sites. The temperature dependence of these shifts again provided evidence for exchange coupling in the oxidized and reduced states. Spectra obtained at intermediate states of oxidation pointed to a rapid inter- or intramolecular electron exchange.

It is needless to say that much effort has gone into producing this admittedly fragmentary picture of the active site of the iron–sulfur proteins and that results from X-ray crystallographic structure determination are now eagerly awaited (cf. Sieker *et al.*, 1972 Watenpaugh *et al.* 1971; Carter *et al.*, 1971). The question may even be asked as to whether the results of all the previous efforts will not be redundant or outdated after the analysis by X-ray diffraction has been completed. Fortunately for those who put in these efforts—unfortunately from the crystallographer's point of view—this is not so; although the X-ray diffraction method will at once present a general and very comprehensive picture, at the present state of the art its resolution in any single detail is limited, while the resonance and magnetic methods are capable of providing more detail in the specific areas of their competence, namely, concerning properties of the metal center (cf. Dunham *et al.*, 1971b; Münck *et al.*, 1972). In addition, some of them are capable of providing information on dynamic aspects and on properties of the proteins in the dissolved state.

IV. FUNCTION OF IRON–SULFUR PROTEINS

According to the historical developments which have led to the discovery of the various families of iron–sulfur proteins, there is no doubt about the function of those that had originally been observed as catalytic "factors," such as the clostridial or plant-type ferredoxins. This is not so for those that have been discovered as colored bands on chromatographic columns, such as the rubredoxins and high-potential iron–sulfur proteins, or as bearers of unusual EPR signals such as some of the iron–sulfur proteins of the mitochondrial electron transport system. It is true that our knowledge of the scope of involvement of ferredoxins in electron transfer reactions is steadily increasing (cf. Buchanan and Arnon, 1970; Thauer *et al.*, 1970), particularly as regards the multitude of processes which microbes are able to catalyze under one condition or another; but it seems tacitly agreed that ferredoxins are of vital functional significance. This is not so for the large quantity of poorly defined nonheme iron and not even for the iron constituents of the established but more complex iron–sulfur proteins, such as the iron–sulfur–flavoproteins and the iron–sulfur proteins in mitochondria. Experiments have not been lacking which aimed at either establishing or disproving their function. Unfortunately, much confusion has resulted from the fact that either authors did not specify or readers did not remember what kind of iron protein was in fact being referred to. In discussing this delicate problem, we must therefore keep in mind quite clearly whether we consider all protein-bound nonheme iron of mitochondria, only that portion which, by chemical analysis, has been shown to be associated with a certain preparation of a specific catalytic activity, say NADH dehydrogenase, or only that portion, which on reduction becomes manifest by an EPR signal of the type of the $g = 1.94$ signal, that is, the portion which can be attributed to an iron–sulfur structure. And here again it must be specified whether negative results were obtained at liquid-nitrogen temperature or at lower temperatures.

Thus it has been repeatedly proposed that nonheme iron is related to oxidative phosphorylation. This may be so, but the earliest experiments in this direction, involving orthophenanthroline (Butow and Racker, 1965), lack specificity and at best aim at whatever may be nonheme iron, but not at some of the better known iron–sulfur structures which are notoriously unresponsive to orthophenanthroline. It has also been reported that the portion of the electron transport system of mitochondria including the pathway from succinate to oxygen can be reconstituted without an additional nonheme iron protein (Yamashita and

Racker, 1969). It may not be readily apparent to the uninitiated that the enzyme succinate dehydrogenase with its full complement of nonheme iron is necessary for this reconstitution. Only the iron–sulfur protein of the cytochrome b–c_1 region (Rieske *et al.*, 1964) was not specifically added; its absence was, however, not positively ascertained by EPR spectroscopy.

On the other hand, it has been clearly shown in rapid kinetic studies with a number of isolated enzymes and types of particulate subcellular fractions that the rate of response of the iron–sulfur structure to specific reductants and oxidants is as rapid as that of other catalytic groups—e.g., flavins—of the same proteins, and is within the limits set by the overall catalytic activity (Beinert and Palmer, 1965a; Bray *et al.*, 1964; Beinert *et al.*, 1965; Rajagopalan *et al.*, 1968; Aleman *et al.*, 1968; DerVartanian *et al.*, 1969). The evidence for a role in electron transfer is therefore as good for the iron–sulfur group as for many other catalysts. The usual dilemma remains, as it also does for other potential catalysts, that it is very difficult to distinguish whether the iron–sulfur group is on the main path or on a rapidly equilibrating side path of electron transfer.

An entirely different approach toward establishing—or disproving—whether iron–sulfur proteins are essential in electron transport or oxidative phosphorylation has recently been taken by several investigators (Imai *et al.*, 1968; Light *et al.*, 1968; Ohnishi *et al.*, 1969; Garland *et al.*, 1970; Garland, 1970; Clegg *et al.*, 1969; Ohnishi, 1970; Clegg and Garland, 1971; Haddock and Garland, 1971; Light and Garland, 1971; Ragan and Garland, 1971; Tottmar and Ragan, 1971). This approach relies on iron starvation, and by its very nature will probably remain limited to microorganisms. It appears that under these conditions of iron shortage the cytochromes are more tenaciously preserved than the iron–sulfur catalysts which we now know how to detect (Imai *et al.*, 1968; Ohnishi *et al.*, 1969). Three independent studies on two different organisms seemed to indicate initially that electron transport and oxidative phosphorylation can proceed practically unimpaired with preparations in which no EPR signals in the $g = 1.94$ region were detectable at liquid-nitrogen temperature. More recent studies, however (Clegg and Garland, 1971; Haddock and Garland, 1971; Light and Garland, 1971; Ragan and Garland, 1971; Tottmar and Ragan, 1971), using iron or sulfur starvation of organisms grown in continuous culture, provide a more detailed analysis. This shows quite convincingly that at least phosphorylation at site 1 (NADH dehydrogenase region) depends on the presence of an iron–sulfur protein or proteins. In addition to this experimental evidence, the following points are worth considering.

First, the exact role of the iron–sulfur proteins associated with the electron transport system is not known, and indeed they may not be essential for functions tested in some of the reported experiments. However, they may have other functions, which we have not yet thought to test for. Second, the EPR signal detectable at liquid-nitrogen temperature in the $g = 1.94$ region was taken as a criterion for the presence or absence of the iron–sulfur structure. It is known that a number of compounds of this family are not detectable at liquid-nitrogen temperature, and it has indeed been reported and confirmed that a number of additional $g = 1.94$ type signals are detectable at lower temperatures in preparations of the microbial (Ohnishi *et al.*, 1970) as well as mammalian electron transport systems (Orme-Johnson *et al.*, 1971). Third, it has been shown some years ago that *Clostridium pasteurianum*, when short of iron, can elaborate a flavoprotein, known as flavodoxin—an expression discouraged by the forthcoming rules on nomenclature (see below)—which can take over the functions of ferredoxin with almost equal efficiency (Knight and Hardy, 1967). This has been confirmed with other microorganisms (Dubourdieu and LeGall, 1970; Mayhew *et al.*, 1969; Vetter and Knappe, 1971), and a similar protein has also been found in algae (Smillie, 1963; Zumft and Spiller, 1971). Nobody has yet suggested, on account of this, that the ferredoxins are redundant, unessential proteins. It should be remembered that the experiments on the role of the iron–sulfur structure in electron transport, referred to above, were done on microorganisms.

Claims that xanthine oxidase can be deprived of its iron and still be functional (Uozumi *et al.*, 1967; Bayer and Voelter, 1966) could not be confirmed in a number of laboratories, including one from which these reports initially emanated (cf. Hart *et al.*, 1970).

Nevertheless, the finding of the "flavodoxins" as legitimate and effective lieutenants of ferredoxins was not without irony, at a time when a great number of scientists were laboring to decipher the structure of these "essential" and ubiquitous compounds. At least it attests to the cleverness of nature, which we now have to admire not only for having invented a host of useful metabolic machines, but also for having set specifications such that there is a "way out" in emergencies. Flavodoxins, incidentially, are the smallest flavoproteins known, and, although only a byproduct of research on iron–sulfur proteins, promise to be of great usefulness in the flavoprotein field. Amino acid sequence data available to date (Tanaka *et al.*, 1971b; Fox and Brown, 1971) indicate that the protein structure of the flavodoxins is unrelated to that of the ferredoxins. The elaboration by microorganisms of flavoproteins such as the flavodoxins and their apparent role stimulate thoughts as to why flavin and iron–sulfur groups are combined in a number of proteins, how this may have come about in evolution, and what it may mean in terms of function. Is the iron–sulfur group in such proteins a remnant which is superseded by the invention of the flavin or vice versa, if flavin preceded the iron–sulfur structure in evolution? Were at one time, or are even now in some cases, iron–sulfur and flavin groups actually interchangeable in function? Is one group some kind of built-in spare part in case difficulties arise, e.g., the protein is maintained to hold both working groups in case either iron or flavin should come in short supply?

V. NOMENCLATURE

Fortunately, the tempers and communicative urges of investigators differ; at one end of the spectrum there are those who have a flair for inventing impressive names for the products of their labor, and at the other there are those who cautiously cling to sober, operationally descriptive, but cumbersome names. Obviously, in a rapidly advancing multifaceted field, such as that of the iron–sulfur proteins, nomenclature quickly grows into a jungle, which is bound to confuse the outsider and newcomer and, if nothing else, irritate the oldtimer.

On the other hand, unfortunately, there is at this stage insufficient information to set forth well-thought-out rules, as one might be able to do ten or twenty years hence. An interim solution must therefore be found, which leans on the historical past and what is by now entrenched, but which at the same time leaves sufficient flexibility to accommodate further developments and allows expansion without a change in its principal structure.

The operational term "nonheme iron proteins" fulfilled a useful role, as long as there was very little information on the nature of the iron complexes present in ferredoxins and the related proteins discussed above. It can now—and this has been generally accepted—be replaced by the more specific term "iron–sulfur proteins" in those cases where this actually applies, namely, where sulfur is a ligand of the iron, as has been shown by X-ray crystallography for rubredoxin, clostridial-type ferredoxins, and high-potential iron proteins, by sulfur hyperfine interaction for the proteins of the plant ferredoxin type, and as can be safely assumed, on the basis of a number of analogies, to be also true for the more complex iron–flavo– or molybdenum–iron–flavoproteins. (The exception should be made though that proteins in which the iron is held by a dominating ligand, such as porphyrin, are not iron–sulfur proteins, even if a sulfur containing group should be an additional ligand.) The term nonheme iron should, therefore, be relegated to those instances in which the distinction between heme and nonheme proteins is the primary aim and a more specific designation of the compounds in question is not available or not sufficiently comprehensive. We would then have, as sketched in Fig. 1, a division of iron proteins into heme-proteins, iron–sulfur proteins, and other classes, yet to be established. Such other classes may eventually be constituted of iron-binding and -transferring proteins such as the transferrins and conalbumin, of the oxygenases (cf. however, Nagami, 1972), and of other iron-containing proteins [e.g., aconitase (Villafranca and Mildvan, 1971), protein B of ribonucleotide reductase (Brown *et al.*, 1968)] when more about the structure of these proteins becomes known. According to the definition given above, hemerythrin (Groskopf *et al.*, 1966; Garbett *et al.*, 1969), the oxygen-carrying pigment of some marine organisms, would not belong to the iron–sulfur proteins, although it is an iron protein and sulfhydryl groups play a role in the structure of this protein. Present evidence does not indicate, however, that the iron of hemerythrin has sulfur ligands.

The question now arises, what naturally given dividing lines can we follow for establishing subclasses of the iron–sulfur proteins, and what names should we chose for them? Concerning the latter question, experience teaches that, if confusion and mixed usage are to be avoided, it is wise to retain existing and generally used names, as long as they

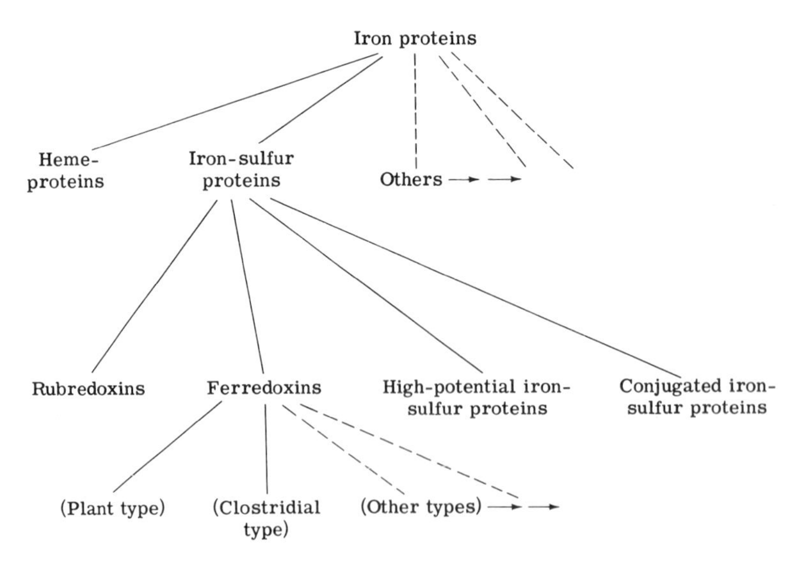

Fig. 1. Tentative scheme for the development of a nomenclature for iron–sulfur proteins.

are acceptable. The name ferredoxin is indeed as good a name as what most compounds in the biochemical arsenal have ended up with, but the coining of misnomers which utilize only the fragment "doxin" or "redoxin" is discouraged. As for reasonably safe and established dividing lines between iron–sulfur proteins, it would seem that the absence or presence of labile sulfur is a prime distinguishing mark. It appears to go parallel with the one-iron center structure—as in rubredoxin—or two- (or more) iron center structure—as in ferredoxins. A division into rubredoxins and ferredoxins would therefore be logical. Since there are sufficiently few proteins which altogether qualify as a rubredoxin, at this juncture, we do not have to concern ourselves as to how this subclass should be limited. This is not so with the ferredoxins, and we have to spell out how far we want to extend this term. Originally this name was given to the eight-iron protein from clostridial (Mortenson *et al.*, 1962) which serves as a catalyst in a number of electron transfer reactions at very low redox potential. After the chemical similarity and functional analogy with the chloroplast factors (see above) had become obvious, it was proposed that the name ferredoxin also be applied to these proteins (Tagawa and Arnon, 1962). This has since been generally accepted, as there was good reason in this proposal and as the name ferredoxin was appealing. It was emphasized at the time the term ferredoxin was first extended to cover bacterial and plant electron transfer factors that electron carriers with this name should characteristically operate at potentials close to the hydrogen electrode and should be able to catalyze photoreduction of pyridine nucleotides. We may ask: Now that more representatives of the class of iron–sulfur proteins are known, should the term ferredoxin be extended farther, and would this be in conflict with previous definitions? The guidelines which have been issued recently by the commission on Biological Nomenclature of the IUPAC-IUB recommend extending it in order to maintain the essence of the first definition, namely, that ferredoxins should operate at low redox potentials, $E_0' < 0$ (although not necessarily as low as clostridial ferredoxin), but the definition that a ferredoxin should be able to catalyze the photoreduction of pyridine nucleotides must be abandoned, if iron–sulfur proteins such as those from *P. putida* (Cushman *et al.*, 1967), *Azotobacter vinelandii* (Shethna *et al.*, 1968), or from adrenals (Kimura, 1968) are to be included under this term. The similarity of the structure of the iron site in all these proteins, as has been evident from EPR, Mössbauer, and magnetic susceptibility studies, would, however, make it logical to include these proteins among the ferredoxins. If the condition of "low redox potential" is maintained, a subclass of iron–sulfur proteins, which do have labile sulfur, is excluded. This is the thus far small, but well-characterized, class of "high-potential" iron–sulfur proteins (Bartsch, 1963; Dus *et al.*, 1967). The notion that these proteins should be grouped together in a separate subclass of iron–sulfur proteins is supported by their entirely different magnetic properties. These proteins are paramagnetic ($S = \frac{1}{2}$) when oxidized and diamagnetic, when reduced (Moss *et al.*, 1969); their EPR and optical spectra are entirely different from those of the ferredoxins (Palmer *et al.*, 1967). Recently, however, it has been proposed (Carter *et al.*, 1972) that the oxidized form of high-potential iron proteins is at a higher oxidation state than that of the ferredoxins, such that reduced, high-potential iron proteins would actually be at the oxidation state of oxidized ferredoxins (see Section III, C).

In addition to the subclass of rubredoxins, ferredoxins, and—for lack of a more appealing name—"high-potential iron proteins," we then have a large and, until further knowledge is available, heterogeneous subclass of conjugated iron–sulfur proteins to which iron-flavoproteins, molybdenum–iron–flavoproteins and molybdenum–iron proteins would belong. Some of these will, of course, with equal justification, be classified among flavo– or molybdenum proteins. Other than this last subclass of complex proteins, which can be readily subdivided according to the additional electron carriers present, the ferredoxins constitute the largest subclass, and we may attempt to follow some natural dividing lines to arrive at groups within this subclass. Here, if we again lean on historical developments, it has been clear and spelled out by many investigators over the years that there is a considerable difference in the properties of ferredoxins such as that from spinach and that from *Clostridium pasteurianum.* The light absorption and EPR spectra differ characteristically, so do the primary structure, the iron content, and electron transfer stoichiometry. It is, therefore, reasonable to consider ferredoxins of the type of spinach ferredoxin as one group and those akin to clostridial ferredoxin as another. For historical reasons, the designations "plant-type" and "bacterial-type" ferredoxins have been in use. To the newcomer, however, this may lead to confusion, since "plant-type" ferredoxins have been isolated from bacteria, even from *Clostridium pasteurianum,* although "bacterial-type" ferredoxins have not been found in plants. It is my feeling that, with a minimum of historical knowledge and consideration, it should be possible to avoid such confusion if the terms "plant-type," not "plant" ferredoxins, and "clostridial-type" instead of "bacterial' are used. This is probably as far as subdivision is feasible and useful today but, without any doubt, we have to be on the lookout for new information (cf. Travis *et al.*, 1971; Orme-Johnson *et al.*, 1972; Shanmugam *et al.*, 1972), according to which additions and revisions will have to be made.

Thus it appears at the time of this writing as if there may be a class of iron–sulfur proteins intermediate between those containing two iron and two labile sulfur atoms and those containing eight of each of these constituents, namely, proteins with one iron–sulfur center and four irons and four labile sulfurs. Although the composition of the iron–sulfur center of these proteins is analogous to that of the high-potential iron–sulfur proteins, they seem to be most closely related to the clostridial-type ferredoxin in their optical and magnetic properties. Focusing on the paramagnetic center and disregarding the size of the protein, one might look at these proteins as something akin to one-half of the clostridial-type ferredoxin.

An appropriate iron–sulfur protein would then be characterized as a rubredoxin, ferredoxin, high-potential, or conjugated iron–sulfur protein; source, function, and, when available, optical and EPR characteristics should be mentioned. For ferredoxins it is useful, although it will not be required at this point, to indicate whether it is of the plant, clostridial, or of a novel type. Distinctions of this latter kind will be formalized at some later date when the primary structures and coordination sites of more ferredoxins are established.

A few investigators have used a notation which rests solely on the head count of irons and labile sulfurs; labile sulfur is designated by S^*. According to this, rubredoxins are 1 Fe proteins, spinach ferredoxin is a 2 Fe–2 S^* protein, clostridial ferredoxin an 8 Fe–8 S^* protein, and the high-potential iron proteins are 4 Fe–4 S^* proteins. This is a convenient shorthand notation in written communications and tables; it probably would be awkward in verbal and particularly in verbal international usage. Unfortunately, this system discards useful information and distinguishing characteristics which have been elaborated through the years. That rubredoxin is an iron–sulfur protein is not apparent from this notation, nor are the striking differences between high-potential iron proteins and ferredoxins, nor the obvious differences between the "plant-type" and "clostridial-type" ferredoxins. As an exclusive system for general use, I find it too conservative, with too much touch of computerlike rigor and too little reliance on the subtle discriminatory power of the human mind.

In closing this section, I think I speak for most investigators active in the field of iron–sulfur proteins or concerned with an orderly developed of biochemical nomenclature, when I make a strong plea to all colleagues to refrain from inventing new names for proteins of this general class and to try following the guidelines suggested by the International Commission [*Eur. J. Biochem.* **35**, 1 (1973)]. It appears that these guidelines are still sufficiently flexible to be able to accommodate even the most unexpected.

ACKNOWLEDGMENTS

I am indebted to Dr. D. L. Drabkin for kindly giving permission to use his data for Table II and for additional suggestions concerning them, to Dr. Karl Dus for supplying information in advance of publication, and to Dr. W. H. Orme-Johnson and Dr. G. Palmer for critical reading of early drafts of this manuscript.

REFERENCES

Aggarwal, S. J., Rao, K. K., and Matsubara, H. (1971). *J. Biochem. (Japan)* **69**, 601.

Aleman, V., Handler, P., Palmer, G., and Beinert, H. (1968). *J. Biol. Chem.* **243**, 2569.

Amoore, J. E. (1963). *Nature (London)* **199**, 38.

Arnon, D. I. (1965a). *Science* **149**, 1460.

Arnon, D. I. (1965b). *In* "Non-Heme Iron Proteins: Role in Energy Conversion" (A. San Pietro, ed.), p. 137. Antioch Press, Yellow Springs, Ohio.

Arnon, D. I., Whatley, F. R., and Allen, M. B. (1957). *Nature (London)* **180**, 1325.

Bachmayer, H., Benson, A. M., Yasunobu, K. T., Garrard, W. T., and Whiteley, H. R. (1968). *Biochemistry* **7**, 986.

Balch, A. L. (1969). *J. Amer. Chem. Soc.* **91**, 6962.

Balch, A. L., Dance, I. G., and Holm, R. H. (1968). *J. Amer. Chem. Soc.* **90**, 1139.

Bartsch, R. G. (1963). *In* "Bacterial Photosynthesis" (H. Gest, A. San Pietro, and L. P. Vernon, eds.), p. 315. Antioch Press, Yellow Springs, Ohio.

Bayer, E., and Voelter, W. (1966). *Biochim. Biophys. Acta* **113**, 632.

Beinert, H. (1965). *In* "Non-Heme Iron Proteins: Role in Energy Conversion" (A. San Pietro, ed.), p. 23. Antioch Press, Yellow Springs, Ohio.

Beinert, H., and Lee, W. (1961). *Biochem. Biophys. Res. Commun.* **5**, 40.

Beinert, H., and Orme-Johnson, W. H. (1969). *Ann. N. Y. Acad. Sci.* **158**, 336.

Beinert, H., and Palmer, G. (1965a). *Advan. Enzymol.* **27**, 105.

Beinert, H., and Palmer, G. (1965b). *In* "Oxidases and Related Redox Systems" (T. E. King, H. S. Mason, and M. Morrison, eds.), Vol. 2, p. 567. Wiley, New York.

Beinert, H., and Sands, R. H. (1960). *Biochem. Biophys. Res. Commun.* **3**, 41.

Beinert. H., Heinen, W., and Palmer, G. (1962). Enzyme Models and Enzyme Structure No. 15, Brookhaven Symposia in Biology, p. 229.

Beinert, H., Palmer, G., Cremona, T., and Singer, T. P. (1965). *J. Biol. Chem.* **240**, 475.

Benson, A. M., Mower, H. F., and Yasunobu, K. T. (1967). *Arch. Biochem. Biophys.* **121,** 563.
Benson, A. M., Tomada, K., Chang, J., Matsueda, G., Lode, E. T., Coon, M. J., and Yasunobu, K. T. (1971). *Biochem. Biophys. Res. Commun.* **42,** 640.
Beutler, E. (1957). *Amer. J. Med. Sci.* **234,** 517.
Beutler, E., and Blaisdell, R. K. (1958). *J. Lab. Med.* **52,** 694.
Boyer, R. F., Lode, E. T., and Coon, M. J. (1971). *Biochem. Biophys. Res. Commun.* **44,** 925.
Bray, R. C., Pettersson, R., and Ehrenberg, A. (1961). *Biochemistry* **81,** 178.
Bray, R. C., Palmer, G., and Beinert, H. (1964). *J. Biol. Chem.* **239,** 2667.
Brintzinger, H., Palmer, G., and Sands, R. H. (1966a). *J. Amer. Chem. Soc.* **88,** 623.
Brintzinger, H., Palmer, G., and Sands, R. H. (1966b). *Proc. Nat. Acad. Sci. U.S.* **55,** 397.
Brown, N. C., Eliasson, R., Reichard, P., and Thelander, L. (1968). *Biochem. Biophys. Res. Commun.* **30,** 522.
Brückmann, G., and Zondek, S. G. (1940). *J. Biol. Chem.* **135,** 23.
Brumby, P. E., Miller, R. W., and Massey, V. (1965). *J. Biol. Chem.* **240,** 2222.
Buchanan, B. B., and Arnon, D. I. (1970). *Advan. Enzymol.* **32,** 119.
Buchanan, B. B., and Rabinowitz, J. C. (1964). *J. Bacteriol.* **88,** 806.
Bulen, W. A., and LeComte, J. R. (1966). *Proc. Nat. Acad. Sci. U.S.* **56,** 979.
Butow, R., and Racker, E. (1965). *In* "Non-Heme Iron Proteins: Role in Energy Conversion" (A. San Pietro, ed.), p. 383. Antioch Press, Yellow Springs, Ohio.
Carter, C W., Jr., Freer, S. T., Xuong, Ng. H., Alden, R. A., and Kraut, J. (1971). *Cold Spring Harbor Symp. Quant. Biol.* **36,** 381.
Carter, C. W., Jr., Kraut, J., Freer, S. T., Alden, R. A., Sieker, L. C., Adman, E., and Jensen, L. H. (1972). *Proc. Nat. Acad. Sci. U.S.* **69,** 3526.
Clegg, R. A., and Garland, P. B. (1971). *Biochem. J.* **124,** 135.
Clegg, R. A., Ragan, C. I., Haddock, B. A., Light, P. A., Garland, P. B., Swann, J. C., and Bray, R. C. (1969). *FEBS Lett.* **5,** 207.
Connelly, N. R., and Dahl, L. F. (1970). *J. Amer. Chem. Soc.* **92,** 7472.
Coucouvanis, D., Lippard, S. J., and Zubieta, J. A. (1969). *J. Amer. Chem. Soc.* **91,** 761.
Crane, F. L., Glenn, J. L., and Green, D. E. (1956). *Biochim. Biophys. Acta* **22,** 475.
Cushman, D. W., Tsai, R. L., and Gunsalus, I. C. (1967). *Biochem. Biophys. Res. Commun.* **26,** 577.
Dalton, H., Morris, J. A., Ward, M. A., and Mortenson, L. E. (1971). *Biochemistry* **10,** 2066.
Davenport, H. E. (1965). *In* "Non-Heme Iron Proteins: Role in Energy Conversion" (A. San Pietro, ed.), p. 115. Antioch Press, Yellow Springs, Ohio.
Davenport, H. E., Hill, R., and Whatley, F. R. (1952). *Proc. Roy. Soc. (London) Ser. B* **139,** 346.
DeRenzo, E. C., Kaleita, E., Heytler, P., Oleson, J. J., Hutchings, B. L., and Williams, J. H. (1953). *J. Amer. Chem. Soc.* **75,** 753.
DerVartanian, D. V., Orme-Johnson, W. H., Hansen, R. E., Beinert, H., Tsai, R. L., Tsibris, J. C. M., Bartholomaus, R. C., and Gunsalus, I. C. (1967). *Biochem. Biophys. Res. Commun.* **26,** 569.
DerVartanian, D. V., Veeger, C., Orme-Johnson, W. H., and Beinert, H. (1969). *Biochim. Biophys. Acta* **191,** 22.

Drabkin, D. L. (1951). *Physiol. Rev.* **31,** 345.
Dubourdieu, M., and LeGall, J. (1970). *Biochem. Biophys. Res. Commun.* **38,** 965.
Dunham, W. R., Bearden, A. J., Salmeen, I., Palmer, G., Sands, R. H., Orme-Johnson, W. H., and Beinert, H. (1971a). *Biochim. Biophys. Acta* **253,** 134.
Dunham, W. R., Palmer, G., Sands, R. H., and Bearden, A. J. (1971b). *Biochim. Biophys. Acta* **253,** 373.
Dus, K., De Klerk, H., Sletten, K., and Bartsch, R. G. (1967). *Biochim. Biophys. Acta* **140,** 291.
Dus, K., Tedro, S., Bartsch, R. G., and Kamen, M. D. (1971). *Biochem. Biophys. Res. Commun.* **43,** 1239.
Eaton, W. A., and Lovenberg, W. (1970). *J. Amer. Chem. Soc.* **92,** 7195.
Eaton, W. A., Lovenberg, W., Palmer, G., Fee, J. A., and Kimura, T. (1971). *Proc. Nat. Acad. Sci. U.S.* **68,** 3015.
Ehrenberg, A. cited in Thornley *et al.* (1966).
Feeney, R. E., and Komatsu, St. K. (1966). *Structure Bonding* **1,** 149.
Fox, J. L., and Brown, J. R. (1971). *Fed. Proc.* **30,** 1242 Abstr.
Friedmann, H. C., and Vennesland, B. (1960). *J. Biol. Chem.* **235,** 1526.
Fritz, J., Anderson, R., Fee, J. A., Palmer, G., Sands, R. H., Orme-Johnson, W. H., Beinert, H., Tsibris, J. A., and Gunsalus, I. C. (1971). *Biochim. Biophys. Acta* **253,** 110.
Garbett, K., Darnall, D. W., Klotz, I. M., and Williams, R. J. P. (1969). *Arch. Biochem. Biophys.* **135,** 419.
Garland, P. B. (1970). *Biochem. J.* **118,** 329.
Garland, P. B., Bray, R. C., Clegg, R. A., Haddock, B. A., Light, P. A., Ragan, C. I., Skyrme, J. E., and Swann, J. C. (1970). *Abstr. Int. Congr. Biochem., 8th, Lucerne, Switzerland.*
Gewitz, H. S., and Völker, W. (1962). *Z. Physiol. Chem.* **330,** 124.
Gibson, J. F., and Bray, R. C. (1968). *Biochim. Biophys. Acta* **153,** 721.
Gibson, J. F., Hall, D. O., Thornley, J. H. M., and Whatley, F. R. (1966). *Proc. Nat. Acad. Sci. U.S.* **56,** 987.
Glickson, J. D., Phillips, W. D., McDonald, C. C., and Poe, M. (1971). *Biochem. Biophys. Res. Commun.* **42,** 271.
Green, D. E. (1956). *In* "Enzymes: Units of Biological Structure and Function," Henry Ford Hospital Int. Symp. (O. H. Gaebler, ed.), p. 465. Academic Press, New York.
Green, D. E., and Wharton, D. C. (1963). *Biochem. Z.* **338,** 335.
Groskopf, W. R., Holleman, J. W., Margoliash, E., and Klotz, I. M. (1966). *Biochemistry* **5,** 3783.
Gubler, C. J., Cartwright, G. E., and Wintrobe, M. M. (1957). *J. Biol. Chem.* **224,** 533.
Guillard, R. D., McKenzie, E. D., Mason, R., Mayhew, S. G., Peel, J. L., and Strangroom, J. E. (1965). *Nature (London)* **208,** 769.
Gunsalus, I. C. (1968). *In* "Biochemie des Sauerstoffs" (B. Hess and H. J. Staudinger, eds.), p. 188. Springer-Verlag, Berlin.
Gutman, M., Singer, T. P., and Beinert, H. (1971). *Biochem. Biophys. Res. Commun.* **44,** 1572.
Haddock, B. A., and Garland, P. B. (1971). *Biochem. J.* **124,** 155.
Hall, D. O., Gibson, J. F., and Whatley, F. R. (1966). *Biochem. Biophys. Res. Commun.* **23,** 81.

Hardy, R. W. F., Knight, Jr., E., McDonald, C. C., and D'Eustachio, A. T. (1965). *In* "Non-Heme Proteins: Role in Energy Conversion" (A. San Pietro, ed.), p. 275. Antioch Press, Yellow Springs, Ohio.
Hart, L. I., McGartoll, M. A., Chapman, H. R., and Bray, R. C. (1970). *Biochem. J.* **116,** 851.
Hayaishi, O. (1962). "The Oxygenases." Academic Press, New York.
Hayaishi, O., and Nozaki, M. (1969). *Science* **164,** 389.
Herriott, J. R., Sieker, L. C., Jensen, L. H., and Lovenberg, W. (1970). *J. Mol. Biol.* **50,** 391.
Herskovitz, T., Averill, B. A., Holm, R. H., Ibers, T. A., Phillips, W. D., and Weiler, J. F. (1972). *Proc. Nat. Acad. Sci. U.S.* **69,** 2437.
Hofmann, T., and Harrison, P. M. (1963). *J. Mol. Biol.* **6,** 256.
Imai, K., Asano, A., and Sato, R. (1968). *J. Biochem.* **63,** 219.
Johnson, C. E., Elstner, E., Gibson, J. F., Benfield, G., Evans, M. C. W., and Hall, D. O. (1968). *Nature (London)* **220,** 1291.
Johnson, C. E., Bray, R. C., Cammack, R., and Hall, D. O. (1969). *Proc. Nat. Acad. Sci. U.S.* **63,** 1234.
Johnson, C. E., Rao, K. K., Cammack, R., and Hall, D. O. (1971). *Biochem. J.* **122,** 257.
Kearney, E. B., and Singer, T. P. (1955). *Biochim. Biophys. Acta* **17,** 596.
Kelly, M., and Lang, G. (1970). *Biochim. Biohpys. Acta* **223,** 86.
Keresztes-Nagy, S., Perini, F., and Margoliash, E. (1969). *J. Biol. Chem.* **244,** 981.
Kimura, T. (1968). *Structure Bonding* **5,** 1.
Kimura, T., Tasaki, A., and Watari, H. (1970). *J. Biol. Chem.* **245,** 4450.
King, T. E. (1964). *Biochem. Biophys. Res. Commun.* **16,** 511.
Knight, Jr., E., and Hardy, R. W. F. (1967). *J. Biol. Chem.* **241,** 2752; **242,** 1370.
Lee, S. S., Travis, J., and Black, Jr., C. C. (1970). *Arch. Biochem. Biophys.* **141,** 676.
LeGall, J., DerVartanian, D. V., Spilker, E., Lee, J. P., and Peck, Jr., H. D. (1971). *Biochim. Biophys. Acta* **234,** 525.
Light, P. A., and Garland, P. B. (1971). *Biochem. J.* **124,** 123.
Light, P. A., Ragan, C. I., Clegg, R. A., and Clegg, P. B. (1968). *FEBS Lett.* **1,** 4.
Loach, P. A., Hadsell, R. M., Sekura, D. L., and Stemer, A. (1970). *Biochemistry* **9,** 3127.
Loew, G. H. and Steinberg, D. A. (1971). *Theor. Chim. Acta (Berlin)* **23,** 239.
Losada, M., Whatley, F. R., and Arnon, D. I. (1961). *Nature (London)* **190,** 606.
Lovenberg, W., and Sobel, B. (1965). *Proc. Nat. Acad. Sci. U.S.* **54,** 193.
Lovenberg, W., Buchanan, B. B., and Rabinowitz, J. C. (1963). *J. Biol. Chem.* **238,** 3899.
Ludewig, S., and Franz, S. W. (1970). *Arch. Biochem. Biophys.* **138,** 397.
Lusty, C. J., Machinist, J., and Singer, T. P. (1965). *J. Biol. Chem.* **240,** 1804.
Mahler, H. R., and Elowe, D. (1953). *J. Amer. Chem. Soc.* **75,** 5769.
Mahler, H. R., and Elowe, D. (1954). *J. Biol. Chem.* **210,** 165.
Massey, V. (1957). *J. Biol. Chem.* **229,** 763.
Massey, V. (1958). *Biochim. Biophys. Acta* **30,** 500.

Matsubara, H., Sasaki, R. M., and Chain, R. K. (1967). *Proc. Nat. Acad. Sci.* **57**, 439.

Matsubara, H., Jukes, T. H., and Cantor, C. R. (1968). Structure, Function, and Evolution in Proteins, Brookhaven Symposia in Biology, No. 21, p. 201.

Mayhew, S. G., Foust, G. D., and Massey, V. (1969). *J. Biol. Chem.* **244**, 903.

McFarlane, W. D. (1932). *Biochem. J.* **26**, 1061.

McFarlane, W. D. (1934). *J. Biol. Chem.* **106**, 245.

Moleski, C., Moss, T. H., Orme-Johnson, W. H., and Tsibris, J. C. M. (1970). *Biochim. Biophys. Acta* **214**, 548.

Mortenson, L. E. (1963). *Ann. Rev. Microbiol.* **17**, 115.

Mortenson, L. E., Valentine, R. C., and Carnahan, J. E. (1962). *Biochem. Biophys. Res. Commun.* **7**, 448.

Moss, T. H., Petering, D., and Palmer, G. (1969). *J. Biol. Chem.* **244**, 2275.

Münck, E., Debrunner, P. G., Tsibris, J. C. M., and Gunsalus, I. C. (1972). *Biochemistry* **11**, 855.

Nagami, K. (1972). *Biochem. Biophys. Res. Commun.* **47**, 803.

Nakos, G., and Mortenson, L. E. (1971a). *Biochemistry* **10**, 455; *Biochim. Biophys. Acta* **229**, 431.

Nakos, G., and Mortenson, L. E. (1971b). *Biochemistry* **10**, 2442.

Neilands, J. B. (1966). *Structure Bonding* **1**, 59.

Newman, D. J., Ihle, J. N., and Dure, L. (1969). *Biochem. Biophys. Res. Commun.* **36**, 947.

Nicholas, D. J. D., Wilson, P. W., Heinen, W., Palmer, G., and Beinert, H. (1962). *Nature (London)* **196**, 433.

O'Brien, I. G., and Gibson, J. F. (1970). *Biochim. Biophys. Acta* **215**, 393.

Ohnishi, T. (1970). *Biochem. Biophys. Res. Commun.* **41**, 344.

Ohnishi, T., Schleyer, H., and Chance, B. (1969). *Biochem. Biophys. Res. Commun.* **36**, 487.

Ohnishi, T., Asakura, T., Wohlrab, H., Yonetani, T., and Chance, B. (1970). *J. Biol. Chem.* **245**, 901.

Orme-Johnson, W. H., and Beinert, H. (1969a). *J. Biol. Chem.* **244**, 6143.

Orme-Johnson, W. H., and Beinert, H. (1969b). *Biochem. Biophys. Res. Commun.* **36**, 337.

Orme-Johnson, W. H., Hansen, R. E., Beinert, H., Tsibris, J. C. M., Bartholomaus, R. C., and Gunsalus, I. C. (1968). *Proc Nat. Acad. Sci.* **60**, 368.

Orme-Johnson, N. R., Orme-Johnson, W. H., Hansen, R. E., Beinert, H., and Hatefi, Y. (1971). *Biochem. Biophys. Res. Commun.* **44**, 446.

Orme-Johnson, W. H., Stombaugh, N. A., and Burris, R. H. (1972). *Fed. Proc.* **31**, 448 Abstr.

Palmer, G., and Massey, V. (1969). *J. Biol. Chem.* **244**, 2614.

Palmer, G., and Sands, R. H. (1966). *J. Biol. Chem.* **241**, 253.

Palmer, G., Sands, R. H., and Mortenson, L. E. (1966). *Biochem. Biophys. Res. Commun.* **23**, 357.

Palmer, G., Brintzinger, H., Estabrook, R. W., and Sands, R. H. (1967). *In* "Magnetic Resonance in Biological Systems." (A. Ehrenberg, B. G. Malmström, and T. Vänngård, eds.), p. 159. Pergamon, Oxford.

Palmer, G., Dunham, W. R., Fee, J. A., Sands, R. H., Iikuza, T., and Yonetoni, T. (1971). *Biochim. Biophys. Acta* **245**, 201.

Peterson, J. A., Basu, D., and Coon, M. J. (1966). *J. Biol. Chem.* **241**, 5162.

Poe, M., Phillips, W. D., McDonald, C. C., and Lovenberg, W. (1970). *Proc. Nat. Acad. Sci. U.S.* **65,** 797.

Poe, M., Phillips, W. D., Glickson, J. D., McDonald, C. C., and San Pietro, A. (1971a). *Proc. Nat. Acad. Sci. U.S.* **68,** 68.

Poe, M., Phillips, W. D., McDonald, C. C., and Orme-Johnson, W. H. (1971b). *Biochem. Biophys. Res. Commun.* **42,** 705.

Pollock, J. R., and Neilands, J. B. (1970). *Biochem. Biophys. Res. Commun.* **38,** 989.

Prelog, V. (1963). *Pure Appl. Chem.* **6,** 327.

Ragan, C. I., and Garland, P. B. (1971). *Biochem. J.* **124,** 171.

Rajagopalan, K. V., and Handler, P. (1964). *J. Biol. Chem.* **239,** 1509.

Rajagopalan, K. V., Fridovich, I., and Handler, P. (1962a). *J. Biol. Chem.* **237,** 922.

Rajagopalan, K. V., Aleman, V., Handler, P., Heinen, W., Palmer, G., and Beinert, H. (1962b). *Biochem. Biophys. Res. Commun.* **8,** 220.

Rajagopalan, K. V., Handler, P., Palmer, G., and Beinert, H. (1968). *J. Biol. Chem.* **243,** 3797.

Reiss-Husson, F., De Klerk, H., Jolchine, G., Jauneau, E., and Kamen, M. D. (1971). *Biochim. Biophys. Acta* **234,** 73.

Richert, D. A., and Westerfeld, W. W. (1953). *J. Biol. Chem.* **203,** 915.

Richert, D. A., and Westerfeld, W. W. (1954). *J. Biol. Chem.* **209,** 179.

Rieske, J. S., MacLennan, D. H., and Coleman, R. (1964). *Biochem. Biophys. Res. Commun.* **15,** 338.

San Pietro, A., and Lang, H. M. (1958). *J. Biol. Chem.* **231,** 211.

San Pietro, A. (ed.) (1965). "Non-Heme Iron Proteins: Role in Energy Conversion," p. 15. Antioch Press, Yellow Springs, Ohio.

San Pietro, A., and Black, C. C., Jr. (1965). *Annu. Rev. Plant Physiol.* **16,** 155.

Schapira, G., and Dreyfus, J.-C. (1948). *Bull. Soc. Chim. Biol.* **30,** 82.

Shanmugam, K. T., Buchanan, B. B., and Arnon, D. I., (1972). *Biochim. Biophys. Acta* **256,** 477.

Shethna, Y. I., Wilson, P. W., Hansen, R. E., and Beinert, H. (1964). *Proc. Nat. Acad. Sci. U.S.* **52,** 1263.

Shethna, Y. I., DerVartanian, D. V., and Beinert, H. (1968). *Biochem. Biophys. Res. Commun.* **31,** 862.

Siegel, L. M., and Kamin, H. (1968). *In* "Flavins and Flavoproteins" (K. Yagi, ed.), p. 15. Univ. Park Press, Baltimore, Maryland.

Sieker, L. C., and Jensen, L. H. (1965). *Biochem. Biophys. Res. Commun.* **20,** 33.

Sieker, L. C., Adman, E., and Jensen, L. H. (1972). *Nature (London)* **235,** 40.

Singer, T. P., and Massey, V. (1957). *Rec. Chem. Progr.* **18,** 201.

Smillie, R. M. (1963). *Plant Physiol.* **37,** 716.

Strahs, G., and Kraut, J. (1968). *J. Mol. Biol.* **35,** 503.

Sugeno, K., and Matsubara, H. (1968). *Biochem. Biophys. Res. Commun.* **32,** 951.

Suzuki, K., and Kimura, T. (1965). *Biochem. Biophys. Res. Commun.* **19,** 340.

Tagawa, K., and Arnon, D. I. (1962). *Nature (London)* **195,** 537.

Tanaka, M., Nakashima, T., Benson, A. M., Mower, H. F., and Yasunobu, K. T. (1966). *Biochemistry* **5,** 1666.

Tanaka, M., Haniu, M., and Yasunobu, K. T. (1970). *Biochem. Biophys. Res. Commun.* **39,** 1182.

Tanaka, M., Haniu, M., Matsueda, G., Yasunobu, K. T., Himes, R. H., Akagi, J. M., Barnes, E. M., and Devanathan, T. (1971a). *J. Biol. Chem.* **246**, 3953.

Tanaka, M., Haniu, M., Yasunobu, K. T., Mayhew, S. G., and Massey, V. (1971b). *Biochem. Biophys. Res. Commun.* **44**, 886.

Tanaka, M., Haniu, M., Yasunobu, K. T., and Kimura, T. (1973). *J. Biol. Chem.* **248**, 1141.

Thauer, R. K., Rupprecht, E., and Jungermann, K. (1970). *FEBS Lett.* **8**, 304.

Thornley, J. H. M., Gibson, J. F., Whatley, F. R., and Hall, D. O. (1966). *Biochem. Biophys. Res. Commun.* **24**, 877.

Toan, T., Fehlhammer, W. P., and Dahl, L. F. (1971). *J. Amer. Chem. Soc.* **94**, 3389.

Tottmar, S. O. C., and Ragan, C. I. (1971). *Biochem. J.*, **124**, 853.

Travis, J., Newman, D. J., LeGall, J., and Peck, H. D., Jr. (1971). *Biochem. Biophys. Res. Commun.*, **45**, 452.

Tsai, R. L., Tsibris, J. C. M., Gunsalus, I. C., Orme-Johnson, W. H., Hansen, R. E., and Beinert, H. as quoted by Tsibris and Woody (1970).

Tsai, R. L., Gunsalus, I. C., and Dus, K. (1971). *Biochem. Biophys. Res. Commun.* **45**, 1300.

Tsibris, J. C. M., Tsai, R. L., Gunsalus, I. C., Orme-Johnson, W. H., Hansen, R. E., and Beinert, H. (1968). *Proc. Nat. Acad. Sci. U.S.* **59**, 959.

Tsibris, J. C. M., and Woody, R. W. (1970). *Coordination Chem. Rev.* **5**, 417.

Tsunoda, J. N., Yasunobu, K. T., and Whiteley, H. R. (1968). *J. Biol. Chem.* **243**, 6262.

Uozumi, M., Hayashikawa, R., and Piette, L. H. (1967). *Arch. Biochem. Biophys.* **119**, 288.

Valentine, R. C., Jackson, R. L., and Wolfe, R. S. (1962). *Biochem. Biophys. Res. Commun.* **7**, 453.

Vetter, H., Jr., and Knappe, J. (1971). Hoppe-Seyler's *Z. Physiol. Chem.* **352**, 433.

Villafranca, J. J., and Mildvan, A. S. (1971). *J. Biol. Chem.* **246**, 772.

Watenpaugh, K. D., Sieker, L. C., Herriott, J. R., and Jensen, L. H. (1971). *Cold Spring Harbor Symp. Quant. Biol.* **36**, 359.

Wei, C. H., and Dahl, L. F. (1965). *Inorg. Chem.* **4**, 1.

Wei, C. H., Wilkes, G. R., Treichel, P. M., and Dahl, L. F. (1966). *Inorg. Chem.* **5**, 900.

Weinstein, B. (1969). *Biochem. Biophys. Res. Commun.* **35**, 109.

Whiteley, H. R., and Woolfolk, C. A. (1962). *Biochem. Biophys. Res. Commun.* **9**, 517.

Yamashita, S., and Racker, E. (1969). *J. Biol. Chem.* **244**, 1220.

Zeylemaker, W. P., DerVartanian, D. V., and Veeger, C. (1965). *Biochim. Biophys. Acta* **99**, 183.

Zumft, W. G., and Spiller, H. (1971). *Biochem. Biophys. Res. Commun.* **45**, 112.

CHAPTER 2

Bacterial Ferredoxins and/or Iron–Sulfur Proteins as Electron Carriers

LEONARD E. MORTENSON and GEORGE NAKOS

I. INTRODUCTION

Bacterial ferredoxin (Fd) was discovered in 1960 at which time it was found to be required as an electron carrier in key oxidation–reduction reactions in *Clostridium pasteurianum* (Mortenson *et al.*, 1962, 1963). Prior to the discovery of bacterial ferredoxin, a plant protein, photosynthetic pyridine nucleotide reductase (now known as plant ferredoxin), was shown to be required for transfer of electrons from the light adsorption–electron excitation event of photosynthesis to $NADP^+$

(San Pietro, 1961). This chapter (1) reviews and discusses the role of bacterial ferredoxins in coupled oxidation–reduction reactions, (2) describes a "new" iron–sulfur protein, hydrogenase, and its role in oxidation–reduction, and (3) describes two additional bacterial iron–sulfur proteins, azoferredoxin (AzoFd or Fe protein) and molybdoferredoxin (MoFd or Mo–Fe protein).

II. DEMONSTRATION OF FERREDOXIN-REQUIRING REACTIONS

Based on the anionic properties of Fd, a method was devised (Mortenson, 1963, 1964a) for demonstrating a requirement for Fd in a reaction. A crude extract of bacterial cells is prepared and passed through a column (usually under anaerobic conditions) of diethylaminoethyl cellulose (DEAE cellulose). The exchange capacity of the DEAE cellulose column must be such that of the total protein introduced only about 5% is adsorbed. Ferredoxin, because of its high affinity for DEAE cellulose, binds with the "5%" protein fraction and can be seen as a brown band at the top of the column. When the protein solution, unadsorbed by DEAE cellulose under these conditions, is analyzed for the activity in which Fd is suspected to play a role, it *usually* will have little or no activity unless Fd is added. The requirement for Fd in many of the activities in Table I was demonstrated in this manner. Since there are other proteins such as the flavodoxins (Knight and Hardy, 1966) that function in many of the roles where Fd functions and these proteins bind to DEAE cellulose less avidly than Fd, *all* activity may not be removed from the crude extract by the DEAE column procedure. Obviously, to rigorously establish the role of Fd or flavodoxin in oxidation–reduction reactions, one must purify both the electron acceptors and the oxidation–reduction catalyst. This has been accomplished in some of the reactions listed in Table I.

Ferredoxin has been found in bacteria of widely different types. The function of some isolated "ferredoxins" is unknown (Table I). In general, it can be said that most organisms whose metabolism involves molecular hydrogen, possess one or more ferredoxins. Most organisms that have no cytochromes contain ferredoxins, although "ferredoxin" is found in some organisms that do contain cytochromes, i.e., *Azotobacter vinelandii* (Shethna *et al.*, 1966; Yoch *et al.*, 1970; DerVartanian *et al.*, 1969). Although most organisms have not been examined for the presence of Fd, it is likely that one or more iron–sulfur proteins that serve as electron carriers will be found in all organisms.

TABLE I
FERREDOXIN-MEDIATED BIOLOGICAL OXIDATION–REDUCTION REACTIONS

Source	Function	References
1. Microbial		
Clostridium	Pyruvate + $P_i \xrightarrow{CoA}$ acetyl phosphate + CO_2 + H_2	Mortenson *et al.* (1962, 1963)
	$2\ H^+ + 2\ e^-(S_2O_4^{2-}) \rightleftarrows H_2$	Mortenson *et al.* (1962); Valentine *et al.* (1963b)
	$H_2 + NADP^+ \rightleftarrows$ NADPH + H^+	Valentine *et al.* (1962b); Whiteley and Woolfolk (1962); Jungermann *et al.* (1969)
	Pyruvate + $NADP^+ \rightarrow$ NADPH	Valentine *et al.* (1962b)
	Pyruvate + $NAD^+ \rightarrow$ NADH	Valentine *et al.* (1963a)
	Pyruvate + urate → xanthine	Valentine *et al.* (1963a)
	$H_2 + NO_2^-$ (or NH_2OH) $\rightarrow NH_3$	Mortenson *et al.* (1962); Valentine and Wolfe (1963); Valentine *et al.* (1963c)
	$N_2 + 3\ H_2 \rightarrow 2\ NH_3$	Mortenson *et al.* (1963); Mortenson (1964b); D'Eustachio and Hardy (1964)
	Acetyl phosphate + CO_2 → pyruvate	Bachofen *et al.* (1964); Andrew and Morris (1965); Raeburn and Rabinowitz (1965)
	Sulfite → sulfide	Akagi (1965)
	Propionyl-CoA + CO_2 → ketobutyrate + CoA	Buchanan (1969)
	6-Hydroxynicotinic acid ⇄ 6-oxo-1,4,5,6-tetrahydronicotinic acid	Holcenberg and Tsai (1969)
	2 NADH $\xrightarrow{\text{Acetyl CoA}}$ 2 NAD^+ + H_2	Thauer *et al.* (1969); Mortenson (1968)
	Formation of one carbon unit from CO_2	Thauer *et al.* (1970)
	CO_2 ⇄ formate	Jungermann *et al.* (1970)
Azotobacter	$N_2 + 3\ H_3 \rightarrow 2\ NH_3$ (?)	Yoch *et al.* (1969); DerVartanian *et al.* (1969); Shethna *et al.* (1966)
Micrococcus	$H_2 + NADP^+ \rightarrow$ NADPH + H^+	Whiteley and Woolfolk (1962)
	Hypoxanthine → xanthine + H_2	Valentine *et al.* (1962a)
	$2\ H^+ + 2\ e^-(S_2O_4^{2-}) \rightleftarrows H_2$	Valentine *et al.* (1962a)
	Pyruvate + $P_i \xrightarrow{CoA}$ acetyl phosphate + H_2 + CO_2	Valentine and Wolfe (1963)

TABLE I (*Continued*)

Source	Function	References
	α-Ketoglutarate $\rightarrow H_2$ + unknown products	Valentine and Wolfe (1963)
	FMN (or FAD) + $H_2 \rightarrow$ $FMNH_2$ (or $FADH_2$)	Whiteley and Woolfolk (1962)
	Riboflavin + $H_2 \rightarrow$ reduced flavin	Whiteley and Woolfolk (1962)
Methanobacillus	Acetaldehyde $\rightarrow$ acetate	Buchanan *et al.* (1964)
Pseudomonas[a]	Methylene hydroxylation	Cushman and Gunsalus (1966); Cushman *et al.* (1967)
Chromatium	Acetyl phosphate + $CO_2 \xrightarrow{CoA}$ pyruvate + P_i	Buchanan *et al.* (1964)
	$H_2 + NAD^+ \rightarrow NADH + H^+$	Weaver *et al.* (1965)
	Not known (high-potential iron proteins)	Bartsch (1963); Dus *et al.* (1967)
	$H_2 + NADP^+ \rightarrow NADPH + H^+$	Buchanan and Bachofen (1968)
	Propionyl-CoA + $CO_2 \rightarrow$ α-ketobutyrate + CoA	Buchanan (1969)
	$N_2 + 3\ H_2 \rightarrow 2\ NH_3$	Yoch and Arnon (1970)
Rhizobium	Mediate electron transfer from illuminated chloroplasts to nitrogenase	Koch *et al.* (1970); Yoch *et al.* (1970)
2. Plants	Photoreduced Fd $\xrightarrow{H_2ase} H_2$	Tagawa and Arnon (1962)
	$NADP^+$ reduction by light	Tagawa and Arnon (1962); Shin and Arnon (1965)
	Photoproduction of O_2	Arnon *et al.* (1964)
	ATP formation (cyclic photophosphorylation)	Tagawa *et al.* (1963a, b)
	ATP formation (noncyclic photophosphorylation)	Arnon *et al.* (1964)
	NO_2^- reduction by light	Ramirez *et al.* (1966); Joy and Hagelman (1966)
	$NO_3^- \rightarrow NO_2^-$	Losada and Pomeque (1966)
	Fructose diphosphatase activation	Buchanan *et al.* (1967)
	Sulfite $\rightarrow$ sulfide	Schmidt and Trest (1969)
3. Animals[b]	Deoxycorticosterone hydroxylation	Omura *et al.* (1965); Suzuki and Kimura (1965); Kimura and Suzuki (1967); Kimura *et al.* (1967)

[a] Putidaredoxin.
[b] Adrenodoxin.

III. PURIFICATION OF FERREDOXIN

There are several methods for purifying Fd (Mortenson *et al.*, 1962; Mortenson, 1964a; Lovenberg *et al.*, 1963). The method one selects depends on what proteins in the crude extract other than Fd one would also like to obtain. If only Fd is desired, the quickest way to prepare it in large scale is to treat a suspension of resuspended dry or wet cells with 50–60% cold acetone (other solvents also work) followed by mixing, centrifuging the precipitated proteins, and collecting the brownish yellow supernatant solution. Some enzymes still can be recovered from the acetone precipitate, if desired, but others are denatured by the acetone. To the resulting 50–60% cold acetone solution, DEAE cellulose (either dry or as a wet slurry) is added in batches with stirring until the Fd is just removed (judged by color). The DEAE with bound Fd (and other contaminants) is easily and rapidly removed from the acetone by first decanting and then filtering. The Fd is then eluted from the DEAE cellulose with a buffered solution of NaCl ($Cl^- > 0.5$ *M*). The remaining steps of Fd purification include ammonium sulfate fractionation and gel filtration. Some enzymes in the acetone precipitate can be recovered. However, if this method destroys other desirable enzymes, one can either use the earlier method (Mortenson *et al.*, 1962), where large amounts of crude extract are passed through DEAE columns and all the protein can be recovered, or a recently reported method where the bulk of the protein of the crude extract is removed by ammonium sulfate precipitation (up to 60% saturation) and then the Fd is removed from the ammonium sulfate solution by passing it through a DEAE column (Mayhew, 1971). The latter method is based on the discovery that proteins can be absorbed on DEAE cellulose from ammonium sulfate solution (Mortenson, 1964a; Lovenberg and Sobel, 1965). The ammonium sulfate precipitate can be resolubilized and other enzymes recovered from it.

Ferredoxin from many sources has been purified and partially characterized (Volume II, Chapter 2). Most purifications used the methods briefly described above or modifications of these methods.

IV. REACTIONS THAT REQUIRE FERREDOXIN

Ferredoxin functions primarily as an electron carrier. It accepts electrons from an enzyme-catalyzed oxidation and in turn becomes oxidized by another enzyme involved in electron transport. A compilation of such reactions in different organisms is given in Table I. These reactions can be divided into three separate groups: (a) reactions in which H_2 is evolved, (b) oxidations coupled via Fd to reduction of other carriers, and (c) reactions where H_2 is oxidized by hydrogenase and the electrons are used via Fd for a variety of reductions.

A. Hydrogen Evolution

Any organism that oxidizes a substrate to produce reduced Fd and contains hydrogenase probably evolves H_2 as a part of its metabolism. The general reaction for this coupled reaction and examples are given in Table II.

Hydrogen evolution by anaerobic bacteria represents a final stage of electron transport where H^+ acts as a terminal electron acceptor. These organisms also excrete various organic compounds that are more reduced than the starting carbon substrate (usually a sugar). This excretion also represents a final stage in electron transport in which electrons from NADH produced during glycolysis are transferred to reductases where they are used to reduce various intermediates or end products of glycolysis (Wood, 1961). An example is the several-step production

TABLE II

EXAMPLES OF HYDROGEN-EVOLVING REACTIONS REQUIRING FERREDOXIN

Model reaction

$$\text{Substrate} + \text{Fd}^{a} \underset{}{\overset{\text{dehydrogenase}}{\rightleftharpoons}} \text{Fd}(2\,e^-) + \text{oxidized products}$$

$$\text{Fd}(2\,e^-) \underset{\pm 2\text{H}^+}{\overset{\text{hydrogenase}}{\rightleftharpoons}} \text{Fd} + \text{H}_2$$

Substrate	Products	Examples of organisms (in which reaction is found)
Pyruvate	Acetyl S $\sim$ CoA, CO_2, H_2	*C. pasteurianum*
		M. lactilyticus
		B. rettgeri
		P. elsdenii
		C. lactoacetophilum
		C. acidi-urici
α-Ketoglutarate	H_2 + unidentified products	*M. lactilyticus*
Hypoxanthine	H_2 + xanthine	*M. lactilyticus*
α-Ketobutyrate	H_2 + unidentified products	*C. pasteurianum*
		M. lactilyticus

[a] Assuming bacterial ferredoxins are two-electron acceptors; if one electron-accepting "plant" ferredoxins are used, 2 Fd would be required.

of butyric acid from acetyl S∼CoA and NADH in certain clostridia (Stadtman and Barker, 1950; Mortenson, 1962, 1963).

An example of an oxidation–reduction that is coupled primarily to H_2 evolution is the "phosphoroclastic" reaction (Fig. 1). In this reaction, pyruvate, a product of glycolysis, is oxidized by the pyruvic acid dehydrogenase system. The reaction is completely dependent on the presence of coenzyme A (CoA) (Wolfe and O'Kane, 1953) and Fd (Mortenson *et al.*, 1962, 1963), and the products are acetyl S∼CoA, CO_2, and reduced Fd. Most of the reduced Fd is oxidized by hydrogenase and the electrons transferred to H^+ to produce H_2 (see Section VI). Smaller amounts of the reduced Fd are used for the reduction to sulfite to sulfide, $NADP^+$ to NADPH, and N_2 to ammonia.

Finally, and very significantly, electrons are transferred from NADH to Fd to produce reduced Fd even though the potential of the half-cell $NADH \rightleftharpoons NAD^+ + 2\,e^- + H^+$ (−320 mV) is more positive than the half-cell, reduced $Fd \rightleftharpoons Fd + 2\,e^-$ (−420 mV). This transfer occurs because during fermentation the ratio of NADH to NAD^+ is high, whereas the ratio of reduced Fd to Fd is low. Ferredoxin reduced by NADH:Fd oxidoreductase then is oxidized by hydrogenase to yield H_2, and the process represents a partial electron transport system in some anaerobes (Mortenson, 1968; Thauer *et al.*, 1969). This process allows ATP to be made at the expense of some of the acetyl S∼CoA rather than all the acetyl S∼CoA being used to produce butyrate (Fig. 1). In *Clostridium pasteurianum* this reaction requires acetyl S∼CoA as an effector; in other organisms (Table I) no effector appears to be required, possibly because the systems examined were not free from CoA.

B. Coupled Reactions

Substrate oxidations can be coupled by the mediation of Fd to the reduction of a variety of biological oxidation–reduction catalysts (coenzymes). Examples of these are the reduction of $NADP^+$, NAD^+, FMN, FAD, and riboflavin (Table I). Some and possibly all of these reduced Fd-coupled reductions are catalyzed by specific reductases (probably flavoproteins). This has been specifically shown for the reduction of $NADP^+$ and NAD^+ (Valentine and Wolfe, 1963), where reduced Fd is oxidized by a flavoprotein which in turn reduces $NADP^+$ and NAD^+ to NADPH and NADH. These reactions could be required by the cell to produce needed NADPH and possibly at times NADH, although NADH is produced in plentiful supply by most organisms during metabolism (fermentation) of the carbon source. These are important reactions in photosynthetic organisms since reduction of Fd coupled to excitation of chlorophyll by light is in turn coupled to the reduction of

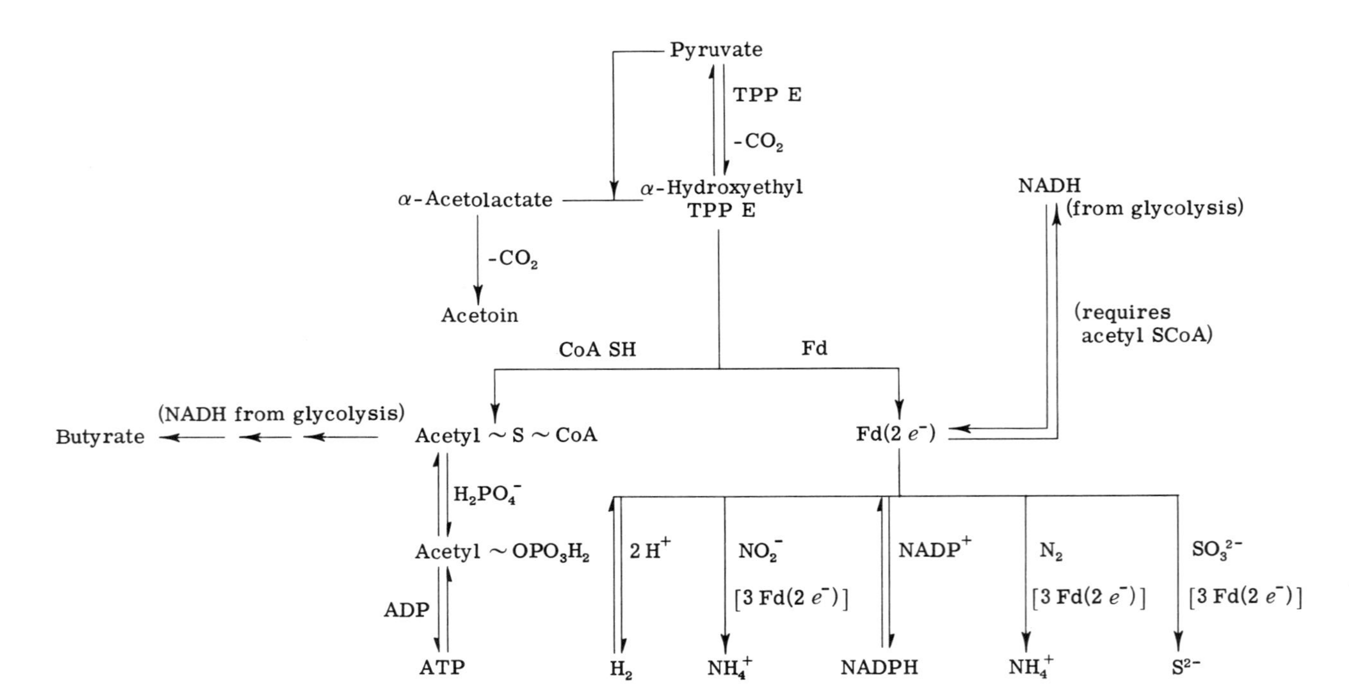

Fig. 1. Some ferredoxin reactions of *C. pasteurianum*. Amount of NADH oxidized in this reaction is equal to the amount of acetyl phosphate formed from acetyl S~CoA (Daesch and Mortenson, 1968; Mortenson, 1968). The remainder of NADH from glycolysis is used for the synthesis of butyrate.

nicotinamide adenine dinucleotides (Tagawa and Arnon, 1962; Shin and Arnon, 1965). The presence of NADH:Fd oxidoreductase in fermentative organisms, where H_2 is evolved, is required not to produce but to oxidize NADH (see Section IV,A) so that the oxidized coenzyme can function again in metabolism to permit continuous and maximum substrate level phosphorylation (Mortenson, 1968; Thauer *et al.*, 1969).

Reduced Fd produced during carbohydrate metabolism also is coupled directly as an electron source to several enzyme-catalyzed reductions. Examples of these are the reduction of N_2 to ammonia by nitrogenase, reduction of sulfite to sulfide by sulfite reductase, and the reduction of nitrite to ammonia by nitrite reductase (Table I).

C. Hydrogen as Reductant

Under conditions where a fermentative organism is growing and producing H_2, the organism has a plentiful supply of reduced Fd and

TABLE III
HYDROGENASE–FERREDOXIN–COUPLED REDUCTIONS

Model Reaction

$$H_2 + \text{Hydrogenase} \overset{\text{Fd}}{\rightleftharpoons} \text{Fd}(2\,e^-) + 2\,H^+$$

$$\text{Fd}(2\,e^-) + \text{substrate} \rightarrow \text{reduced substrate} + \text{Fd}$$

Substrate[a]	Product	Organism
NO_2^-	NH_4^+	*C. pasteurianum*
NH_2OH	NH_4^+	*C. pasteurianum*
		M. lactilyticus
$NADP^+$	NADPH	*C. pasteurianum*
		M. lactilyticus
NAD^+	NADH	*C. acidi-urici*
FMN, FAD	$FMNH_2$, $FADH_2$	*M. lactilyticus*
Riboflavin	Riboflavin H_2	*M. lactilyticus*
Inorganic anions	Reduced anion	*M. lactilyticus*
Pyruvate	Lactate	*M. lactilyticus*
Uric acid	Hypoxanthine	*M. lactilyticus*
SO_2^{2-} [b]	S^{2-}	*C. pasteurianum*
		C. thermoaceticum
N_2	NH_4^+	*C. pasteurianum*

[a] In some of these reactions there may be an intermediate reaction whereby Fd$(2\,e^-)$ is oxidized by an oxidoreductase and the electrons transferred to another carrier such as NAD^+. The NADH (or other carrier) produced would then be oxidized and substrate reduced.

[b] Probably the same enzyme catalyzing $NO_2^- \rightarrow NH_4^+$.

does not need to produce reduced Fd via the hydrogen–hydrogenase system. However, in some situations where substrate is low and H_2 is available, reduced Fd, needed for cellular reduction, could be produced from H_2 gas. At any rate, in cell-free extracts many biochemical reactions requiring reduced Fd have been supported with H_2 as the initial reductant. Examples of these are given in Table III. It has recently been shown, for example, that the reductant (reduced Fd) for N_2 fixation and acetylene reduction by pure nitrogenase can be supplied by pure hydrogenase, H_2 and Fd (G. Nakos and L. E. Mortenson, unpublished results).

V. NUMBER OF ELECTRONS CARRIED BY FERREDOXINS

Early studies established that Fd was an electron carrier but did not establish how many electrons were accepted by Fd (Mortenson *et al.*, 1962, 1963). Tagawa and Arnon (1962 and later in 1968) suggested that clostridial Fd was a one-electron acceptor. Later it was established that spinach Fd was a one-electron acceptor (Whatley *et al.*, 1963; Horio and San Pietro, 1964). However, in a comprehensive study, Sobel and Lovenberg (1966) established that (1) the ferredoxin from *Clostridium pasteurianum* was a two-electron acceptor and not a one-electron acceptor, (2) that oxidized Fd had about 50% ferrous iron (four Fe^{2+} and four Fe^{3+}) and this changed to about 75% after reduction (six Fe^{2+} and two Fe^{3+}), and (3) that free SH groups did not appear during reduction. Several different lines of evidence were used to establish that clostridial Fd was a two-electron acceptor: (a) reduction with NADPH coupled to a reduced Fd:NADP oxidoreductase, (b) release and measurement of iron before and after reduction, (c) reduction by the hydrogen–hydrogenase couple, (d) potentiometric titration with sodium dithionite, and (e) reduction of Fd by light and chloroplasts to which $NADP^+$ was added anaerobically and the amount of NADPH produced determined.

Similar techniques were used to determine that ferredoxin from *Anacystis nidulans* and spinach Fd are one-electron acceptors, whereas ferredoxin from *Clostridium welchii* and *Chromatium* are two-electron carriers (Evans *et al.*, 1968).

Ferredoxins reduced by either dithionite or coupled enzymatic reactions exhibit electron paramagnetic resonance spectra but only at low temperatures (Beinert and Lee, 1961; Palmer and Sands, 1966; Palmer *et al.*, 1966). Orme-Johnson and Beinert (1969c) devised a technique for titration of ferredoxins (or other anaerobic titrations) with the use of a solid dithionite. With this technique they titrated various ferredoxins to their maximum electron paramagnetic resonance (EPR) signal, then determined the amount of dithionite required and by double integration of the resulting EPR signal, the number of electrons involved in the signal. With six different "ferredoxins" containing two iron atoms and two labile sulfide groups and one "ferredoxin" with four atoms each of iron and labile sulfide, one-electron was required for maximum reduction (Orme-Johnson and Beinert, 1969a).

The same technique was used to titrate the Fd of *Clostridium pasteurianum* which had previously been shown to have a complex EPR signal (Palmer *et al.*, 1966) and to be a two-electron acceptor (Sobel and Lovenberg, 1966) or possibly a one-electron acceptor (Tagawa and Arnon, 1968). It was found that there are two EPR signals that appear at different points during the titration and that all spectral changes that occurred with increasing dithionite ceased when the equivalent of two electrons had been added. This elegant experiment of Orme-Johnson and Beinert (1969b) both confirmed and extended the results that showed clostridial Fd to be a two-electron acceptor and at the same time explained the complexity of the EPR signal.

From the above facts, one can now conclude with reasonable confidence that each molecule of the bacterial-type ferredoxins that contains eight iron atoms and function in the range of —400 mV, accepts the two electrons removed from pyruvate during pyruvate oxidation and also the two electrons from the hydrogen–hydrogenase couple during H_2 oxidation (Fig. 1). Whether or not this type of bacterial Fd also functions in one-electron transfer reactions remains to be established. The ferredoxins that contain two iron atoms appear to function only as single electron carriers. A discussion of these ferredoxins will be included with other chapters of this treatise.

VI. HIGH MOLECULAR WEIGHT IRON–SULFUR PROTEINS INVOLVED IN ELECTRON TRANSPORT

Table IV lists several high molecular weight iron–sulfur proteins that have been shown to be involved in oxidation–reduction reactions. Some of these will be discussed in Chapters 9 and 10 of this volume. They have been included here so that one can compare them with the reactions in Table I. In some cases one might consider the iron sulfide structure of these proteins to represent a "built-in Fd." Three of these larger molecular weight iron–sulfur proteins, molybdoferredoxin, azoferredoxin,

TABLE IV

HIGH MOLECULAR WEIGHT IRON–SULFUR PROTEINS INVOLVED IN BIOLOGICAL OXIDATION–REDUCTION REACTIONS

Protein	Function	Molecular weight	Source	References
NADH dehydrogenase	Electron transport Oxidative phosphorylation	80,000	Hog heart	Mahler *et al.* (1952) Mahler and Elowe (1954); Singer (1963)
Xanthine oxidase	Xanthine + $O_2 \rightarrow$ urate, oxidizes a wide variety of purines and aldehydes	300,000	Bovine milk	Aleman *et al.* (1965)
			Mammalian liver	DeRenzo (1956); Hart and Bray (1967)
			Chicken liver	Rajagopalan and Handler (1967)
			Micrococcus lactilyticus	Smith *et al.* (1967)
Succinate dehydrogenase	Electron transport oxidative phosphorylation	200,000	Bovine heart	Singer *et al.* (1957); Kearney (1960)
			Hog heart	Zeylemaker *et al.* (1965)
			Yeast	Singer *et al.* (1957)
Aldehyde oxidase	Aldehyde $\rightarrow O_2$	300,000	Hog liver	Rajagopalan *et al.* (1962)
Dihydroorotate dehydrogenase	Orotate $\rightarrow$ dihydroorotate	115,000	*Zymobacterium oroticum*	Friedmann and Vennesland (1960); Aleman and Handler (1967)
Ubiquinone-cytochrome c reductase	Electron transport oxidative phosphorylation	26,000	Beef heart mitochondria	Rieske *et al.* (1964); Coleman *et al.* (1964)
Azoferredoxin (Fe protein)	N_2 fixation	55,000 (dimer)	*Clostridium pasteurianum*	Mortenson *et al.* (1967); Moustafa and Mortenson (1969); Taylor (1969); Nakos and Mortenson (1971a); Vandecasteele and Burris (1970)
		?	*Azotobacter vinelandii*	Bulen and LeComte (1966); Moustafa (1970)
		?	*Azotobacter chroococcum*	Kelly (1969)

		51,000	Soybean nodules	Klucas *et al.* (1968); Bergersen and Turner (1970)
		?	*Bacillus polymyxa*	Witz and Wilson (1967)
		66,000 (dimer)	*Klebsiella pneumoniae*	Kelly and Lang (1970); Postgate (personal communication)
Molybdoferredoxin (Fe–Mo protein)	N_2 fixation	220,000 (tetramer) 110,000 (dimer)	*Clostridium pasteurianum*	Mortenson *et al.* (1967); Taylor (1969); Vandecasteele and Burris (1970); Nakos and Mortenson (1971b); Dalton *et al.* (1971); recent unpublished data
		270,000 (dimer)	*Azotobacter vinelandii*	Bulen and LeComte (1966); Burns *et al.* (1970)
		?	*Azotobacter chroococcum*	Kelly (1969)
		182,000	Soybean nodules	Klucas *et al.* (1968); Bergersen and Turner (1970)
		?	*Bacillus polymyxa*	Witz and Wilson (1967)
		220,000 (tetramer) 110,000 (dimer)	*Klebsiella pneumoniae*	Kelly and Lang (1970)
Hydrogenase	Sulfate reduction ?	45,000	*Desulfovibrio vulgaris*	Haschke and Campbell (1971)
		60,000	*D. vulgaris*	LeGall *et al.* (1971)
	$2H^+ + 2e^- \rightleftarrows H_2$ (Reduced Fd $+ 2H^+ \rightleftarrows H_2$)	60,000 (dimer)	*Clostridium pasteurianum*	Nakos and Mortenson (1971c,d)

and hydrogenase, will be discussed in more detail in Sections VII and VIII.

VII. HYDROGENASE

Hydrogenase (hydrogen:ferredoxin oxidoreductase EC 1.12.1.1) is involved in hydrogen metabolism and electron transport of a large number of microorganisms (Gray and Gest, 1965), of certain plants (Renwick *et al.*, 1964), and probably of animals (Kuruta, 1962). Hydrogenase from *Clostridium pasteurianum* couples the oxidation of hydrogen gas (H_2 uptake) to the reduction of ferredoxin and certain dyes (Mortenson *et al.*, 1962; Valentine *et al.*, 1962a) and couples the oxidation of reduced Fd to the reduction of protons (H_2 evolution). Hydrogenase was recently described as an iron–sulfur protein with some properties similar to other iron–sulfur proteins (Nakos and Mortenson, 1971c,d).

A. Purification

Purification procedures for hydrogenase have been described for the bacteria, *Clostridium butylicum* (Peck and Gest, 1957), *Desulfovibrio desulfuricans* (Sadana and Jaganathan, 1956; Sadana and Morey, 1961; Riklis and Rittenberg, 1961; Yagi *et al.*, 1968), *Desulfovibrio vulgaris* (Haschke and Campbell, 1971), and *Proteus vulgaris* (Schengrurd and Krasna, 1969). Partial purification of hydrogenase from *Clostridium pasteurianum* has also been reported (Shug *et al.*, 1954; Yagi *et al.*, 1968; Kleiner and Burris, 1970).

Recently, Nakos and Mortenson (1971c) described a purification procedure for hydrogenase from *C. pasteurianum*. In this method all the purification steps are performed in an air-free atmosphere under H_2 gas. Cells of *C. pasteurianum* are grown with either N_2 or NH_3 as the sole source of nitrogen, and cell-free extracts are prepared according to Mortenson (1964b). The crude extract is treated with 3% (by weight of the protein) of protamine sulfate to remove nucleic acids and nucleoproteins. The supernatant solution from the protamine sulfate treatment is heated for 10 minutes at 55°C in a water bath. Insoluble proteins are removed by centrifugation, and the supernatant solution is loaded on a DEAE-cellulose column packed and equilibrated anaerobically with 0.05 *M* tris-HCl pH 8 buffer. The absorbed proteins are eluted with a linear KCl gradient, and the protein solution from the fractions containing hydrogenase activity is concentrated and passed through a Sephadex G-100 column. From the two bands resolved on elution with 0.05 *M* tris-HCl pH 8 buffer the slower moving band contains all the hydrogenase activity. Finally, the protein from the slower band is concentrated and further fractionated on a longer Sephadex G-100 column, and, of the two closely running protein bands seen, the first band contains all of the hydrogenase activity.

At this stage of purification, hydrogenase has a specific activity, in the hydrogen evolution assay, of about 70 μmoles/mg protein/minute, and it shows only one protein band on analytical polyacrylamide disc gel electrophoreis with a coincident activity band and one peak in the analytical ultracentrifuge. It contains four iron atoms and four "acid-labile" sulfide groups per molecule.

Ackrell *et al.* (1966), Kidman *et al.* (1969), and Kleiner and Burris (1970) have reported that crude extracts from *C. pasteurianum* or partially purified hydrogenase from the organism shows three to five hydrogenase activity bands after electrophoresis on polyacrylamide gels. These bands were interpreted to be "isoenzymes" of hydrogenase. Subsequent studies with a combination of crude extracts and purified enzyme suggest that the activity bands seen on disc gels of crude extracts are artifacts. Addition of pure hydrogenase (one band in disc gel electrophoresis) to crude extracts from which hydrogenase has been removed and which show no activity bands on disc gel, results in multiple bands similar to those seen in a crude extract (Nakos and Mortenson, 1971d).

B. Properties

1. pH Optimum, Heat, and Oxygen Sensitivity

The optimum pH in the methyl viologen assay (H_2 evolution) is about 7.8, whereas that in the methylene blue assay is about 7.0. Pure hydrogenase is stable for several months on storage at 0°–4°C under H_2 or in liquid nitrogen. Hydrogenase is completely inactivated when heated at 75°C for 1–2 minutes under a hydrogen atmosphere. Exposure of the enzyme to air (O_2) for 50–60 minutes results in its complete inactivation (Nakos and Mortenson, 1971c).

2. Absorption Spectra

The absorption spectrum of hydrogenase (as isolated) is shown in Fig. 2. The molar extinction coefficients at 280 nm and 400 nm are 24.5×10^3 and 8.2×10^3, respectively. When the enzyme is exposed to air (O_2) for about $\frac{1}{2}$ hour, little change in the absorption at 400 nm was observed. Longer treatment in air increased the absorption, and a shoulder appeared at 315 nm. Treatment of the isolated enzyme

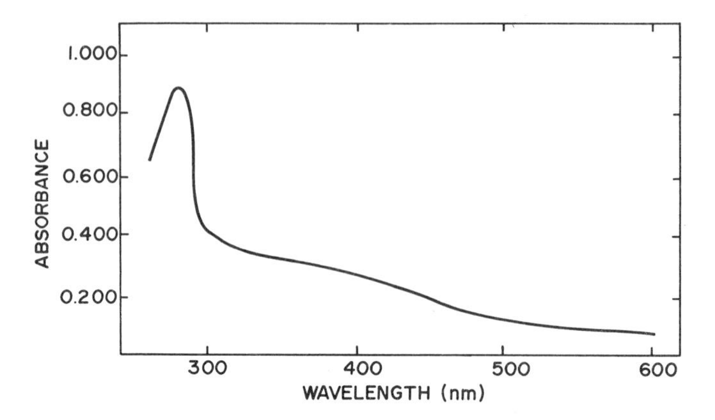

Fig. 2. The absorption spectrum of hydrogenase under anaerobic conditions. Protein concentration is 2.2 mg/ml. The solvent was 0.05 *M* tris-HCl, pH 8. The spectrum was taken in a Cary-14 recording spectrophotometer (from Nakos and Mortenson, 1971d).

with a 40 mole excess of dithionite caused a 10% reduction in its millimolar extinction coefficient at 400 nm. Sodium mersalyl treatment or dialysis against sodium dodecyl sulfate or urea drastically decreased the absorption of hydrogenase in the visible region. This is accompanied by loss of its iron, "acid-labile" sulfide and enzymatic activity (Nakos and Mortenson, 1971d).

3. Electron Paramagnetic Resonance (EPR) Spectra

The EPR spectrum of hydrogenase (as isolated) is shown in Fig. 3A. This signal of the $g = 1.94$ type disappears when the enzyme is exposed to air (Fig. 3B) and a new but smaller signal at about $g = 2.01$ appears. As seen in Fig. 3C, the original signal is almost completely restored when the air-oxidized hydrogenase (Fig. 3B) is incubated with either dithionite or H_2 (G. Nakos, L. E. Mortenson, and G. Palmer, previously unpublished).

4. Molecular Weight and Subunit Structure

The molecular weight of hydrogenase from *C. pasteurianum* is 60,000 as estimated by gel filtration on Sephadex columns (Yagi *et al.*, 1968; Kidman *et al.*, 1969; Nakos and Mortenson, 1971c). Molecular weights of 90,000 and 45,000 for hydrogenase from *Desulfovibrio desulfuricans* (Riklis and Rittenberg, 1961) and *D. vulgaris* (Haschke and Campbell, 1971), respectively, have also been reported. Hydrogenase dissociates into two subunits of identical size upon treatment of the protein with 1.0% sodium dodecyl sulfate plus 1.0% 2-mercaptoethanol followed by carboxymethylation. The molecular weight of the subunits estimated by the method of Shapiro *et al.* (1967) was found to be 30,000 (Nakos and Mortenson, 1971d). Electrophoresis of native hydrogenase on polyacrylamide gels made 8.0 *M* in urea does not dissociate hydrogenase into subunits, i.e., only one band is seen. This band on the urea gel corresponds to the native enzyme and shows hydrogenase activity when tested with the method of Ackrell *et al.* (1966). However, when the native enzyme is first dialyzed against 8.0 *M* urea containing 0.1 *M* 2-mercaptoethanol for 12 hours and then electrophoresed on polyacrylamide gel made 8.0 *M* in urea, a partial dissociation of hydrogenase into subunits is observed (Nakos and Mortenson, 1971d). Most of the enzyme still remains as the dimer and only one protein band appears in addition to the native enzyme band. Since disc gel electrophoresis in urea does not eliminate protein charge as a separating characteristic and yet there is only one dissociating band (monomer), the two subunits of equal size also seem to have identical isoelectric points.

5. Elemental Analysis, Amino Acid Analysis, and Metal Composition

The elemental composition of the hydrogenase sample from *C. pasteurianum* is: C, 48.5%; H, 7.4%; N, 14.4%; and S, 1.2%. The amino acid composition of hydrogenase from *C. pasteurianum* is shown in Table V. A mean residue weight of 197 and a partial specific volume ($\bar{v}$) of 0.75 (McMeekine *et al.*, 1949) were calculated. The determination of the partial specific volume does not take into account the iron and

TABLE V
Amino Acid Composition of Hydrogenase from *C. pasteurianum*[a]

Amino acid	Moles/mole[b]	Amino acid	Moles/mole[b]
Aspartic acid	53	Methionine	14
Threonine	38	Isoleucine	54
Serine	28	Leucine	50
Glutamic acid	23	Tyrosine	22
Proline	18	Phenylalanine	24
Glycine	36	Lysine	50
Alanine	50	Histidine	14
Valine	58	Arginine	16
½ Cystine	4	Tryptophan	0

[a] From Nakos and Mortenson (1971d).
[b] A molecular weight of 60,000 is used for the native enzyme. This value was obtained from gel filtration (Nakos and Mortenson, 1971c) and from sodium dodecyl sulfate–polyacrylamide gel electrophoresis (Nakos and Mortenson, 1971d).

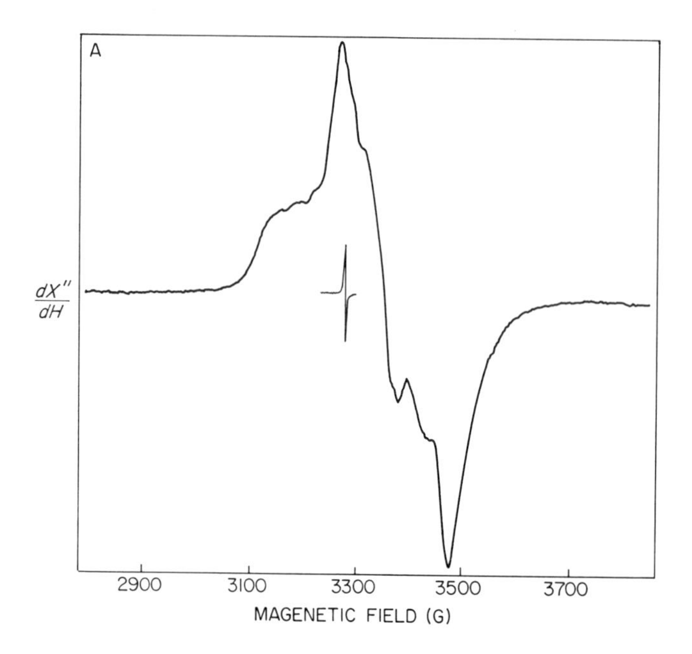

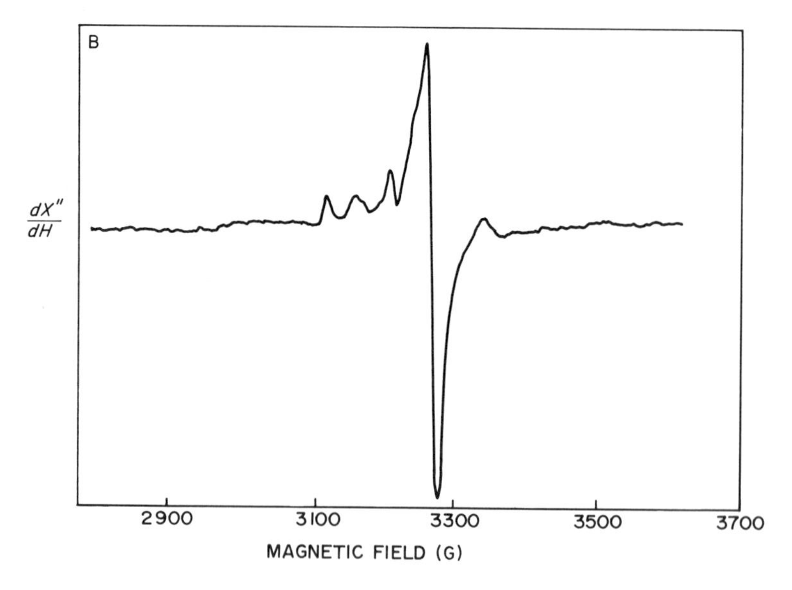

Fig. 3. (A) Electron paramagnetic resonance spectrum of 1 m*M* hydrogenase as isolated. The spectrum was obtained at 25°K in a Varian V-45000 EPR spectrometer operating at 9.197 GHz. Field modulation frequency, 100kHz; amplitude, 6 G; power, 3 mW; scan rate, 400 G/minute; time constant, 0.3 second. A free radical field marker ($g = 2.003$) is also recorded. Data obtained in collaboration with G. Palmer (from Nakos and Mortenson, 1971d, and previously unpublished data). (B) EPR spectrum after oxidation for 5 minutes in air. Conditions the same as (A) except that the gain was 100 instead of 32.

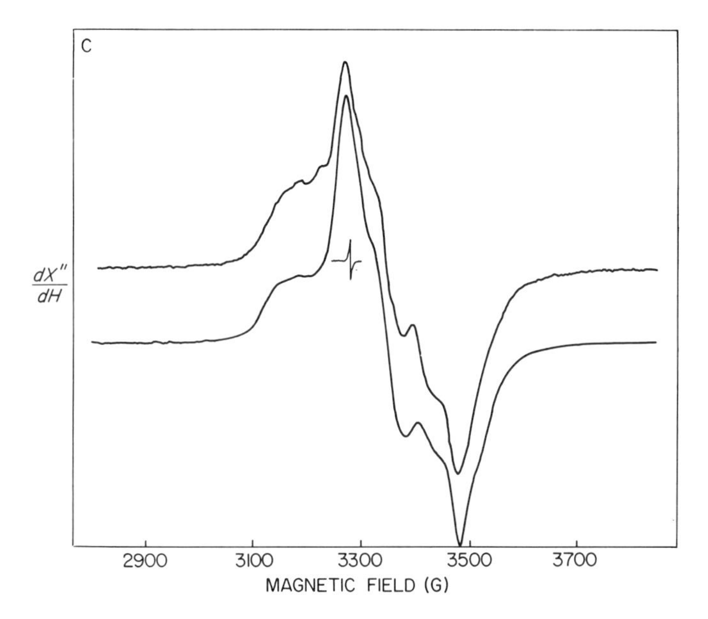

Fig. 3. (C) EPR spectrum of Fig. 3 (B) after equilibration with H_2 (top spectrum) and $Na_2S_2O_4$ under N_2 (bottom spectrum) Conditions were as Fig. 3A except that the gain was 40.

the "acid-labile" sulfide present in the native protein. Hydrogenase contains a relatively high number of acidic and neutral amino acids as well as lysine. This is consistent with the fact that the isoelectric point of hydrogenase is about 5.0 (Nakos and Mortenson, 1971c) and that hydrogenase is weakly adsorbed on DEAE cellulose. Active hydrogenase (specific activity = 70 μmoles of H_2 evolved/minute/mg protein) contains four iron atoms per molecule. Analysis for other metals show trace amounts of Zn and nonmeasurable amounts of K, Mg, Ca, Co, and Mo. The dithiol reagent for Mo gives a zero molybdenum content for hydrogenase. This is in disagreement with two previous reports (Shug *et al.*, 1954; Kleiner and Burris, 1970) that molybdenum is associated with hydrogenase and plays a role in its enzymatic activity.

6. Reaction with Sulfhydryl Reagents

Clostridium pasteurianum hydrogenase was titrated with *p*-hydroxymercuribenzoate (PMB) according to the procedure of Boyer (1954). Based on the absorbance change at 250 nm, approximately 12.5 moles of PMB react with 1.0 mole of the protein. Since 2 moles of PMB react with 1.0 mole of sodium sulfide (Lovenberg *et al.*, 1963) and since hydrogenase contains 4 moles each of "acid-labile" sulfide (Nakos and

TABLE VI
A SUMMARY OF SOME PHYSICAL AND CHEMICAL PROPERTIES OF HYDROGENASE FROM *C. pasteurianum*[a]

Molecular weight	60,000
Polypeptide chains[b]	2
Isoelectric point	5
"Acid-labile" sulfide (moles/mole)	4
Iron atoms (moles/mole)	4 (or 2)[c]
$\frac{1}{2}$ Cystine (moles/mole)	4
Methionine (moles/mole)	14
—SH equivalents (moles/mole)[d]	12
Total sulfur (%)	1.2 ± 0.01
ϵ_{mM}[e]	8.2
Turnover number	3,000[f]/15,000[g]
$\bar{v}$	0.75

[a] From Nakos and Mortenson (1971d).
[b] Two at 30,00 each (appear to be identical).
[c] *o*-Phenanthroline treatment removes two Fe atoms without loss of activity. Trace amounts of Zn and nondetectable amounts of K, Ca, Mg, Cu, and Mo were found.
[d] The number of PMB titratable groups (SH = 1, S^2 = 2)
[e] In liters cm^{-1} $mmole^{-1}$, λ_{max} at 400 nm.
[f] In the methyl viologen assay.
[g] In the methylene blue assay (benzyl viologen reduction falls between these two values).

Mortenson, 1971c) and half-cystine residues (Table V) per mole of protein, one would expect 12.0 moles of PMB to react with 1.0 mole of hydrogenase. These results indicate that all the sulfur in hydrogenase (except that in the methionine) occurs in a form that can react with PMB without prior reduction and is, therefore, not present as disulfide. Based on a molecular weight of 60,000 for hydrogenase and the 4 moles each of half-cysteine and "acid-labile" sulfide plus the 14 moles of methionine, the total sulfur content of hydrogenase should be 1.18%. This agrees with the value of 1.20% found from the elemental analysis.

Table VI summarizes some physical and chemical properties of hydrogenase.

7. EFFECTS OF IRON CHELATING AGENTS AND MERCURIALS ON HYDROGENASE

Hydrogenase, in the presence of dithionite, reacts with excess *o*-phenanthroline to give the characteristic color of ferrous iron *o*-phenanthroline complex (λ_{max} = 512 nm). The amount of color produced is equivalent to 2 moles of iron per mole of original protein, and yet *o*-phenanthroline added to hydrogenase (original protein) in four- to fortyfold mole excess has no effect on its activity.

When the hydrogenase–*o*-phenanthroline mixture is passed through a Sephadex G-25 column, two colored bands are resolved. The first band is hydrogenase (brown protein), whereas the second band corresponds to the ferrous *o*-phenanthroline complex (red). This hydrogenase has a specific activity equal to that of the enzyme originally isolated. However, it now contains only two iron atoms per molecule. These two iron atoms do not react further with either *o*-phenanthroline or with Tiron. The millimolar extinction coefficient of the two-iron hydrogenase at 400 nm is about 8.2, the same as the native enzyme. Upon anaerobic titration with PMB, it still shows the presence of 12 —SH equivalents. When the two-iron hydrogenase is treated with a thirty- to fiftyfold mole excess of sodium mersalyl [3 to 4 mole excess over the titratable (—SH plus $2 \times S^{2-}$)] for 30 minutes, all of its remaining iron is released and reacts with Tiron but not with *o*-phenanthroline.

The fact that hydrogenase is always isolated with four iron atoms suggests that the "extra" two iron atoms are required for other functions, perhaps participation in structural features of the enzyme unrelated to its enzymatic activity. The presence of 12 titratable —SH equivalents after the removal of the two iron atoms indicates that none of the "acid-labile" sulfide groups was released by the *o*-phenanthroline treatment. The millimolar extinction at 400 nm of 8.2 per two iron atoms or 4.1 per iron atom is in reasonable agreement with the extinction for other iron–sulfur proteins (Hong and Rabinowitz, 1970). The EPR spectrum is similar to the untreated hydrogenase.

Hydrogenase when treated anaerobically for 20 minutes with 24-fold mole excess sodium mersalyl (twofold over titratable —SH *equivalents*) lost about 80% of its activity. Addition of excess 2-mercaptoethanol after the 20-minute period did not restore activity. Mersalyl inhibits hydrogenase only through reaction of its mercury with the —SH and/or S^{2-} groups, since if mersalyl is first complexed with 2-mercaptoethanol and then added to hydrogenase, no inhibition is observed. In a separate experiment, hydrogenase again was treated with 24 mole excess mersalyl; this product was treated with dithionite and *o*-phenanthroline, and the mixture was fractionated on a Sephadex G-25 column. The equivalent of less than one iron atom per molecule remained with the protein (it is extremely difficult to remove all iron from iron–sulfur proteins). In all the protein fractions, 4 moles of mercury were found associated with each mole of protein. This suggests that all the "acid-labile" sulfide was removed from the protein by the mersalyl treatment and that mersalyl was bound only to the four —SH groups of apoprotein.

8. SUMMARY

Hydrogenase is an iron–sulfur protein of the "ferredoxin" type. The minimum content of iron and sulfide for maximum activity is two and four, respectively. This represents an unusual case where there appears to be twice as much sulfide as iron. Both iron atoms of the "two-iron" hydrogenase assayed as ferric iron since they complexed anaerobically with Tiron. Since activity, sulfide and iron are lost when hydrogenase is inactivated by mersalyl, and since the iron sulfide "1.94" EPR signal, present in isolated hydrogenase, is lost on oxidation and is restored on treatment with H_2 or dithionite, the iron sulfide–sulfhydryl chromophore appears to function in the activity of hydrogenase. Krasna and Rittenberg (1954) have presented evidence that hydrogenase probably combines a proton and with two electrons to form a hydrided enzyme $H{:}E^-$ which then reacts with H^+ to yield $H_2 + E$. This suggests that clostridial hydrogenase may function through an iron–sulfur hydride or iron hydride and that one of the two "iron–sulfur complexes" is hydrided, whereas the other carries an electron only.

VIII. NITROGENASE: A COMPLEX OF TWO IRON–SULFUR PROTEINS

A. Structure and Properties of Nitrogenase Components

Two protein components are required for N_2 fixation by systems from all bacteria examined (see Table IV). Invariably one contains iron and "acid-labile" sulfide and the other contains molybdenum, iron, and "acid-labile" sulfide. The iron protein (azoferredoxin or Fe protein) is the smaller of the two and when isolated from *C. pasteurianum* it is found to exist in solution as a dimer of molecular weight 55,000 with subunits of 27,500 each. Each subunit appears to be identical and contains two iron atoms, two "acid-labile" sulfide groups, and six "free" cysteine residues. Apparently the chromophore is a complex of all these groups, since when treated with metal chelating agents or sulfide and sulfhydryl complexing mercurials, activity and color are lost. The Fe protein is extremely sensitive to oxygen and loses all activity after exposure to air for about a minute.

The Mo–Fe protein isolated from *Azotobacter* (called molybdoferredoxin when from clostridia) appears to be a dimer of molecular weight 270,000 (Table IV), whereas when isolated from *C. pasteurianum*, its minimum molecular weight is 110,200 and it appears to consist of two dissimilar subunits of molecular weight 59,500 and 50,700. The C-terminal amino acid of one is leucine and of the other alanine. The Mo–Fe protein from *Clostridium* (110,200 minimum molecular weight) contains one molybdenum atom, about twelve Fe atoms, twelve sulfide groups, and fifteen titratable SH groups. The Mo–Fe protein from *Azotobacter* contains two molybdenum atoms/270,000. Reports on the iron content of *Azotobacter* Mo–Fe protein, vary from 17 to greater than 30 if based on a molecular weight of 270,000 (a dimer).

The combination of Mo–Fe and Fe proteins is required for all activities attributed to nitrogenase. When isolated in pure solution, the Fe protein (AzoFd) of *C. pasteurianum* is present as a dimer and the Mo–Fe protein as a tetramer. If in the assay for nitrogen fixation or acetylene reduction AzoFd is kept constant and MoFd added in increasing concentrations, initially MoFd is assayed maximally but at the higher concentrations not only is MoFd not measured optimally but there is actually a decrease in activity. This suggests that the 1:1 MoFd:AzoFd adduct is inactive and that only the 1:2 species is functional. In the opposite combination with MoFd constant and AzoFd added in increasing concentration, a sigmoidal response is obtained. Again, this would be expected since initially the MoFd:AzoFd ratio would be high.

B. Reduction of Dinitrogen; Reductant-Dependent ATP Utilization

Nitrogen fixation by isolated pure components of nitrogenase requires as supporting substrates, magnesium ATP and a low-potential reductant (reduced Fd or sodium dithionite). Because of these requirements and the fact that little ATP is consumed by nitrogenase in the absence of reductant, it has been suggested that nitrogenase acts as a reductant requiring ATPase (Mortenson, 1964c; Hardy and Knight, 1966). In addition, it was shown in extracts of *Azotobacter vinelandii* (Bulen *et al.*, 1965) that in the absence of the substrate nitrogen and to some extent even in its presence, H^+ is reduced and molecular H_2 is evolved. This H_2 evolution is absolutely dependent on ATP and has been called ATP-dependent H_2 evolution. Early studies showed that 5 moles of ATP were required for each H_2 evolved. This discovery was confirmed and extended to other organisms (Burns, 1965; Hardy and Knight, 1966; Mortenson, 1966). Later with pure components of clostridial nitrogenase, it was shown that both components, AzoFd and MoFd, were required for H_2 evolution (Mortenson, 1965, 1966) and that 4 moles of ATP were consumed for each mole of H_2 evolved. This was also shown for the *Azotobacter* system but 5 ATP/H_2 evolved was still observed (Bulen and LeComte, 1966). In the clostridial system some ATP was "hydrolyzed" in the absence of reductant and the rate of reductant-independent ATPase was pH dependent, whereas the total ATPase activity was only slightly affected by pH (Bui and Mortenson, 1970). When the reductant-independent ATP utilization was subtracted from the reductant-dependent ATP utilization at all points from pH 6.0 to 7.0 the ratio of ATP used per H_2 evolved was constant and equal to two (Jeng *et al.,* 1970). This suggested that 2 moles of ATP was needed for the transfer of 2 moles of electrons to 2 moles of H^+ (and presumably 6 moles of ATP would be required for the reduction of 1 mole of N_2 to 2 moles of ammonia). A similar suggestion has been made for the *Azotobacter* system (Silverstein and Bulen, 1970). Although *Azotobacter* has only a slight reductant-independent ATPase, these authors have presented an hypothesis supported by extensive and carefully controlled kinetic data, showing how the utilization of only two molecules of ATP could be coupled to the reduction of two molecules of H^+.

Experiments to determine how electrons flow from reductant to N_2 with components from *C. pasteurianum* have shown by spectral measurements in the range of iron–sulfur adsorption (400 nm) that reduced AzoFd is oxidized by the addition of MoFd and magnesium ATP (M. L. Walker and L. E. Mortenson, unpublished). Such experiments are extremely difficult to run, because one has to rigorously exclude oxygen and they must be thoroughly controlled. At this time we consider them to be preliminary. For the oxidation of reduced AzoFd and concomitant change and decrease of its EPR spectra (complex signal in the $g = 1.94$ region), one also must add MoFd and magnesium ATP. This also results in a rapid decrease in the EPR spectrum of MoFd; all g values, 4.28, 3.81, and 2.01, decrease proportionately (unpublished results). Like the experiments with hydrogenase (see previous section), these experiments suggest a role for the iron–sulfur chromophore in electron transport. Unlike hydrogenase, the electron transport required for nitrogenase activity requires ATP and looks like an ATP-dependent reverse electron flow (termed "electron activation"). The rationale for suggesting that the iron–sulfur chromophore is involved stems, of course, from the knowledge that the $g = 1.94$ EPR signal results from the iron chromophore and the absorption spectral changes are in the 300–400 nm range, which is the wavelength range of maximal adsorption shown for iron–sulfur proteins (see Volume II, Chapter 1 of this treatise).

REFERENCES

Ackrell, B. A. C., Asato, R. N., and Mower, H. F. (1966). *J. Bacteriol.* **92,** 828.
Akagi, J. M. (1965). *Biochem. Biophys. Res. Commun.* **21,** 72.
Aleman, V., and Handler, P. (1967). *J. Biol. Chem.* **242,** 4087.

Aleman, V., Smith, S. T., Rajagopalan, K. V., and Handler, P. (1965). *In* "Non-Heme Iron Proteins" (A. San Pietro, ed.), pp. 327–348. Antioch Press, Yellow Springs, Ohio.

Andrew, I. G., and Morris, J. G. (1965). *Biochim. Biophys. Acta* **97,** 176.

Arnon, D. I., Tsujimoto, H. Y., and McSwain, B. D. (1964). *Proc. Nat. Acad. Sci. U.S.* **51,** 1274.

Arnon, D. I., Tsujimoto, H. Y., and McSwain, B. D. (1967). *Nature (London)* **214,** 562.

Bachofen, R., Buchanan, B. B., and Arnon, D. I. (1964). *Proc. Nat. Acad. Sci. U.S.* **51,** 690.

Bartsch, R. G. (1963). *In* "Bacterial Photosynthesis" (H. Gest, A. San Pietro, and L. P. Vernon, eds.) p. 315. Antioch Press, Yellow Springs, Ohio.

Beinert, H., and Lee, W. (1961). *Biochem. Biophys. Res. Commun.* **5,** 40.

Bergersen, F. J. and Turner, G. L. (1970). *Biochim. Biophys. Acta* **214,** 28.

Boyer, P. D. (1954). *J. Amer. Chem. Soc.* **76,** 4331.

Buchanan, B. B. (1969). *J. Biol. Chem.* **244,** 4218.

Buchanan, B. B., and Bachofen, R. (1968). *Biochim. Biophys. Acta* **162,** 607.

Buchanan, B. B., Rabinowitz, R., and Arnon, D. I. (1964). *Proc. Nat. Acad. Sci. U.S.* **52,** 839.

Buchanan, B. B., Kalberer, P. P., and Arnon, D. I. (1967). *Biochem. Biophys. Res. Commun.* **29,** 74.

Bui, P. T., and Mortenson, L. E. (1969). *Biochemistry* **8,** 2462.

Bulen, W. A., and LeComte, J. R. (1966). *Proc. Nat. Acad. Sci. U.S.* **56,** 979.

Bulen, W. A., Burns, R. C., and LeComte, J. R. (1965). *Proc. Nat. Acad. Sci. U.S.* **53,** 532.

Burns, R. C. (1965). *In* "Non-Heme Iron Proteins: Role in Energy Conversion" (A. San Pietro, ed.), pp. 289–297. Antioch Press, Yellow Springs, Ohio.

Burns, R. C., Holstein, R. D., and Hardy, R. W. F. (1970). *Biochem. Biophys. Res. Commun.* **39,** 90.

Coleman, R., Rieske, J. S., and Wharton, D. (1964). *Biochem. Biophys. Res. Commun.* **15,** 345.

Cushman, D. W., and Gunsalus, I. C. (1966). *Bacteriol. Proc.* p. 86.

Cushman, D. W., Tsai, R. L., and Gunsalus, I. C. (1967). *Biochem. Biophys. Res. Commun.* **26,** 577.

Daesch, G., and Mortenson, L. E. (1968). *J. Bacteriol.* **96,** 346.

Dalton, H., Morris, J. A., Ward, M. A., and Mortenson, L. E. (1971). *Biochemistry* **10,** 2066.

D'Eustachio, A. J., and Hardy, R. W. F. (1964). *Biochem. Biophys. Res. Commun.* **15,** 319.

DeRenzo, E. C. (1956). *Advan. Enzymol.* **17,** 293.

DerVartanian, D. V., Shethna, Y. I., and Beinert, H. (1969). *Biochim. Biophys. Acta* **194,** 548.

Dus, K., DeKlerk, H., Sletten, K., and Bartsch, R. G. (1967). *Biochim. Biophys. Acta* **140,** 291.

Evans, M. C. W., Hall, D. O., Bothe, H., and Whatley, F. R. (1968). *Biochem. J.* **110,** 485.

Friedmann, H. C., and Vennesland, B. (1960). *J. Biol. Chem.* **235,** 1526.

Gray, C. T., and Gest, H. (1965). *Science* **148,** 180.

Hardy, R. W. F., and Knight, E., Jr. (1966). *Biochim. Biophys. Acta* **132,** 520.

Hart, L. I., and Bray, R. C. (1967). *Biochim. Biophys. Acta* **146,** 611.

Haschke, R. H., and Campbell, L. L. (1971). *J. Bacteriol.* **105**, 249.
Holcenberg, J. S., and Tsai, R. L. (1969). *J. Biol. Chem.* **244**, 1211.
Hong, J. S., and Rabinowitz, J. C. (1970). *J. Biol. Chem.* **245**, 6574.
Horio, T., and San Pietro, A. (1964). *Proc. Nat. Acad. Sci. U.S.* **51**, 1226.
Jeng, D. Y., Morris, J. A., and Mortenson, L. E. (1970). *J. Biol. Chem.* **245**, 2809.
Joy, R. W., and Hagelman, R. H. (1966). *Biochem. J.* **100**, 263.
Jungermann, K., Thauer, R. K., Rupprecht, E., Ohrloff, C., and Decker, K. (1969). *FEBS Lett.* **3**, 144.
Jungermann, K., Kirchniawy, H., and Thauer, R. K. (1970). *Biochem. Biophys. Res. Commun.* **41**, 682.
Kearney, E. B. (1960). *J. Biol. Chem.* **235**, 865.
Kelly, M. (1969). *Biochim. Biophys. Acta* **171**, 9.
Kelly, M., and Lang, G. (1970). *Biochim. Biophys. Acta* **223**, 86.
Kidman, A. D., Yanagihara, R., and Asato, R. N. (1969). *Biochim. Biophys. Acta* **191**, 170.
Kimura, T., and Suzuki, K. (1967). *J. Biol. Chem.* **242**, 485.
Kimura, T., Suzuki, K., Omura, T., Cooper, D. Y., and Estabrook, R. W. (1967). *Proc. Int. Congr. Biochem., Tokyo III, 7th* p. 561 (Abstr.).
Kleiner, D., and Burris, R. H. (1970). *Biochim. Biophys. Acta* **212**, 417.
Klucas, R. W., Koch, B., Russell, S. A., and Evans, H. J. (1968). *Plant Physiol.* **43**, 1906.
Knight, E., Jr., and Hardy, R. W. F. (1966). *Biochim. Biophys. Acta* **113**, 626.
Koch, B., Wong, P., Russell, S. A., Howard, R., and Evans, H. J. (1970). *Biochem. J.* **118**, 773.
Krasna, A. I., and Rittenberg, D. (1954). *J. Amer. Chem. Soc.* **76**, 3015.
Kuruta, Y. (1962). *Exp. Cell Res.* **28**, 424.
LeGall, G., DeVartanian, D. V., Spilker, E., Lee, Jin-Po, and Peck, H. D., Jr. (1971). *Biochem. Biophys. Acta* **234**, 525.
Losada, M., and Pomeque, A. (1966). *Biochim. Biophys. Acta* **126**, 578.
Lovenberg, W., and Sobel, B. E. (1965). *Proc. Nat. Acad. Sci. U.S.* **54**, 193.
Lovenberg, W., Buchanan, B. B., and Rabinowitz, J. C. (1963). *J. Biol. Chem.* **238**, 3899.
Mahler, H. R., and Elowe, D. G. (1954). *J. Biol. Chem.* **210**, 165.
Mahler, H. R., Sarkar, N. K., Vernon, L. P., and Alberty, R. A. (1952). *J. Biol. Chem.* **199**, 585.
McMeekin, T. C., Groves, M. L., and Hipp, N. J. (1949). *J. Amer. Chem. Soc.* **71**, 3298.
Mayhew, S. G. (1971). *Anal. Biochem.* **42**, 191.
Mortenson, L. E. (1962). *In* "The Bacteria" (R. Stanier and I. C. Gunsalus, eds.) Vol. III, p. 119. Academic Press, New York.
Mortenson, L. E. (1963). *Ann. Rev. Microbiol.* **17**, 115.
Mortenson, L. E. (1964a). *Biochim. Biophys. Acta* **81**, 71.
Mortenson, L. E. (1964b). *Biochim. Biophys. Acta* **81**, 473.
Mortenson, L. E. (1964c). *Proc. Nat. Acad. Sci. U.S.* **52**, 272.
Mortenson, L. E. (1965). *In* "Non-Heme Iron Proteins: Role in Energy Conversion" (A. San Pietro, ed.), pp. 243–259. Antioch Press, Yellow Springs, Ohio.
Mortenson, L. E. (1966). *Biochim. Biophys. Acta* **127**, 18.
Mortenson, L. E. (1968). *Survey Progr. Chem.* **4**, 127.

Mortenson, L. E., Valentine, R. C., and Carnahan, J. E. (1962). *Biochem. Biophys. Res. Commun.* **7,** 448.
Mortenson, L. E., Valentine, R. C., and Carnahan, J. E. (1963). *J. Biol. Chem.* **238,** 794.
Mortenson, L. E., Morris, J. A., and Jeng, D. Y. (1967). *Biochim. Biophys. Acta* **141,** 516.
Moustafa, E. (1970). *Biochim. Biophys. Acta* **206,** 178.
Moustafa, E., and Mortenson, L. E. (1969). *Biochim. Biophys. Acta* **172,** 106.
Nakos, G., and Mortenson, L. E. (1971a). *Biochemistry* **10,** 455.
Nakos, G., and Mortenson, L. E. (1971b). *Biochim. Biophys. Acta* **229,** 431.
Nakos, G., and Mortenson, L. E. (1971c). *Biochim. Biophys. Acta* **227,** 576.
Nakos, G., and Mortenson, L. E. (1971d). *Biochemistry* **10,** 2442.
Omura, T., Sanders, E., Cooper, D. Y., Rosenthal, O., and Estabrook, R. W. (1965). *In* "Non-Heme Iron Proteins: Role in Energy Conversion" (A. San Pietro, ed.), p. 401. Antioch Press, Yellow Springs, Ohio.
Orme-Johnson, W. H., and Beinert, H. (1969a). *J. Biol. Chem.* **244,** 6143.
Orme-Johnson, W. H., and Beinert, H. (1969b). *Biochem. Biophys. Res. Commun.* **36,** 337.
Orme-Johnson, W. H., and Beinert, H. (1969c). *Anal. Biochem.* **32,** 425.
Palmer, G., and Sands, R. H. (1966). *J. Biol. Chem.* **241,** 253.
Palmer, G., Sands, R. H., and Mortenson, L. E. (1966). *Biochem. Biophys. Res. Commun.* **23,** 357.
Peck, H. D., Jr., and Gest, H. J. (1957). *J. Bacteriol.* **73,** 569.
Raeburn, S., and Rabinowitz, J. C. (1965). *Biochem. Biophys. Res. Commun.* **18,** 303.
Rajagopalan, K. V., and Handler, P. (1967). *J. Biol. Chem.* **242,** 4097.
Rajagopalan, K. V., Fridovich, I., and Handler, P. (1962). *J. Biol. Chem.* **237,** 922.
Ramirez, J. M., DelCampo, F. F., Pomeque, A., and Losada, M. (1966). *Biochim. Biophys. Acta* **118,** 58.
Renwick, G. M., Giumarro, C., and Siegel, S. M. (1964). *Plant Physiol.* **39,** 303.
Rieske, J. S., MacLennan, D. H., and Coleman, R. (1964). *Biochem. Biophys. Res. Commun.* **15,** 338.
Riklis, E., and Rittenberg, D. (1961). *J. Biol. Chem.* **236,** 2526.
Sadana, J. C., and Jaganathan, V. C. (1956). *Biochim. Biophys. Acta* **19,** 440.
Sadana, J. C., and Morey, A. V. (1961). *Biochim. Biophys. Acta* **50,** 153.
San Pietro, A. (1961). *In* "Light and Life" (W. E. McElroy and B. Glass, eds.) p. 631. Johns Hopkins Univ. Press, Baltimore, Maryland.
Schengrurd, C., and Krasna, A. I. (1969). *Biochim. Biophys. Acta* **185,** 332.
Schmidt, A., and Trebst, A. (1969). *Biochim. Biophys. Acta* **180,** 529.
Shapiro, A. L., Vinuela, E., and Maizel, J. V. (1967). *Biochem. Biophys. Res. Commun.* **28,** 185.
Shethna, Y. I., Wilson, P. W., and Beinert, H. (1966). *Biochim. Biophys. Acta* **113,** 225.
Shin, M., and Arnon, D. I. (1965). *J. Biol. Chem.* **240,** 1405.
Shug, A. L., Wilson, P. W., Green, D. E., and Mahler, H. R. (1954). *J. Amer. Chem. Soc.* **76,** 3355.
Silverstein, R., and Bulen, W. A. (1970). *Biochemistry* **9,** 3809.
Singer, T. P. (1963). *In* "The Enzymes" (P. D. Boyer, H. Lardy, and K. Myrback, eds.) Vol. 7, p. 345. Academic Press, New York.

Singer, T. P., Kearney, E. B., and Massey, V. (1957). *Advan. Enzymol.* **18,** 65.
Smith, S. T., Rajagopalan, K. V., and Handler, P. (1967). *J. Biol. Chem.* **242,** 4108.
Sobel, B. E., and Lovenberg, W. (1966). *Biochemistry* **5,** 6.
Stadtman, E. R., and Barker, H. A. (1950). *J. Biol. Chem.* **184,** 769.
Suzuki, K. and Kimura, T. (1965). *Biochem. Biophys. Res. Commun.* **19,** 340.
Tagawa, K., and Arnon, D. I. (1962). *Nature (London)* **195,** 537.
Tagawa, K., and Arnon, D. I. (1968). *Biochim. Biophys. Acta* **153,** 602.
Tagawa, K., Tsujimoto, H. Y., and Arnon, D. I. (1963a). *Proc. Nat. Acad. Sci. U.S.* **49,** 567.
Tagawa, K., Tsujimoto, H. Y., and Arnon, D. I. (1963b). *Proc. Nat. Acad. Sci. U.S.* **50,** 544.
Taylor, K. B. (1969). *J. Biol. Chem.* **244, 171.**
Thauer, R. K., Jungermann, K., Rupprecht, E., and Decker, K. (1969). *FEBS Lett.* **4,** 108.
Thauer, R. K., Rupprecht, E., and Jungermann, K. (1970). *FEBS Lett.* **8,** 304.
Valentine, R. C. and Wolfe, R. S. (1963). *J. Bacteriol.* **85,** 1114.
Valentine, R. C., Jackson, R. L., and Wolfe, R. S. (1962a). *Biochem. Biophys. Res. Commun.* **7,** 453.
Valentine, R. C., Brill, W. J., and Wolfe, R. S. (1962b). *Proc. Nat. Acad. Sci. U.S.* **48,** 1856.
Valentine, R. C., Brill, W. J., and Sager, R. D. (1963a). *Biochem. Biophys. Res. Commun.* **12,** 315.
Valentine, R. C., Mortenson, L. E., and Carnahan, J. E. (1963b). *J. Biol. Chem.* **238,** 1141.
Valentine, R. C., Mortenson, L. E., Mower, H. F., Jackson, R. L., and Wolfe, R. S. (1963c). *J. Biol. Chem.* **238,** 856.
Vandecasteele, J. P., and Burris, R. H. (1970). *J. Bacteriol.* **101, 794.**
Weaver, P., Tinker, K. and Valentine, R. C. (1965). *Biochem. Biophys. Res. Commun.* **21,** 195.
Whatley, F. R., Tagawa, K. and Arnon, D. I. (1963). *Proc. Nat. Acad. Sci. U.S.* **49,** 266.
Whiteley, H. R., and Woolfolk, C. A. (1962). *Biochem. Biophys. Res. Commun.* **9,** 517.
Witz, D. F., and Wilson, P. W. (1967). *Bacteriol. Proc.* p. 112.
Wolfe, R. S., and O'Kane, D. J. (1953). *J. Biol. Chem.* **205,** 755.
Wood, W. A. (1961). *In* "The Bacteria" (I. C. Gunsalus and R. Y. Stanier eds.), Vol. II, pp. 59–149. Academic Press, New York.
Yagi, T., Honya, M., and Tamiya, N. (1968). *Biochim. Biophys. Acta* **153,** 699.
Yoch, D. C., and Arnon, D. I. (1970). *Biochim. Biophys. Acta* **197,** 180.
Yoch, D. C., Benemann, J. R., Valentine, R. C., and Arnon, D. I. (1969). *Proc. Nat. Acad. Sci. U.S.* **64,** 1404.
Yoch, D. C., Benemann, J. R., Arnon, D. I., Valentine, R. C., and Russell, S. A. (1970). *Biochem. Biophys. Res. Commun.* **38,** 838.
Zeylemaker, W. P., DerVartanian, D. V., and Veeger, C. (1965). *Biochim. Biophys. Acta* **99,** 183.

CHAPTER 3

Comparative Biochemistry of Iron-Sulfur Proteins and Dinitrogen Fixation

R. W. F. HARDY and R. C. BURNS

The biochemistry of N_2 fixation and iron–sulfur proteins have a familial relationship. Not only did research on N_2 fixation uncover the relatively simple iron–sulfur proteins, the ferredoxins, but also nitrogenase, the N_2-fixing enzyme, is itself composed of unique iron–sulfur proteins. In fact, there are at least 60×10^6 kg of nitrogenase in the earth's biosphere or about 10^6 kg of the Fe–S prosthetic group performing possibly the most unique and essential reaction of any iron–sulfur protein. In addition, nonbiological models containing iron and sulfur are effective for N_2 fixation.

This chapter will summarize knowledge of iron–sulfur proteins and models in N_2 fixation. Sufficient information on nitrogenases now exists from 16 different organisms to permit a comprehensive presentation of nitrogenase biochemistry organized on a comparative basis. Other recent noncomparative reviews of the biochemistry (Bergersen, 1969, 1971; Burns and Hardy, 1973; Burris, 1969, 1971; Evans, 1969; Evans and Russell, 1971; Hardy and Burns, 1968; Hardy and Knight, 1968; Hardy *et al.*, 1971; Mortenson, 1962, 1968; Postgate, 1970, 1971a,b; Silver, 1971; Yates, 1971a) and biochemical methods (Burns and Hardy, 1972a; Bulen and LeComte; 1972; Evans *et al.*, 1972) as well as inorganic chemistry (Allen, 1971; Allen and Bottomley, 1968; Borodko and Shilov, 1969; Chatt, 1969, 1970; Chatt and Richards, 1971; Ferguson and Love, 1970; Henrici-Olivé and Olivé, 1969; Kuchynka, 1969; Leigh, 1971; van Tamelen, 1970, 1971) and inorganic biochemistry (Shilov and Lichtenshtein, 1971; Hardy *et al.*, 1973; Murray and Smith, 1968) of N_2 fixation are available. It is hoped that upon reading this, the specialist will be stimulated to further experimentation by associations, inconsistencies, or omissions made obvious by the comparative approach. It is also hoped that the comparative approach will orient the nonspecialist to the general relationships of the iron–sulfur proteins of N_2 fixation. These relationships were not apparent to at least one reviewer (Smith, 1966) of the only previous book devoted to iron–sulfur proteins, who wrote: "Section 3 deals with nitrogen fixation. Although one can see the connexion with non-heme iron proteins in general, and with ferredoxin in particular, one cannot help feeling that this section was largely off the main line of interest of the symposium. Equally this difficult field still seems to lack sufficient information of a kind to stimulate the nonspecialist reader. The papers report various experiments on aspects of nitrogen fixation with little relationship between them."

I. RELATIONSHIP OF IRON–SULFUR PROTEINS AND DINITROGEN FIXATION

Specific relationships between the older science of N_2 fixation and the younger science of iron–sulfur proteins are (1) nutritional requirement for additional iron for growth of organisms on N_2 versus fixed N; (2) isolation of the first iron–sulfur protein, clostridial ferredoxin, as a by-product of fractionation experiments with nitrogenase preparations; (3) demonstration of ferredoxins as electron donors for nitrogenase; (4) discovery of the flavodoxins, a new class of natural electron transfer proteins which can replace ferredoxins; (5) fractionation of nitrogenase into two iron-containing proteins called Mo–Fe protein and Fe protein; (6) purification and characterization of the Mo–Fe and Fe proteins as unique iron–sulfur proteins; (7) abiological N_2 fixation by molybdo–iron–thiol models.

The discovery of clostridial ferredoxin, the first purified protein recognized as an iron–sulfur protein, may be considered a by-product of research on N_2 fixation. Nutritional experiments during the physiological era of N_2-fixation research suggested that organisms grown on N_2 as the sole source of N required more Fe in the media than those grown on fixed N (Wilson, 1968; Carnahan and Castle, 1958). Accordingly, fractions of crude cell-free N_2-fixing extracts of *Clostridium pasteurianum* from DEAE cellulose columns were analyzed for iron. A dark-brown, acidic protein fraction was observed with remarkably high iron content. This protein was found to undergo oxidation and reduction and was suggested to function in electron transfer between pyruvate and H_2 or pyruvate and nitrogenase (Hardy and Knight, 1968). Later an iron-rich fraction obtained in a similar manner was shown to restore pyruvate oxidation and H_2 evolution to DEAE cellulose treated extracts of *C. pasteurianum* (Mortenson *et al.*, 1962, 1963). Because of its high iron content and activity as a redox protein, the protein was named ferredoxin. Subsequently ferredoxin was shown to transfer electrons from either pyruvate or H_2 to nitrogenase (D'Eustachio and Hardy, 1964; Hardy and D'Eustachio, 1964; Mortenson, 1964a,b).

The ferredoxin nomenclature has been extended to include iron–sulfur proteins such as plant ferredoxins that function as electron-transferring proteins. In fact, the name ferredoxin has been popularized in a recent best-selling novel, "The Couples," (Updike, 1966), where a surprisingly acceptable definition was used:

> What's ferredoxin?
> A protein. An electron carrier with a very low redox potential.

Flavodoxins, flavin mononucleotide-containing electron-transferring proteins that function as ferredoxin replacements, were also spin-offs from N_2-fixation research. Flavodoxin was first obtained from *C. pasteurianum* which was grown on media containing low levels of iron in an attempt to obtain apoferredoxin (Knight and Hardy, 1966a; Knight *et al.*, 1966). These bacteria produced little ferredoxin or apoferredoxin, but instead produced flavodoxin, a protein containing FMN and no iron. The demonstration that flavodoxin can replace ferredoxin shows that the simple iron–sulfur proteins of the ferredoxin class are not physiologically unique, natural electron-transferring proteins.

Nitrogenase, the enzyme that catalyzes N_2 fixation, has been fractionated into two proteins (Bulen and LeComte, 1966; Mortenson, 1965; Mortenson *et al.*, 1967). Both proteins contain Fe, explaining the nutritional requirement for additional iron during growth on N_2; only one of the proteins contains Mo, the other metal required in additional amounts for growth on N_2. These proteins are designated by a variety of terms (Hardy *et al.*, 1973). The most common are based on their metal content. The Mo- and Fe-containing protein is called Mo–Fe protein, and the Fe-containing protein is called Fe protein. A proposed designation, azofermo and azofer, respectively, is based on their metal composition as well as their unique ability to react with dinitrogen.

The Mo–Fe and Fe proteins of nitrogenases from at least three sources have been purified to homogeneity (Eady *et al.*, 1972; Burns *et al.*, 1970; Dalton *et al.*, 1971; Moustafa and Mortenson, 1969). All are iron–sulfur proteins, but they differ from other iron–sulfur proteins both in chemical and physical characteristics, as well as in the biological activity of their enzymatically active complex, nitrogenase.

Chemical models for nitrogenase have been actively sought. Recent evidence suggests that complexes of molybdenum and iron with thiol ligands may catalyze N_2 fixation under ambient conditions (Schrauzer *et al.*, 1971). Thus, iron and sulfur are components of (1) nitrogenase, (2) electron donors to nitrogenase, and (3) models of nitrogenase.

II. DEFINITION OF NITROGENASE

A uniform definition may be used to describe all known nitrogenases (Hardy *et al.*, 1973). This definition holds that nitrogenases are complexes of Mo–Fe–S and Fe–S proteins, whose syntheses are repressed by fixed nitrogen, and whose activity couples ATP hydrolysis to electron transfer (reductant-dependent ATP utilization) from ferredoxins or flavodoxins for reduction of

N_2 to NH_3 (dinitrogen fixation)
N_3^- to $N_2 + NH_3$
N_2O to $N_2 + H_2O$
$RCCH$ to $RCHCH_2$
RCN to $RCH_3 + NH_3$
RNC to RNH_2 + alkanes and alkenes
and/or $2\ H_3O^+$ to $H_2 + 2\ H_2O$ (ATP-dependent H_2 evolution)

with H_2 a competitive inhibitor specifically of N_2 reduction, and CO an inhibitor of all reductions except that of H_3O^+. This definition is presented schematically in Fig. 1.

III. IRON–SULFUR PROTEINS AS REDUCTANTS FOR NITROGENASE

The electrons required in the nitrogenase reaction (Fig. 1) can be provided by an artificial reductant or by an electron-transferring agent coupled to a substrate source of electrons. Hydrosulfite is the sole artificial reductant known to react directly with nitrogenase and is effective with all known nitrogenases (Table I). Consequently, it is used routinely in almost all experimental work with nitrogenase except investigations concerning the natural electron-transferring agent or source of electrons.

Only two classes of naturally occurring agents are known to transfer electrons to nitrogenase. They are the ferredoxins and flavodoxins. We define ferredoxins broadly as iron–sulfur, electron-transferring proteins and flavodoxins as FMN-containing, electron-transferring proteins. The insistence that candidate proteins must function in the phosphoroclastic reaction and/or in light-dependent chloroplast $NADP^+$ reduction has led to a confusing expansion of nomenclature with the introduction of terms such as azotoflavin (Benemann *et al.*, 1969) and bacteroid nonheme iron protein (Koch *et al.*, 1970). We prefer for purposes of clarity to call these proteins simply *Azotobacter* flavodoxin, as recently sug-

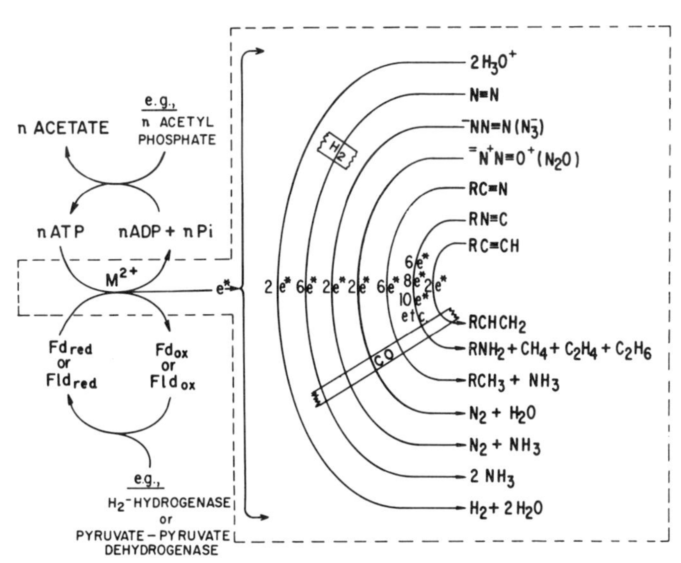

Fig. 1. Schematic diagram of the nitrogenase reaction (enclosed in dotted area) indicating substrates (showing triple bond cleaved or reduced), products, inhibitors, and natural electron donors with electron and phosphagen sources as for *C. pasteurianum*.

TABLE I
IRON–SULFUR PROTEINS AND NITROGENASE REDUCTANTS

Organism	Electron-transferring agents[a]		Electron-donor systems
	Ferredoxins	Other	
C. pasteurianum	*Clostridium* Fd	*Clostridium* Fld MV	Pyruvate → Fd or Fld or MV (D'Eustachio and Hardy, 1964; Hardy and D'Eustachio, 1964; Mortenson, 1964a; Knight and Hardy, 1966; Knight *et al.*, 1966) Formate → → Fd (Mortenson, 1966) $KBH_4 \rightarrow H_2 \rightarrow$ Fd or Fld or MV (D'Eustachio and Hardy, 1964; Mortenson, 1964a) NADH → → Fd (D'Eustachio and Hardy, 1964) $Na_2S_2O_4$ (Burns, 1965; Hardy *et al.*, 1965)
B. polymyxa	*Bacillus* Fd		Pyruvate → Fd (Shethna *et al.*, 1971) Formate → → Fd (Fisher and Wilson, 1970) $Na_2S_2O_4$ (Witz *et al.*, 1967)
K. pneumoniae	*Clostridium* Fd	BV MV	$H_2 \rightarrow$ *Clostridium* H_2ase and Fd (Parejko and Wilson, 1971) $H_2 \rightarrow$ *Desulfovibrio* H_2ase and BV or MV (Eady *et al.*, 1972) $Na_2S_2O_4$ (Mahl and Wilson, 1968)
C. ethylicum	*Chloropseudomonas* Fd *Clostridium* Fd *Chromatium* Fd		Pyruvate → Fd (Evans and Smith, 1971) Ascorbate → DCPIP → illuminated heated chloroplast → Fd (Evans and Smith, 1971)
Chromatium	*Clostridium* Fd *Chromatium* Fd	MV BV	$H_2 \rightarrow$ MV or BV (Winter and Arnon, 1970) $H_2 \rightarrow$ *Clostridium* H_2ase and Fd (Winter and Arnon, 1970) Ascorbate → DCPIP → illuminated heated chloroplast → Fd[b] (Yoch and Arnon, 1970) $Na_2S_2O_4$ (Winter and Arnon, 1970)
R. rubrum			$Na_2S_2O_4$ (Bulen *et al.*, 1965a; Burns and Bulen, 1966)
A. cylindrica	Plant Fd *Anabaena* Fd *Clostridium* Fd	Fld	G-6-P or isocitrate → $NADP^+$ → Fd (Bothe, 1970, 1971; Smith *et al.*, 1971a) Ascorbate → DCPIP → illuminated heated chloroplast → Fd or Fld (Bothe, 1970; Smith *et al.*, 1971a) $H_2 \rightarrow$ *Clostridium* H_2ase and Fd (Haystead and Stewart, 1972) $Na_2S_2O_4$ (Smith and Evans, 1970)

Gloeocapsa			$Na_2S_2O_4$ (Gallon *et al.*, 1972)
P. boryanum			$Na_2S_2O_4$ (Haystead *et al.*, 1970)
A. chroococcum		BV	G-6-P → → BV (Yates and Daniel, 1970) NADH → BV (Yates, 1971b) $Na_2S_2O_4$ (Kelly, 1966)
A. vinelandii	*Azotobacter* Fd *Clostridium* Fd	*Azotobacter* Fld BV MV	NADPH → *Azotobacter* Fd and Fld or *Clostridium* Fd (Benemann *et al.*, 1971 Ascorbate → DCPIP → illuminated heated chloroplast → *Azotobacter* Fd or Fld (Benemann *et al.*, 1969) β-Hydroxybutyrate → NAD^+ → BV or MV (Klucas and Evans, 1968) H_2 → *Clostridium* H_2ase and Fd (Bulen *et al.*, 1964) $Na_2S_2O_4$ (Bulen *et al.*, 1965a)
M. flavum		BV	NADH → BV (lactate, pyruvate, H_2, or G-6-P also active) (Biggins and Postgate, 1971) $Na_2S_2O_4$ (Biggins and Postgate, 1969)
Dolichos lab-lab			$Na_2S_2O_4$ (Kennedy, 1970)
Lupinus sp.			$Na_2S_2O_4$ (Evans, 1969; Manorik *et al.*, 1970)
O. sativus			$Na_2S_2O_4$ (Evans, 1969)
G. max bacteroids	*Glycine max* bacteroid Fd *Azotobacter* Fd *Clostridium* Fd	BV MV FMN FAD *Azotobacter* Fld	H_2 → *Clostridium* H_2ase and Fd (Wong *et al.*, 1971) β-Hydroxybutyrate → NAD^+ → BV or MV or FMN or FAD[c] (Klucas and Evans, 1968) G-6-P → $NADP^+$ → *Azotobacter* Fld and bacteroid Fld or Flavin (Wong *et al.*, 1971) Ascorbate → DCPIP → illuminated heated chloroplast → bacteroid Fd or *Azotobacter* Fld (Wong *et al.*, 1971; Yoch *et al.*, 1970) $Na_2S_2O_4$ (Koch *et al.*, 1967b; Bergersen and Turner, 1970)

[a] Abbreviations include: DCPIP, dichlorophenolindophenol; Fd, ferrodoxin; Fld, flavodoxin; BV, benzylviologen; MV, methylviologen; NAD^+, nicotinamide adenine dinucleotide; $NADP^+$, nicotinamide adenine dinucleotide phosphate; FMN, flavinmononucleotide; FAD, flavin adenine dinucleotide; G-6-P, glucose 6-phosphate; H_2ase, hydrogenase;

[b] Plant Fd inactive.

[c] Clostridial Fd inactive.

gested based on a reexamination of functions (van Lin and Bothe, 1972), and *Glycine max* bacteroid ferredoxin.

Coupled ferredoxin–nitrogenase systems have been isolated from *Clostridium, Bacillus, Chloropseudomonas, Chromatium, Anabaena, Azotobacter,* and *Glycine max* bacteroids (Table I). In some cases, such as *Clostridium* and *Bacillus,* the ferredoxin component will restore nitrogenase activity to a ferredoxin-depleted extract with either H_2 or pyruvate as a physiological electron donor. In other cases, such as *Azotobacter* or *Glycine max* bacteroids, the respective ferredoxin was essential for (or stimulated) nitrogenase activity with an artificial electron donor such as the ascorbate, DCPIP, and illuminated chloroplast system described below.

Clostridium ferredoxin couples readily to nitrogenases of other species and has been used with nitrogenases from six different organisms, and *Azotobacter* ferredoxin couples with bacteroid nitrogenase. This suggests a lack of species specificity for this coupling. Spinach ferredoxin, however, is inactive with almost all nitrogenases. These observations indicate that nitrogenases require a ferredoxin which contains more than 2 Fe and 2 S^{2-} atoms per molecule and is, therefore, presumably capable of transferring one or two electrons per molecule. Further characterization of the recently isolated ferredoxins from *Bacillus* (Orme-Johnson *et al.,* 1972), *Azotobacter* (Yoch and Arnon, 1972), and *Glycine max* bacteroids (Koch *et al.,* 1970) will enable a more definitive comparison of these ferredoxins with *Clostridium* ferredoxin and differentiation from plant ferredoxins.

Flavodoxins have been isolated from N_2-fixing organisms, *Clostridium* (Knight and Hardy, 1966; Knight *et al.,* 1966; Knight and Hardy, 1967) and *Azotobacter* (Benemann *et al.,* 1969; van Lin and Bothe, 1972), and are suggested to occur in *Glycine max* bacteroids (Wong *et al.,* 1971). For electron transfer to nitrogenase they are interchangeable with their respective ferredoxins. Flavodoxins contain a single FMN prosthetic group and no iron. Similarities of flavodoxin and ferredoxin suggest that the requirements for electron transfer to nitrogenase may include (1) a redox potential ≤ -0.20 V, (2) a low isoelectric point, and (3) ability to transfer one or two electrons per molecule. An Fe–S prosthetic group is obviously not essential. For several nitrogenases, nonbiological electron-transferring agents with a redox potential of ≤ -0.35 V, such as methyl viologen and benzyl viologen, are effective replacements for ferredoxin or flavodoxin.

All of the electron carriers require a substrate source of electrons. Sources effective with different nitrogenases are tabulated in Table I.

The physiological source of electrons for nitrogenase has been identified only in *Clostridium* and possibly *Bacillus* where pyruvate metabolized via the phosphoroclastic reaction reduces ferredoxin or flavodoxin or methyl viologen. An alternate natural (but probably not physiological) source of electrons in *Clostridium* is H_2, which reduces ferredoxin via hydrogenase, and this system is effective with several nitrogenases. A useful experimental source of electrons is ascorbate, DCPIP, and an illuminated spinach chloroplast preparation modified by heating to eliminate O_2 evolution (Yoch and Arnon, 1970). This artificial system has been used with nitrogenases from photosynthetic and nonphotosynthetic organisms, and it provided the assay system used for the isolation of *Azotobacter* ferredoxin (Yoch *et al.*, 1969) and flavodoxin (Benemann *et al.*, 1969) and *Glycine max* bacteroid ferredoxin (Yoch *et al.*, 1970; Koch *et al.*, 1970). Other substrate sources of electrons include glucose 6-phosphate or NADPH in *Anabaena* (Bothe, 1970; Smith *et al.*, 1971a), *Azotobacter* (Benemann *et al.*, 1971), and *Glycine max* bacteroids (Wong *et al.*, 1971) and β-hydroxybutyrate in *Azotobacter* (Klucas and Evans, 1968) and *Glycine max* bacteroids (Klucas and Evans, 1968). The unusual requirement for both ferredoxin and flavodoxin with NADPH (Benemann *et al.*, 1971; Wong *et al.*, 1971), coupled with the low activities (or simply stimulations of activity) observed with these systems, suggests that the physiological electron source for these nitrogenases has not yet been found.

IV. EXTRACTION, FRACTIONATION, AND PURIFICATION OF NITROGENASE

A. Sources

Nitrogenase, the enzyme that catalyzes the reduction of N_2 to NH_3, has been extracted from sixteen different organisms. These sources may be conveniently grouped by physiological types, proceeding from strict anaerobes to strict aerobes, as shown at top of p. 74.

Nitrogenase production by various source organisms can be maximized by employing special growth conditions. As already indicated, additional Mo and Fe is required in the growth medium because of the Mo and Fe content of nitrogenase. Nitrogenase is repressed by fixed forms of N such as NH_3, but fixed N has never been shown to mutate N_2-fixing bacteria into non-N_2-fixing bacteria, as the following widely distributed

Anaerobic bacteria	*Gloeocapsa* (Gallon *et al.*, 1972)
Clostridium pasteurianum (Carnahan *et al.*, 1960)	*Plectonema boryanum* (Haystead *et al.*, 1970)
Facultative anaerobes	Aerobic bacteria
Bacillus polymyxa (Witz *et al.*, 1967; Grau and Wilson, 1963; Detroy *et al.*, 1968)	*Azotobacter chroococcum* (Kelly, 1966)
Klebsiella pneumoniae (Parejko and Wilson, 1971; Mahl and Wilson, 1968; Detroy *et al.*, 1968)	*A. vinelandii* (Bulen *et al.*, 1964, 1965a)
Photosynthetic bacteria	*Mycobacterium flavum* (Biggins and Postgate, 1969, 1971)
Chloropseudomonas ethylicum (Evans and Smith, 1971)	Legume bacteroids
Chromatium (Winter and Arnon, 1970; Winter and Ober, 1971)	*Dolichos lab-lab* (Kennedy, 1970a)
Rhodospirillum rubrum (Burns and Bulen, 1966; Munson and Burris, 1969)	*Glycine max* (Koch *et al.*, 1967a, b)
Blue-green algae	*Lupinus* sp. (Evans, 1969; Starchenkov *et al.*, 1971; Kretovich *et al.*, 1970; Manorik *et al.*, 1971)
Anabaena cylindrica (Smith and Evans, 1970; Haystead *et al.*, 1970)	*Ornithopus sativus* (Evans, 1969)
	Nonlegume symbionts
	None

but misleading statement indicates: "Under the impact of a heavy use of inorganic nitrogen fertilizer, the nitrogen-fixing bacteria originally living in the soil may not survive, or, if they do, may mutate into non-fixing forms" (Commoner, 1971). Organisms cultured for extraction of nitrogenase are generally grown on N-free media or media containing low levels of N and 0.8–1.0 atm of N_2. Replacement of N_2 with Ar during final growth in order to maximize derepression of nitrogenase is used with blue-green algae (Haystead and Stewart, 1972; Smith and Evans, 1970); while low, but nonrepressive, concentrations of NH_3 produce higher growth rates and yields of nitrogenase with other organisms, such as *R. rubrum* (Munson and Burris, 1969) or *C. pasteurianum* (Daesch and Mortenson, 1968).

B. Extraction

Nitrogenases have been successfully extracted with a variety of techniques, and there is little specificity in method of extraction. Recoveries in some cases are excellent with acellular activities of about 10 nmoles N_2 reduced/minute/mg of extract protein for *Azotobacter* and *Clos-*

tridium; these approach cellular activities. Since nitrogenase accounts for 1–5% of the protein of N_2-fixing organisms, substantial amounts of nitrogenase are obtained in these crude extracts. On the other hand, algal and *R. rubrum* extracts, for example, contain only 10–20% of the cellular activity.

The first effective cell-free nitrogenase preparations were obtained by rupture of frozen *Clostridium* cells in a Hughes press or by autolysis of dried cells of the same organism under anaerobic conditions (Carnahan *et al.,* 1960). Subsequently, other methods including lysis with lysozyme and DNase, sonication, or rupture with a pressure cell have been effectively applied to most classes of N_2-fixing organisms (Table II).

Special methodology was developed for effective extraction of nitrogenase from legume bacteroids (Koch *et al.,* 1967a,b). Development of this methodology was a product of an attempt to isolate a ferredoxin from bacteroids. Variable amounts of a dark-brown material were obtained in the fractions originally associated with ferredoxin. Subsequently the material was attributed to reactions involving the abundant polyphenols released from the plant cells during maceration. Polyvinylpyrrolidone and ascorbate, a combination already found useful for various isolations from plant material, was used to remove nitrogenase-inactivating reactants during maceration of the nodules. The bacteroids obtained by this procedure yielded the first high-activity cell-free preparations of nodular nitrogenase. Polyvinylpyrrolidone and ascorbate are now generally used for preparation of various legume bacteroids prior to nitrogenase extraction. This technique may have general application for the isolation of nitrogenase from other N_2-fixing plant systems, such as nonlegume nodules and phylloplane associations.

Most extraction procedures produce a soluble O_2-sensitive nitrogenase. However, certain methods of extraction yield specific non-nitrogenase material which protects nitrogenase from O_2. For example, *Azotobacter* nitrogenase, obtained by cell rupture in a pressure cell, is O_2 insensitive (Bulen *et al.,* 1965a) and sediments after 3 hours at 180,000 *g,* while that obtained through osmotic lysis is O_2 sensitive (Oppenheim *et al.,* 1970) and is not precipitated by a similar centrifugation. The stability and centrifugal properties of the *Azotobacter* nitrogenase produced by pressure cell treatment is attributed to its attachment to membranes. Electron micrographs of *A. vinelandii* grown on N_2 versus fixed N (Oppenheim and Marcus, 1970) have been used to support the concept that nitrogenase is localized on internal cellular membranes. Sedimentation characteristics of *Chromatium* crude extracts (Winter and Ober, 1971) suggest that this nitrogenase is in a lipid particle with a density of 1.02 gm/ml calculated from sedimentation rate in a sucrose gradient. Triton X-100 solubilized the particle.

C. Purification

The nitrogenase in crude extracts can be purified to yield preparations with specific activities of up to 100 nmoles N_2 reduced/minute/mg protein. Unfractionated nitrogenase has not been purified to homogeneity, although a specific activity of about 200 has been calculated for it on the basis of the activities of the purified individual components (Hardy *et al.*, 1973).

Steps used in purification vary to a small extent with the source of the nitrogenase (Table II), but to a greater extent with the specific research group. In general, purification methods usually include combinations of one or more of the following steps: (1) centrifugation, (2) heat treatment at 50°–60°C to precipitate nonnitrogenase protein, (3) protamine sulfate addition in two steps to remove nucleic acids and then to precipitate nitrogenase, and (4) anion exchange chromatography with DEAE cellulose. Other techniques applied only occasionally include solvent fractionation, precipitation with polypropylene glycol in place of protamine sulfate, ammonium sulfate precipitation, and gel filtration. The similarity of the various nitrogenases is demonstrated by the successful use of heat treatment and protamine sulfate precipitation for purification of nitrogenases from five different organisms. Except for the crude extracts of a few organisms, nitrogenases are O_2 sensitive and also cold labile (see Section V,C). Consequently, purifications must be performed under anaerobic conditions and preferably at room temperature. Purified nitrogenases stored under these conditions are stable for several days or more.

Unfractionated nitrogenases are recommended over recombinations of fractionated components of nitrogenase for investigations of reaction characteristics (see Section V,B). Different recombining ratios show different characteristics for reactions such as some substrate reductions or ATP hydrolysis. However, the inhomogeneity of current nitrogenase preparations make them unsuitable for chemical and physical characterization.

D. Fractionation

Fractionation of nitrogenase into its two component proteins, the Mo–Fe protein and Fe protein, and subsequent purification of the indi-

TABLE II

EXTRACTION, FRACTIONATION, AND PURIFICATION OF NITROGENASE AND ITS MO–FE AND FE PROTEINS

Organism	Extraction	Nitrogenase purification steps	Fractionation and purification of Mo–Fe and Fe proteins
C. pasteurianum	Hughes press[a] lysozyme and DNase[b-e]	protamine sulfate[b,f] alcohol and acetone[g]	DEAE → 0.065 M $MgCl_2$ → Mo–Fe protein └→ 0.09 M $MgCl_2$ → Fe protein → Sephadex G-100[h] cold storage at 0–1°C for 7–10 days → Mo–Fe protein[i]
	autolysis of dried cells[a]		heat → Fe protein[j] 2–2.5% protamine sulfate (P.S.) → Fe protein → P.S. → Sephadex G-100 → P.S.[b,k] 5–12% P.S. → Mo–Fe protein → 4–6% P.S. → 2–5% P.S. → Sephadex G-100[b,k] Sephadex[l]
B. polymyxa	pressure cell[e] lysozyme and DNase[d,n]	heat → DEAE → 0.5 M NaCl[n]	DEAE → 0.3 M NaCl → Mo–Fe protein └→ 0.5 M NaCl → Fe protein[e,o]
K. pneumoniae	pressure cell[e,p] lysozyme and DNase[p]	Heat → DEAE → 0.1 M $MgCl_2$[p] DEAE[q]	DEAE → 0.2–0.3 M NaCl → Mo–Fe protein → Sephadex G-200 └→ 0.5 M NaCl or 0.09 M $MgCl_2$ → Fe protein[e,q,r]
C. ethylicum	sonication[r]		DEAE → 0.28 M NaCl → Mo–Fe protein └→ 0.6 M NaCl → Fe protein[s]
Chromatium	sonication[t] pressure cell[t]	centrifugation[u] sucrose gradient[u]	
R. rubrum	sonication[v,w] pressure cell[v]	centrifugation[v] Sephadex G-25[v]	
A. cylindrica	sonication[r,x,y] pressure cell[aa]	DEAE → 0.15 M $MgCl_2$[z]	DEAE → 0.28 M NaCl → Mo–Fe protein └→ 0.6 M NaCl → Fe protein[s]
Gloeocapsa	pressure cell[bb]		
P. boryanum	sonication[y]		
A. chroococcum	pressure cell[cc] sonication	protamine sulfate[dd]	DEAE → 0.04 M $MgCl_2$ → Mo–Fe protein └→ 0.09 M $MgCl_2$ → Fe protein[o,dd]

TABLE II (*Continued*)

Organism	Extraction	Nitrogenase purification steps	Fractionation and purification of Mo–Fe and Fe proteins
A. vinelandii	pressure cell[ee,ff] lysozyme[gg] mechanical disruption[gg] osmotic lysis[hh] phage lysis[ii]	protamine sulfate → $MgCl_2$ → Sephadex G-100[jj] cetyltrimethylammonium bromide[kk] protamine sulfate → heat → protamine sulfate[ll] centrifugation[ff,mm] ammonium sulfate[nn] sucrose gradient[oo] acetone extraction[pp]	DEAE → 0.25 *M* NaCl or 0.035 *M* $MgCl_2$ → ↳Mo–Fe protein → crystallizes at 0.08 *M* NaCl 0.35 *M* NaCl or 0.09 *M* $MgCl_2$ → Fe protein[qq,ll,rr] DEAE → 0.035 *M* $MgCl_2$ ↳ 0.09 *M* $MgCl_2$ → Fe protein → heated at 60°C 15′ → protamine sulfate[kk] DEAE → 0.17 *M* NaCl → Bio-gel P-150 ↳ 0.33 *M* NaCl → DEAE → 0.24 *M* NaCl → 0.35 *M* NaCl[ss] Sephadex G-150[gg] cold → Mo–Fe protein[kk]
M. flavum	sonication[tt] pressure cell[uu]		DEAE[vv]
Dolichos lab-lab	maceration with PVP[ww] Pressure cell[ww]		
Lupinus sp.	maceration with PVP[xx] pressure cell[xx] sonication[yy]	protamine sulfate[zz] Heat → DEAE → 0.5 *M* NaCl[1]	Sephadex G-100[1]
O. sativus	maceration with PVP[xx] pressure cell[xx]		
G. max	anaerobic pressure cell and DNase[2] maceration with PVP[3,4] pressure cell[3,4]	polyacrylamide gel[3] protamine sulfate[5] polypropylene glycol[6] heat[7]	DEAE → 0.035 *M* $MgCl_2$ → Mo–Fe protein ↳ 0.100 *M* $MgCl_2$ → Fe protein[5] Sephadex G-200[7]

[a] Carnahan *et al.* (1960).
[b] Moustafa and Mortenson (1969).
[c] Fisher and Wilson (1970).
[d] Witz *et al.* (1967).
[e] Detroy *et al.* (1968).
[f] Mortenson (1966).
[g] Winter and Burris (1968).
[h] Vandecasteele and Burris (1970).
[i] Moustafa and Mortenson (1968).
[j] Hardy *et al.* (1968).
[k] Dalton *et al.* (1971).
[l] Mortenson *et al.* (1967).
[m] Witz and Wilson (1967).
[n] Grau and Wilson (1963).
[o] Kelly (1969b).
[p] Parejko and Wilson (1971).
[q] Eady *et al.* (1972).
[r] Evans and Smith (1971).
[s] Smith *et al.* (1971b).
[t] Winter and Arnon (1970).
[u] Winter and Ober (1971).
[v] Burns and Bulen (1966).
[w] Munson and Burris (1969).
[x] Smith and Evans (1970).
[y] Haystead *et al.* (1970).
[z] Haystead and Stewart (1972).
[aa] Smith and Evans (1971).
[bb] Gallon *et al.* (1972).
[cc] Kelly (1966).
[dd] Kelly (1969a).
[ee] Bulen *et al.* (1965a).
[ff] Bulen *et al.* (1964).
[gg] Yakovlev *et al.* (1965).
[hh] Oppenheim *et al.* (1970).
[ii] Nimek and Wilson (1963).
[jj] Silverstein and Bulen (1970).
[kk] Moustafa (1970).
[ll] Burns *et al.* (1970).
[mm] Hardy and Knight (1966).
[nn] Ganelin *et al.* (1969).
[oo] Gvozdev *et al.* (1969).
[pp] Novikov *et al.* (1968).
[qq] Bulen and LeComte(1966).
[rr] Kelly *et al.* (1967).
[ss] Kajiyama *et al.* (1969).
[tt] Biggins and Postgate (1969).
[uu] Biggins and Postgate (1971).
[vv] Biggins *et al.* (1971).
[ww] Kennedy (1970a).
[xx] Evans (1969).
[yy] Kennedy (1970b).
[zz] Manorik *et al.* (1971).
[1] Peive *et al.* (1971).
[2] Bergersen and Turner (1968).
[3] Koch *et al.* (1967a).
[4] Koch *et al.* (1967b).
[5] Klucas *et al.* (1968).
[6] Evans *et al.* (1972).
[7] Bergersen and Turner (1970).

vidual proteins to homogeneity has been accomplished, in contrast to the partial purification of the unfractionated nitrogenase. In fact, all the definitive chemical and physical characteristics of nitrogenase have been obtained with homogeneous preparations of the Mo–Fe and Fe proteins. Such preparations have been described for *Clostridium* and *Klebsiella* Mo–Fe and Fe protein and for *Azotobacter* Mo–Fe protein; in the latter case methodology has been discovered for routine crystallization.

The highest reported specific activities of the individual proteins are 675 nmoles N_2 reduced/minute/mg for *Clostridium* (Moustafa and Mortenson, 1969) Fe protein and 375–380 nmoles N_2 reduced/minute/mg for *Azotobacter* (Burns *et al.*, 1970) and *Klebsiella* (Eady *et al.*, 1972) Mo–Fe protein. These specific activities are determined by addition of a limiting amount of the Fe or Mo–Fe protein to an optimum amount of its Mo–Fe or Fe protein partner.

The most widely applied method for fractionation of various nitrogenases into their Mo–Fe and Fe proteins uses stepwise elution from DEAE cellulose columns with increasing concentration of either NaCl or $MgCl_2$ (Table II). At least nine different nitrogenases have been successfully fractionated with this method. The similarities of nitrogenases from different sources is further substantiated by their consistent response to the similar concentrations of salt used for elution of the fractions. The Mo–Fe protein is eluted with 0.25–0.30 *M* NaCl or 0.035–0.065 *M* $MgCl_2$, and the Fe protein is eluted with 0.35–0.6 *M* NaCl or 0.09–0.10 *M* $MgCl_2$. As an alternative, gel filtration with Sephadex G-100, G-150, and G-200 has been used to fractionate nitrogenases from four different sources. Successive protamine sulfate additions are reported to fractionate *Clostridium* nitrogenase; the Fe protein is precipitated by the first addition and the Mo–Fe protein is precipitated by the second addition. Temperature treatments can be used to selectively destroy either protein. The cold lability of the Fe protein permits it to be specifically inactivated with no loss of the Mo–Fe protein. The differential heat sensitivity of the proteins has been applied for selective destruction of the Mo–Fe protein activity.

So far, successful purification of the nitrogenase fractions has been limited to *Clostridium*, *Klebsiella*, and *Azotobacter* Mo–Fe and Fe proteins. Two methods will be described for the *Clostridium* proteins. In one method (Vandecasteele and Burris, 1970), *Clostridium* Mo–Fe protein is rechromatographed on DEAE cellulose to yield Mo–Fe protein with a specific activity of 345 nmoles N_2 reduced/minute/mg protein, the highest reported specific activity for Mo–Fe protein from this organism. The Fe protein is chromatographed on Sephadex G-100 to yield

TABLE III
CHARACTERISTICS OF PURIFIED MO–FE AND FE PROTEINS OF VARIOUS NITROGENASES

	Mo–Fe Protein							Fe Protein					
	Molecular weight (daltons)	Subunit weight (daltons)	Fe (gm atom/mole)	Mo (gm atom/mole)	S (gm atom/mole)	CySH (residues/mole)	Specific activity (nmoles N_2 reduced/minute/mg protein)	Molecular weight (daltons)	Subunit weight (daltons)	Fe (gm atom/mole)	S (gm atom/mole)	CySH (residues/mole)	Specific activity (nmoles N_2 reduced/minute/mg protein)
C. pasteurianum	168,000	50,700	14	1	11–16	21–23	230[c,d]	55,000	27,500	4	4	11	675[l,m,n]
		59,500											
	220,000[a]			2[a]									
	160,000		15	1			345[e]						345[e]
K. pneumoniae	218,000	51,000	17	1	17	17	380[f]	66,800	34,600	4	4	11	275[f]
		60,000											
A. vinelandii	270–300 × 10^3	70,000	32–38	2	26–28	41[b]	375[g,h]						532[o]
G. max	182,000		8.7	0.9			[i]	51,000		0.9			[i]
	132,000		15	0.6	14		[j,k]			50 ng atom/mg			[j]

[a] Most recent modification for homogeneous clostridial Mo–Fe protein is 220,000 daltons with 2 units of each subunit and 2 atoms of Mo (Huang *et al.*, 1973).
[b] Reported as ½ cystine.
[c] Dalton *et al.* (1971).
[d] Nakos and Mortenson (1971a).
[e] Vandecasteele and Burris (1970).
[f] Eady *et al.* (1972).
[g] Burns and Hardy (1970, 1972a).
[h] Burns *et al.* (1970)
[i] Bergersen and Turner (1970).
[j] Klucas *et al.* (1968).
[k] Israel *et al.*, (1972).
[l] Moustafa and Mortenson (1969).
[m] Nakos and Mortenson (1971b).
[n] Jeng *et al.* (1969b).
[o] Moustafa (1970).

Fe protein with specific activity of 460 nmoles N_2 reduced/minute/mg protein. In the other method (Mortenson, 1972; Dalton *et al.*, 1971; Moustafa and Mortenson, 1969), the Mo–Fe protein fraction is successively fractionated with three protamine sulfate precipitations and then chromatographed on a Sephadex column. A major drawback is the necessity to determine the amount of protamine at each step for each preparation. The Fe protein is precipitated with protamine sulfate, chromatographed on Sephadex G-100, and reprecipitated with protamine sulfate. The specific activity of the Mo–Fe protein obtained by this method is somewhat less, about 300 nmoles N_2 reduced/minute/mg protein, while that of the Fe protein is 675 nmoles N_2 reduced/minute/mg protein, the highest specific activity reported for any Fe protein. Homogeneity of the Mo–Fe protein is indicated by the occurrence of a single component after polyacrylamide disc gel electrophoresis. Absence of tryptophan and ultracentrifugal and disc gel electrophoretic examinations also indicate homogeneity for the Fe protein.

Relationship between purity and chemical and physical characteristics of the *Clostridium* Mo–Fe protein has shown considerable variability (Mortenson *et al.*, 1967; Dalton *et al.*, 1971; Dalton and Mortenson, 1970). However, a more recent paper (Dalton *et al.*, 1971) contains the following qualifying statement in the introduction, "a careful investigation of the molecular weight of Mo–Fd (Mo–Fe protein) was undertaken since values previously reported were only estimations which ranged from 100,000 to 200,000" (see Table III for a more careful investigation).

Methods have been recently described for the preparation of homogeneous Mo–Fe and Fe proteins from *Klebsiella* (Eady *et al.*, 1972). Crude extract is centrifuged and the supernatant fractionated on DEAE cellulose to provide the Mo–Fe and Fe proteins. The Mo–Fe protein is chromatographed on Sephadex G-200 to yield a homogeneous preparation of this component based on electrophoretic and sedimentation behavior. The component would not crystallize by the method described below for the corresponding *Azotobacter* protein. Overall yield was reported as 135%. The purified *Klebsiella* Mo–Fe protein had a specific activity of 380 nmoles N_2 reduced/minute/mg protein. The Fe protein was also chromatographed on Sephadex G-200, but two steps were required to separate native from O_2-damaged protein. This protein was indicated to be homogeneous on the basis of sedimentation and electrophoretic behavior. Overall yield was 20%, and the homogeneous protein had a specific activity of 275 nmoles N_2-reduced/minute/mg protein, a value substantially less than that found with purified *Clostridium* or partially purified *Azotobacter* Fe protein (Table III).

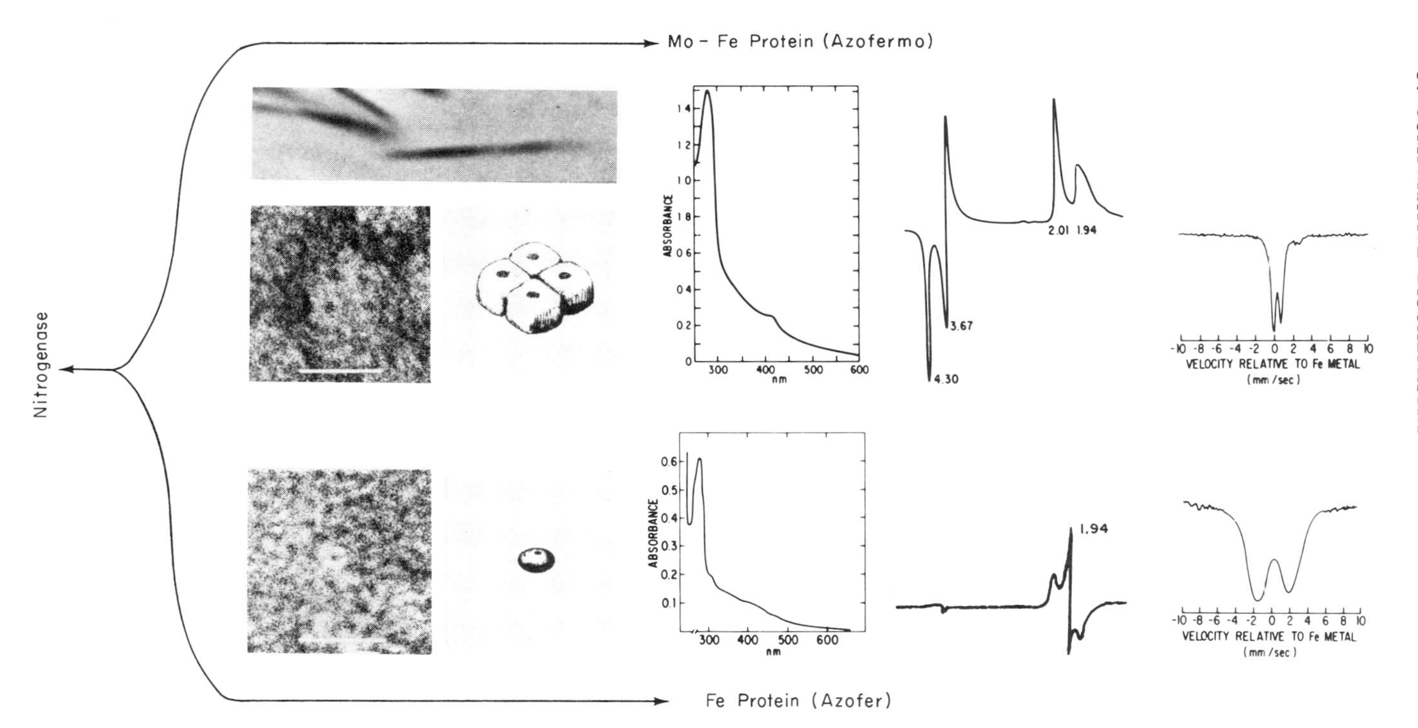

Fig. 2. Nitrogenase and characteristics of its component proteins, Mo–Fe and Fe protein. Light micrograph of crystalline *Azotobacter* Mo–Fe protein (Burns and Hardy, 1972a; Burns *et al.*, 1970), electron micrographs of negatively stained *Azotobacter* Mo–Fe and Fe proteins and models of each protein (Stasny *et al.*, 1971, 1973), UV–visible spectra of *Azotobacter* Mo–Fe protein (Burns and Hardy, 1972a; Burns *et al.*, 1970) and *Clostridium* Fe protein (Moustafa and Mortenson, 1969), ESR spectra of *Azotobacter* Mo–Fe protein (Burns and Hardy, 1972a; Burns *et al.*, 1970) and *Clostridium* Fe protein (Zumft *et al.*, 1973) and Mössbauer spectra of *Azotobacter* Mo–Fe (Burns and Hardy, 1972a; Weiher *et al.*, 1973) and Fe protein (Weiher *et al.*, 1973). All spectra except ESR are of native protein prepared under anaerobic conditions; ESR spectra are of native protein reduced with $Na_2S_2O_4$.

Homogeneous preparations of the Mo–Fe but not Fe protein of *Azotobacter* have been described (Burns and Hardy, 1972a; Burns *et al.*, 1970). The Mo–Fe protein fraction from *Azotobacter* crystallizes as white needles about 1–4 × 30–60 μm (Fig. 2). Crystallization is caused by reduction of the salt concentration from 0.25 to 0.08 *M* NaCl. The protein can be resolubilized to a brown solution and recrystallized as white needles. Crystallization of this protein is unusual in two aspects. The protein changes color and is crystallized by decreasing rather than increasing ionic strength. Crystallization of other Mo–Fe proteins has not yet been reported, although one would expect this method to be effective since ion-exchange chromatography indicates similar charge characteristics. Homogeneity of the *Azotobacter* Mo–Fe protein is indicated by constant specific activity and constant amino acid analyses through several recrystallizations and by sedimentation velocity pattern and single band on polyacrylamide gel electrophoresis (Stasny *et al.*, 1973). The specific activity of the recrystallized *Azotobacter* Mo–Fe protein is 375, similar to the activity reported for *Klebsiella* Mo–Fe protein.

The Fe protein of *Azotobacter* has not been purified to homogeneity. However, a method (Moustafa, 1970) utilizing cetyltrimethylammonium bromide in place of protamine sulfate for precipitation of nitrogenase followed by fractionation on DEAE cellulose, heat treatment, and protamine sulfate precipitation is reported to produce an Fe protein with a specific activity of 535 nmoles N_2 reduced/minute/mg protein. Further reports utilizing this technique are needed.

The purification procedure developed for the *Clostridium* and *Azotobacter* Mo–Fe protein appear to be able to provide luxuriant quantities of purified protein in only a few days. Purified protein should no longer be a limitation for characterization studies such as those described in the next section. Thus the report (Dalton *et al.*, 1971) for *Clostridium* states that: "This method yields about 2 g of pure protein and can be prepared within 2 days from 1 kg of dried cells," while that (Burns *et al.*, 1970) for *Azotobacter* emphasizes that: "The procedure is striking in its simplicity, requiring little time or specialized equipment. Exceptionally high recoveries permit the production of half-gram amounts of crystals by a single operator within several days of crude extract preparation."

V. CHARACTERISTICS OF NITROGENASE AND ITS COMPONENTS

The characteristics of nitrogenase and its Mo–Fe protein and Fe protein are summarized in this section. Topics are macromolecular organization, molecular weight and subunits, recombination, inactivation, chemical composition, and spectral characteristics.

A. Quaternary Structure, Molecular Weight, and Subunits

The quaternary structure of *Azotobacter* nitrogenase, Mo–Fe protein, and Fe protein have been investigated. Definitive results are reported for the Mo–Fe protein, while preliminary results are available for the Fe protein. A structure for nitrogenase, the enzymatically active complex, has not yet been identified.

The Mo–Fe and Fe proteins have been examined by electron microscopy. Results obtained with the resolubilized crystalline *Azotobacter* Mo–Fe protein are most meaningful since this protein was homogeneous and of high specific activity (Stasny *et al.*, 1971, 1973). Three different techniques were used to substantiate these conclusions. Electron micrographs from negative staining and shadow casting revealed a single type of square structure, $90 \times 90 \times 40$ Å, (Fig. 2) for which a molecular weight of 270,000 daltons was calculated, in close agreement with the values of 270,000–300,000 daltons based on sedimentation equilibrium and Mo content. This structural type was further established by the specific reaction of a ferritin conjugated immunoglobulin fraction of anti-Mo–Fe protein with the 90×90 Å particle. A tetrad subunit structure of $45 \times 45 \times 40$ Å particles for the Mo–Fe protein was indicated in high magnifications and was further suggested by the interaction of anti-Mo–Fe protein with each corner of the square. In addition, centers of increased density, possibly representing areas of concentration of Fe and Mo, were found within each of the substructures. The molecular weight of each subunit of the tetrad is calculated as 66,000 daltons, and polyacrylamide gel electrophoresis of Mo–Fe protein in dilute sodium dodecyl sulfate gave a single band for which an approximate size of 70,000 daltons was calculated. Another study (Slepko *et al.*, 1971) of *Azotobacter* Mo–Fe protein used a fraction with only 15% of the specific activity of that used for the above report. A tetrad was also observed in electron micrographs of this negatively stained Mo–Fe protein fraction. Dimensions of the structure were also 90×90 Å; however, a considerable proportion of the tetrads were dimeric so that a height of 90 Å and a cubic structure was proposed.

Electron micrographs of the *Azotobacter* Fe protein are less definitive because of the unavailability of homogeneous preparations (Stasny *et al.*, 1971, 1973). The basic structural subunit is reported as discoidal or ellipsoidal in shape with approximate dimensions of 35×40 Å and an estimated molecular weight of about 25,000 daltons in one study (Fig. 2), and spherical with dimensions of 35 ± 5 Å with an estimated molecular weight of 20,000 daltons in another study. This structural unit appears to repeat itself in dimers, tetramers, and hexagons; the latter appear to aggregate in a "stacked disk" conformation.

In another report (Gvozdev *et al.*, 1971) nitrogenase with a molecular weight of 340,000 ± 60,000 daltons by sedimentation analysis was examined by electron microscopy. The nitrogenase was composed of a high molecular weight and a low molecular weight complex. The high molecular weight complex was suggested to consist of eight subunits of molecular weight 35,000 ± 6000 daltons each with an overall cubic arrangement, while the low molecular weight complex was suggested to consist of four subunits of molecular weight 20,000 ± 6000 daltons each. Possible arrangements of the two types of subunits in the nitrogenase molecule were proposed.

Molecular weights of purified *Clostridium* (Dalton *et al.*, 1971), *Klebsiella* (Eady *et al.*, 1972), and *Azotobacter* (Burns *et al.*, 1970) Mo–Fe proteins have been determined by sedimentation or gel filtration (Table III). Homogeneous *Clostridium* Mo–Fe protein has a molecular weight of 160,000–220,000 daltons. Treatment with sodium dodecyl sulfate or with sodium mersalyl produces two subunits calculated to be 59,500 and 50,700 daltons by polyacrylamide gel electrophoresis (Nakos and Mortenson, 1971a). Microdensitometer tracings of the two electrophoretic bands from the Mo–Fe protein were first interpreted as two of the heavier subunits and one of the lighter but later as two of each. *Klebsiella* Mo–Fe protein, homogeneous by the criteria of disc electrophoresis and ultracentrifugation, has a molecular weight of 218,000 daltons. Treatment with β-mercaptoethanol and sodium dodecyl sulfate produces two types of subunits in equivalent amounts with molecular weights of 50,000 and 60,000 daltons. Homogeneous *Azotobacter* Mo–Fe protein has a molecular weight of 270,000–300,000 daltons on the basis of ultracentrifugation, Mo content, and electron microscopy. Mild treatment with sodium dodecyl sulfate produces a single type of subunit of 70,000 daltons, while stronger treatments yield two similar but not identical subunits of molecular weight of about 40,000 daltons. Exchange of only a few amino acids, e.g., serine for threonine, was observed between the subunits. No preparation of either the *Clostridium* or *Azotobacter* subunits are known to contain metals; this markedly limits the utility of these subunits for characterization of nitrogenase. The molecular weight of the less purified Mo–Fe protein from *Glycine max* is 182,000 daltons (Bergersen and Turner, 1970) calculated from gel filtration and 132,000 daltons by ultracentrifugation (Israel *et al.*, 1972). An unusual component of 93,000 daltons has been described as one component of a three-component

Azotobacter nitrogenase (Kajiyama *et al.*, 1969); this atypical result requires further clarification.

The molecular weight of purified *Clostridium* Fe protein is 55,000 daltons based on gel filtration (Nakos and Mortenson, 1971b). This value is an upward revision of a previous report of 40,000 (Moustafa and Mortenson, 1969; Jeng *et al.*, 1969b). The *Clostridium* Fe protein is dissociated by sodium dodecyl sulfate to give two identical subunits of 27,500 daltons each. The similarity of the subunits is based on single C- and N-terminal amino acids. Metal and sulfide are lost during dissociation, as with the Mo–Fe protein. The purified Fe protein from *Klebsiella* (Eady *et al.*, 1972) has a molecular weight of 66,800 daltons. Treatment with β-mercaptoethanol and sodium dodecyl sulfate produced a single type of subunit of molecular weight 34,600 daltons. The partially purified Fe protein from *Glycine max* bacteroids (Bergersen and Turner, 1970) has a molecular weight of 51,000 daltons based on gel filtration.

B. Recombination of Mo–Fe and Fe Protein

Active nitrogenase can be reconstituted by addition of purified Mo–Fe and Fe proteins from the same, and, in many cases, different source organisms. Requirements for additional components (Taylor, 1969) have not been substantiated (Jeng *et al.*, 1969a). The recombining ratio has been determined for homologous recombinations of *Clostridium* (Vandecasteele and Burris, 1970; Jeng and Mortenson, 1969), *Bacillus* (Kelly, 1969b), *Klebsiella* (Kelly, 1969b), *Mycobacterium* (Biggins *et al.*, 1971), *Azotobacter* (Kelly, 1969a), and *G. max* bacteroid (Murphy and Koch, 1971) components. In general, the optimum w/w ratio of Mo–Fe to Fe protein is 2–3 for N_2 fixation, C_2H_2 reduction, or reductant-dependent ATP utilization. Higher ratios of up to 6.5 have been reported for CN^- or CH_3NC reduction (Kelly, 1969a,b; Biggins *et al.*, 1971), and an altered ratio also favors non-reductant-dependent ATP utilization (Kelly, 1969a; Bui and Mortenson, 1969; Jeng *et al.*, 1970). Recently the ratio of ATP utilized per two electrons transferred has been shown to be markedly affected by the ratio of the two components with excess Mo–Fe protein increasing the ratio from 4 to 20 (Ljones *et al.*, 1972). The molecular weight of nitrogenase has never been reported, but the recombining ratio suggests the combination of about 150,000 daltons of Mo–Fe protein (1 Mo:16 Fe:16 S) with 50,000 daltons of Fe protein (4 Fe:4 S) to yield a nitrogenase complex of about 200,000 daltons (1 Mo:20 Fe:20 S) or multiples thereof. These ratios can be accepted only with some reservation at present because the composition of fractions is not always well defined.

TABLE IV
INTERCHANGEABILITY OF NITROGENASE COMPONENTS FROM DIFFERENT PHYSIOLOGICAL TYPES OF ORGANISMS[a]

Fe protein source \ Mo–Fe protein source	Anaerobes	Facultative anaerobes	Photosynthetic bacteria	Algae	Aerobic bacteria	Symbionts
Anaerobes		− +	−		− −	−
Facultative	−	+	+		+ + +	+
anaerobes	−	+	+		+ + +	
Photosynthetic						
bacteria				+		
Algae			−			
Aerobic	−	+ + −	+		+	+
bacteria		+ + +			+	
Symbionts					+	

[a] + and − indicates result of nitrogenase assay for a specific recombination from Detroy *et al.*, 1968; Hardy *et al.*, 1968; Kelly, 1969b; Smith *et al.*, 1971b; Biggins *et al.*, 1971; Murphy and Koch, 1971.

The interchangeability of Mo–Fe and Fe proteins from different sources has been examined with over thirty combinations (Table IV). The results, in general, suggest similarities of the nitrogenase components. Positive cross reactions with activities from 8 to 142% of the homologous combination were obtained in 23 of the 25 cases involving components from facultative anaerobes, photosynthetic bacteria, algae, aerobes, and symbionts while only one positive cross reaction was observed in the nine cases where either the Mo–Fe or Fe component isolated from an anaerobic organism was combined with a component from any other group. It is concluded that a high degree of interchange is possible with nitrogenase components from other than strict anaerobes. Nitrogenase reactions tested in recombination experiments included N_2 fixation and reduction of alternate substrates. The degree of interchange was, in general, similar for both N_2 fixation and reduction of alternate substrates. In some combinations the C_2H_6/C_2H_4 ratio of products from isonitrile reduction varied in a manner that was correlated with the source of the Mo–Fe protein and thereby suggested the reduction site for at least isonitrile was located in this protein (Kelly, 1969b).

Genetic studies with *Azotobacter* have produced mutants lacking one or both of the proteins of nitrogenase (Fisher and Brill, 1969; Sorger and Trofimenkoff, 1970). Such bacteria are, of course, incapable of fixing N_2. Recently nitrogenase activity has been transferred by transduction (Streicher *et al.*, 1971) or conjugation (Dixon and Postgate, 1971) to non-N_2-fixing mutants of *Klebsiella,* and the not unexpected extension of these conjugation experiments has led to the transfer of N_2-fixing activity to *Escherichia coli,* an organism heretofore unable to fix N_2 (Dixon and Postgate, 1972). Exciting developments are anticipated in this virgin area.

A sigmoidal relationship has frequently been observed between activity and nitrogenase concentration (Bulen *et al.*, 1965a; Hardy *et al.*, 1968; Yates, 1970; Sorger, 1971). Recent results show that the sigmoidicity can be eliminated by increasing concentrations of ATP (Burns and Hardy, 1972b). It is suggested that ATP functions as an activator in formation of the enzymically active form of nitrogenase through either combination of the component proteins or conformational changes of the nitrogenase complex (Kennedy, 1970b; Silverstein and Bulen, 1970; Burns and Hardy, 1972b). This development may necessitate the reevaluation of earlier kinetic data as well as function of added ferredoxins or flavodoxins where stimulation, not restoration, of activity is observed (Section III). Another recent development provides the first evidence which attributes a metabolic regulator role of ammonia on nitrogenase activity (L'vov *et al.*, 1971).

C. Inactivation

All purified nitrogenases are extremely sensitive to O_2. The sensitivity towards O_2 far exceeds that of most other iron–sulfur proteins, with up to 70–75% loss of activity after 10 minutes of exposure to air. The Fe protein is more O_2 sensitive than the Mo–Fe protein. For example, one-half of the activity of *Klebsiella* Fe protein is lost after a 45-second exposure to air, while a 10-minute exposure is required for similar loss of activity by the Mo–Fe protein (Eady *et al.*, 1972). The ability of oxidized Fe protein to inhibit nitrogenase activity (Biggins and Kelly, 1970) suggests that oxidized Fe protein can complex with Mo–Fe protein but forms an inactive nitrogenase. Some N_2-fixing extracts such as those from *Azotobacter* (Bulen and LeComte, 1966; Bulen *et al.*, 1965b) are initially insensitive to O_2 but after a few steps of purification become sensitive and necessitate the use of anaerobic conditions. On the basis of the ubiquitous O_2 sensitivity of purified nitrogenase, it is not surprising that three groups of bacteria—anaerobic, facultative anaerobic, and photosynthetic—fix N_2 only under anaerobic conditions. Two other groups—blue-green algae and aerobic bacteria—are considerably more active in fixation of N_2 at less than atmospheric pO_2 (Biggins and Postgate, 1969; Dalton and Postgate, 1969; Stewart and Lex, 1970). In fact, growth of algae under anaerobic conditions was used to obtain the first reproducible N_2-fixing algal extracts (Smith and Evans, 1970). Algal associations such as lichens are substantially more active than the free algae (Millbank, 1972). This enhancement may be attributed to the relative anaerobicity of the algae within the lichen.

Two regulatory processes, enhanced respiration and conformational change, have been proposed to protect the nitrogenase of the aerobe, *Azotobacter*, from excessive oxygenation. Conformation protection is suggested to operate via a reversible conformational change of nitrogenase, which protects it from O_2 damage, but also prevents N_2 fixation (Drozd and Postgate, 1970).

Nitrogenase in legume symbionts does not show O_2 sensitivity until the pO_2 reaches about 0.5 atm (Bergersen, 1970). Protection from O_2 in the legume symbiont can be attributed to the specialized cellular organization of the nodule which may function to exclude O_2 from nitrogenase by either a metabolic or membrane (Parshall and Hardy, 1973) effect. In legume symbionts nitrogenase is located in the pleiomorphic *Rhizobium*, called a bacteroid, which is enclosed within vesicles within the host plant cells of the nodule.

Most nitrogenases also exhibit cold lability. Inactivation by cold was first observed with crude extracts from *C. pasteurianum* where storage for 24 hours at 0°C produced loss of activity (Dua and Burris, 1963). Some stabilization of activity toward cold is provided by 10% ethanol or acetone (Moustafa and Mortenson, 1969). Cold lability has now been reported for nitrogenases from *C. pasteurianum* (Moustafa and Mortenson, 1969; Dua and Burris, 1963), *A. cylindrica* (Haystead *et al.*, 1970), *P. boryanum* (Haystead *et al.*, 1970), and *M. flavum* (Biggins and Postgate, 1971; Biggins *et al.*, 1971). Reports of cold lability of *G. max* bacteroid nitrogenase have varied with the laboratory (Evans *et al.*, 1972; Bergersen and Turner, 1970), and purified *Klebsiella* Mo–Fe protein does not show cold lability (Eady *et al.*, 1972). Crude extracts of *Chromatium* (Winter and Ober, 1971; Winter and Arnon, 1970) and *Azotobacter* (Bulen and LeComte, 1966; Bulen *et al.*, 1965b) do not exhibit cold lability just as they do not show O_2 sensitivity. However, purified extracts do exhibit cold lability. The cold lability of nitrogenase is associated with the Fe but not the Mo–Fe protein. The effect of cold inactivation is different from that of oxidation on the basis of spectral, metal, and sulfur changes.

The Fe protein, which is more sensitive to cold than O_2, is also more rapidly inactivated by NO than is the Mo–Fe protein (Kelly and Lang,

1970). Treatment of the *Klebsiella* Fe and Mo–Fe protein with 1.5% NO for 5 minutes completely destroys activity of the Fe protein but does not affect the activity of the Mo–Fe protein. The effect of NO and O_2 is suggested to be different, since crude *Azotobacter* nitrogenase is insensitive to O_2 but irreversibly inhibited by NO.

D. Chemical Composition

The Mo–Fe and Fe proteins of nitrogenase have been examined with respect to nature of metals, metal content, labile sulfide content, and cysteine content. In addition, end group analysis and amino acid composition have been reported in some cases. No amino acid sequences are available yet.

Highly purified *Clostridium* and *Klebsiella* and crystalline *Azotobacter* Mo–Fe protein have been extensively analyzed (Burns and Hardy, 1972a; Burns *et al.*, 1970; Dalton *et al.*, 1971; Eady *et al.*, 1972). Molybdenum and iron are the only metals consistently present in stoichiometric amounts; a single report of the presence of Zn in an atypical *Azotobacter* fraction remains unexplained (Kajiyama *et al.*, 1969). The *Clostridium* protein contains 1 Mo, 14–15 Fe, 11–16 labile sulfide (S^{2-}), and 21–23 cysteine (CySH) per molecule of 160,000–170,000 daltons (Dalton *et al.*, 1971; Vandecasteele and Burris, 1970; Nakos and Mortenson, 1971a). However, a recent report supports an earlier estimate of 2 Mo per protein of molecular weight of about 200,000 daltons (Huang *et al.*, 1973). Titration of *Clostridium* Mo–Fe protein with mersalyl detected 55 titratable SH equivalents composed of 16 S^{2-} groups or 32 SH equivalents and 23 sulfhydryl groups. The *Azotobacter* protein contains 2 Mo, 32–38 Fe, 26–28 S^{2-}, and 40 one-half cystines per 270,000–300,000 daltons (Burns and Hardy, 1972a; Burns *et al.*, 1970), while the *Klebsiella* Mo–Fe protein contains 1 Mo, 17 Fe, 17 S^{2-}, and 17 CySH per molecule of 218,000 daltons (Eady *et al.*, 1972). The latter results indicate a substantial increase in Fe over previous reports on *Klebsiella* Mo–Fe protein (Kelly and Lang, 1970). The range of average ratios of 1:14–16:11–17:17–23 for Mo:Fe:S^{2-}:CySH for the three purified Mo–Fe proteins indicates their similarity. Values reported for Mo and Fe content of impure Mo–Fe proteins from *G. max* bacteroids are in general agreement with those above (Bergersen and Turner, 1970; Klucas *et al.*, 1968).

The amino acid composition of *Azotobacter* (Burns and Hardy, 1972a), *G. max* bacteroid (Israel *et al.*, 1972), and *Klebsiella* (Eady *et al.*, 1972) Mo–Fe protein has been reported (Table V); they contain a full complement of 18 amino acids. There is a high proportion of acidic amino acids, >20% of the residues, and these are about twice as prevalent as

TABLE V
AMINO ACID COMPOSITION OF NITROGENASE PROTEINS

	Mo–Fe protein			Fe protein
	A. vinelandii[a]	*G. max* bacteroid[b]	*K. pneumoniae*[c]	*K. pneumoniae*[d]
Aspartic	249	120	199	60
Threonine	115	59	99	36
Serine	134	72	104	24
Glutamic	250	118	195	84
Proline	101	59	87	18
Glycine	206	112	144	60
Alanine	169	107	150	60
Valine	173	83	118	42
Methionine	86	26	72	36
Isoleucine	134	74	95	48
Leucine	190	89	173	42
Tyrosine	79	43	65	18
Phenylalanine	102	53	91	12
Histidine	55	35	46	6
Lysine	177	80	97	36
Arginine	108	62	97	24
Tryptophan	50	19	53	0
Half-cystine	41	15	36	18

[a] Residues per 270,000 MW (Burns and Hardy, 1972a).
[b] Residues per 132,000 MW (Israel *et al.*, 1972).
[c] Residues per 218,000 MW (Eady *et al.*, 1972).
[d] Residues per 68,000 MW (Eady *et al.*, 1972).

basic residues, as expected. Although the complete amino acid composition of *Clostridium* Mo–Fe protein (Dalton *et al.*, 1971; Nakos and Mortenson, 1971a) has not been reported, the percent content of methionine and cysteine are similar to the *Azotobacter* protein.

Highly purified *Clostridium* (Dalton *et al.*, 1971) and *Klebsiella* (Eady *et al.*, 1972) Fe protein have been analyzed. *Clostridium* Fe protein contains 4 Fe, 4 S^{2-}, and 11 CySH per 55,000 daltons, and *Klebsiella*, 4 Fe, 4 S^{2-}, and 11 CySH per 67,000 daltons. Other purified Fe proteins have been found to contain only Fe, but are not of sufficient purity to justify serious comparison of their Fe or S^{2-} content. Both the *Clostridium* and *Klebsiella* Fe proteins lack tryptophan; the *Klebsiella* protein contains the other amino acids with the expected preponderance of the acidic amino acids (Table V).

E. Spectral Characteristics

Nitrogenase and its component proteins have been examined with a variety of spectral techniques including determination of UV, visible and circular dichroism, ESR, and Mössbauer spectra. In comparison with other iron–sulfur proteins, such as the ferredoxins, similarities and unique differences are noted.

The UV and visible spectra of the Mo–Fe and/or Fe proteins from *Clostridium* (Dalton *et al.*, 1971; Moustafa and Mortenson, 1969), *Klebsiella* (Eady *et al.*, 1972), *Azotobacter* (Burns and Hardy, 1972a; Burns *et al.*, 1970), and *Glycine max* bacteroids (Bergersen and Turner, 1970; Klucas *et al.*, 1968) and the circular dichroism spectra of both proteins from *Klebsiella* (Eady *et al.*, 1972) have been reported. Both proteins give brown solutions, and as in the case of other iron–sulfur proteins with more than 2 Fe and 2 S^{2-} per molecule, there are no prominent absorption maxima and only broad absorption from about 300 to 600 nm, e.g., *Azotobacter* Mo–Fe protein (Fig. 2). As with other iron–sulfur proteins, a substantial amount of the absorption from 260–280 nm is contributed by the iron–sulfur groups. The molar extinction coefficients at 280 nm are about 250×10^3, 320×10^3, and 470×10^3 for *Clostridium, Klebsiella,* and *Azotobacter* Mo–Fe proteins, respectively; however, the molar extinction coefficient per iron is between $15–19 \times 10^3$ for each protein. The molar extinction coefficient for *Klebsiella* Fe protein is 112×10^3 at 265 nm or 28×10^3 per iron. Addition of mersalyl to *Clostridium* Fe protein (Fig. 2) eliminates essentially all of the absorption above 350 nm and decreases the absorption at 280 nm by at least one-third. This effect of mersalyl is similar to its effect on ferredoxin. Mersalyl also decolorizes the Mo–Fe protein. Reduction of the native *Azotobacter* Mo–Fe protein produces a minor change with a red shift in the absorption at 412–420 nm and the development of weak maxima at 525 and 557 nm, while the ultraviolet spectra of the oxidized and reduced *Klebsiella* protein were identical. Other minor spectral changes in the area 300–400 nm have been produced with pH changes from 5.6 to 7.2 in the case of *Clostridium* Mo–Fe protein.

Spectral comparison of *Azotobacter* Mo–Fe protein with an analogous protein from *Azotobacter* V-nitrogenase (see Section VI) is informative (Burns and Hardy, 1971). This latter protein is devoid of S^{2-} and Mo and lacks 60% of the iron of Mo–Fe protein; however, the extinction coefficient at 280 nm is substantially reduced. It may be suggested that Mo and S^{2-} play no specific role in the visible absorption spectrum of the Mo–Fe protein.

The circular dichroism spectra of oxidized or reduced *Klebsiella*

Mo–Fe protein were similar, consisting mainly of a negative trough at 220 nm with $[\phi]_{220} = -11{,}600^\circ$ cm^2 dmole^{-1}, while the reduced Fe protein also showed a negative trough at 220 nm with $[\phi]_{220} = -10{,}000^\circ$ cm^2 dmole^{-1}. In contrast to other iron–sulfur proteins (Tsibris and Woody, 1970), the nitrogenase proteins exhibited no circular dichroism in the visible region. An α-helical content of 38 and 33% for the Mo–Fe and Fe protein has been calculated from the circular dichroism results.

Electron spin resonance spectra have been obtained for nitrogenase and/or its component proteins from *C. pasteurianum* (Dalton *et al.*, 1971; Jeng *et al.*, 1969b; Zumft *et al.*, 1973), *C. ethylicum* (Evans *et al.*, 1971), *K. pneumoniae* (Eady *et al.*, 1972), and *A. vinelandii* (Burns and Hardy, 1972a; Burns *et al.*, 1970; Novikov *et al.*, 1968; Davis *et al.*, 1972; Matkhanov *et al.*, 1969; Ivleva *et al.*, 1969a,b; Syrtsova *et al.*, 1969, 1971; Frolov *et al.*, 1971). Purified *Clostridium, Klebsiella,* and *Azotobacter* Mo–Fe protein in the native state possess resonances at g values of 2.01, 3.7, and 4.3 when examined at temperatures below about 30°K (Burns and Hardy, 1972a; Burns *et al.*, 1970; Dalton *et al.*, 1971; Eady *et al.*, 1972). Reduction enhances the intensity of the resonances at g values of 3.7 and 4.3, while oxidation of the protein destroys the resonance at a g value of 3.7 and also destroys biological activity. The protein analogue of Mo–Fe protein from V-nitrogenase does not have resonance at 3.7 in the native or reduced state. It is not yet clear if a resonance occurs at a g value of 1.94 in reduced Mo–Fe proteins as observed in other reduced iron–sulfur proteins (Burns and Hardy, 1972a; Davis *et al.*, 1972). The resonance at a g value of 3.7 has been pointed out as unique to the Mo–Fe protein of nitrogenase (Hardy *et al.*, 1971). It has been variously suggested that the latter resonance arises from high-spin Fe^{2+} or intermediate-spin Fe^{3+} which has a function unique to N_2 fixation. This resonance at 3.7 occurs in such an unusual area that it can be used to diagnose the presence of Mo–Fe protein in intact bacterial cells (Davis *et al.*, 1972), and it should provide a powerful probe to follow the Mo–Fe protein in living organisms.

The Fe proteins of *Clostridium, Chloropseudomonas,* and *Azotobacter* have been examined also for ESR. Resonance at a g value of 1.94 has been reported for *Klebsiella* (Eady *et al.*, 1972) and *Clostridium* (Zumft *et al.*, 1973); this resonance disappeared upon inactivation by oxygen. Oxidation of the *Clostridium* Fe protein destroys activity and also produces a signal at a g value of about 2 (Jeng *et al.*, 1969b). ATP changes the spectrum of *Clostridium* Fe protein from g values of 2.06, 1.94, and 1.87 to 2.04 and 1.93.

The ESR technique and the characteristic iron–sulfur protein resonance at a g value of 1.94 have been intensively utilized for attempted characterization of the iron–sulfur structure and reactions of nitrogenase. For example, one report suggests that the resonance at a g value of 1.94 in reduced *Chloropseudomonas* Mo–Fe protein is altered in intensity by a combination of ATP and a reducible substrate of nitrogenase, N_2 (Evans *et al.*, 1971). Other results with N_2-fixing extracts of *Azotobacter* also suggest an effect of N_2, ATP, and an inhibitor of N_2 fixation, CO, on the resonance at g values of 1.93 and 2.01 (Matkhanov *et al.*, 1969; Ivleva *et al.*, 1969a). The effect of NO, *p*-chloromericuribenzoate, and diazobenzenesulfoxylate on the ESR spectra have also been examined (Ivleva *et al.*, 1969b). Results of these ESR studies have led to a molecular orbital diagram and sketch of a proposed iron–sulfur structure of nitrogenase (Syrtsova *et al.*, 1969). Unfortunately, most of this sophisticated spectral work has been done with preparations whose activity, when reported, was only about 10% of that of homogeneous preparations or whose $Fe:S^{2-}$ ratio was about 3:1 versus the characteristic 1:1 of highly active preparations. The occurrence of several other iron–sulfur proteins besides nitrogenase in extracts of N_2-fixing bacteria such as *Azotobacter* render data on impure nitrogenase or its fractions nondefinitive. It is suggested that future work in this area should employ high-purity preparations and include the unique resonance at 3.67 as well as that at a g value of 1.94 as in recent reports (Zumft *et al.*, 1973; Mortenson *et al.*, 1973).

No resonances attributable to the Mo of the Mo–Fe protein have been observed in either native or reduced preparations. The absence of resonance may be due to the absence of Mo in the correct valence state, masking by nearby iron resonance, or spin coupling. One report indicates that denaturation with acid and reduction produces a resonance at a g value of 1.98 which is consistent with Mo (Ivleva *et al.*, 1969a).

Mössbauer spectra of nitrogenase or its component proteins from *Azotobacter* (Hardy *et al.*, 1971; Burns and Hardy, 1972a; Burns *et al.*, 1970; Novikov *et al.*, 1968; Syrtsova *et al.*, 1969; 1971; Frovlov *et al.*, 1971; Moshkovskii *et al.*, 1967; Ivanov *et al.*, 1968; Weiher *et al.*, 1973) and *Klebsiella* (Eady *et al.*, 1972; Kelly and Lang, 1970) have been recorded. In all cases the [^{57}Fe]nitrogenases have been obtained by isolation from bacteria grown on ^{57}Fe-enriched media, rather than exchange of nitrogenase iron. *Azotobacter* Mo–Fe protein (Burns and Hardy, 1972a; Burns *et al.*, 1970; Weiher *et al.*, 1973) in the native state yields a doublet with an isomer shift of 0.39 mm/second and a quadrupole split of 0.84 mm/second indicative of high-spin Fe^{3+} or Fe^{+} and has been suggested to possibly represent the unique portion of the iron involved in N_2 complexation. Reduction with hydrosulfite converts up to 50% of the iron to a form representative of high-spin Fe^{2+}

or Fe^+. Mössbauer spectra of native *Azotobacter* Fe protein show a single dipole with isomer shift of 0.64 mm/second and quadrupolar split of 0.84 mm/second. Reduction with $Na_2S_2O_4$ gives conversion to a dipole with isomer shift of 1.58 mm/second and quadrupole split of 2.95 mm/second.

Mössbauer spectroscopy has been used for an intensive analysis of each of the Mo–Fe and Fe proteins from *Klebsiella* as well as their combination supplemented with various substrates (Kelly and Lang, 1970; Postgate, 1971c). Unfortunately, it is impossible to evaluate the quality of the nitrogenase preparations used since values of specific activity were not reported. Recent results (Eady *et al.*, 1972) on homogeneous *Klebsiella* proteins negate most of the earlier results as indicated by the authors: "Purification of Kp_2 led to considerable changes in the Mössbauer spectra compared with those reported by Kelly and Lang (1970)." The Mössbauer spectrum of reduced *Klebsiella* Mo–Fe protein consisted of three doublets of $\delta = 0.65$, 0.60, and 0.35 mm/second and $\Delta\epsilon = 3.05$, 0.8, and 0.7 mm/second at 4.2°K, while that of the Fe protein exhibited a doublet at 77°K with $\delta = 0.45$ mm/second and $\Delta\epsilon = 1.1$ mm/second which broadened into a multiplet at 4.2°K.

Other nonspectral characteristics including heat of solution and magnetic susceptibility of *Azotobacter* Mo–Fe protein have been recorded (Burns and Hardy, 1972a; Burns *et al.*, 1970). The heat of solution is —5.5 kcal/mole, and the protein solubility is highly responsive to ionic strength. The magnetic susceptibility measured by both NMR and Farady cage techniques is about three Bohr magnetons per Fe.

Spectral studies of nitrogenase continue to define the characteristics of the static components. These characteristics show similarities to, as well as unique differences from, other iron–sulfur proteins. There is a need for additional spectral work on nitrogenase and its components utilizing only preparations of the highest specific activity. The spectral techniques appear to be useful for definition of the dynamics of nitrogenase activity. Again, high-purity preparations are mandatory. The iron-based spectral techniques—ESR and Mössbauer—would be facilitated by nitrogenase preparations with lower iron contents. Nitrogenase produced by *Clostridium* grown on low levels of iron, such as used for flavodoxin isolation, should be examined for lower iron content. The size of the nitrogenase components has excluded application of the high-resolution PMR technique. Production of metal-containing subunits capable of being reconstituted into an effective nitrogenase would provide useful species for characterization. Exchange or replacement of Fe, as well as exchange or replacement of sulfur with, for example, selenium may also provide useful information. The role of metals will be discussed in the next section, including replacement of Mo with V.

VI. METALS AND BIOLOGICAL DINITROGEN FIXATION

This section will summarize evidence for a functional role of metals in biological N_2 fixation. Examples of this evidence include the following.

1. Nutritional requirement for additional Fe and Mo for growth on N_2 versus fixed N.

2. Inhibition of nitrogenase reaction by chelating agents. *Azotobacter* nitrogenase is inhibited by *o*-phenanthroline, α,α'-dipyridyl, and tiron. Thiol reagents also inhibit activity (Hardy and Knight, 1966; Bulen *et al.*, 1965b).

3. Occurrence of metals in both proteins and correlation of removal of metals with loss of activity. Isolated *Azotobacter* V–Fe protein is deficient in metal and S^{2-} (Burns and Hardy, 1971). Treatment of *Clostridium* Mo–Fe protein with sodium mersalyl removes all of the Mo and about 60% of the iron and destroys activity (Dalton and Mortenson, 1970). Dialysis at pH 9.0 or treatment with *o*-phenanthroline or 8-hydroxyquinoline is reported to remove the Mo but not the Fe of *Azotobacter* nitrogenase (Gvozdev *et al.*, 1969). α,α'-Dipyridyl under anaerobic conditions reacts with 50–60% of the iron of *Clostridium* Mo–Fe protein. Preliminary evidence (Ward *et al.*, 1971) may suggest that iron can be removed by α,α'-dipyridyl in a stepwise manner. Removal of up to three iron atoms from a preparation of Mo–Fe protein containing 14 iron atoms does not affect activity, while removal of seven iron atoms destroys non-reductant-dependent ATP utilization.

4. Presence of unique iron in nitrogenase based on ESR and Mössbauer spectra and changes in these spectra produced by reductants effective for nitrogenase activity (see Section V,E).

5. Nature of substrates: N_2, N_3^-, N_2O, RCN, RNC, and RCCH; and nature of inhibitors: CO, NO, and H_2; and similarities of their affinities for nitrogenase and transition metal complexes (Hardy *et al.*, 1971, 1973). Utilization of $^{14}CN^-$ and [^{14}C]ATP binding to define the role of each of the Mo–Fe and Fe proteins has produced equivocal results (Biggins and Kelly, 1970; Bui and Mortenson, 1968); however, specific binding of $^{14}CN^-$ by the nitrogenase complex is reported to involve ATP and $S_2O_4^{2-}$ and to be inhibited by CO (Biggins and Kelly, 1970).

6. Similarity of nitrogenase-catalyzed reactions to transition metal catalyzed reactions such as N_2 reduction, N_3^- reduction, N_2O reduction, RCN reduction, CH_3NC insertion reactions, and C_2H_2 reduction (Hardy *et al.*, 1971, 1973). Section VII describes Fe- and/or Mo-containing models that effectively mimic the nitrogenase reaction, including N_2 reduction in protonic media and ATP-stimulated H_2 evolution.

7. Catalysis of D_2–H_3O^+ exchange by N_2-fixing nitrogenase and a

metal-containing nitrogenase model (Jackson *et al.*, 1968; Parshall, 1967; Turner and Bergersen, 1969). Nitrogenase catalyzes an exchange between D_2 and H_3O^+. The exchange is dependent upon a specific substrate, N_2, and in addition requires conditions for the reduction of N_2 to NH_3. It is suggested that protons of diazene or hydrazine complexed to Fe and/or Mo are responsible for the exchange. Support for this mechanism is provided by a model of the proposed transition metal-complexed diazene and hydrazine intermediates of N_2 fixation. These models catalyze exchange between D_2 and protonic media.

8. Altered kinetics of the nitrogenase reaction when Mo is replaced by V (Burns and Hardy, 1971; Burns *et al.*, 1971; Fuchsman and Hardy, 1972; McKenna *et al.*, 1970). Replacement of iron with, for example, ruthenium has not been successful.

Comparison of the kinetic characteristics of vanadium-nitrogenase with molybdenum-nitrogenase provides the first direct evidence defining a functional role of metals in the nitrogenase reaction. It has long been known that some N_2-fixing bacteria can utilize V to satisfy their nutritional requirement for Mo. A nitrogenase, designated V-nitrogenase was isolated from *Azotobacter* grown on N_2 in medium supplemented with V in place of Mo. V-Nitrogenase was similar to Mo-nitrogenase in purification procedure, requirements for activity, reactions catalyzed, and biphasic Arrhenius plot. V-Nitrogenase was less active and less stable to heat and in storage than Mo-nitrogenase. A protein was isolated from V-nitrogenase utilizing procedures effective for Mo–Fe protein. This protein was similar to Mo–Fe protein in UV–visible spectra, solubility properties, antigenicity, and amino acid analyses but contained only traces of V and Mo, no S^{2-}, and about 40% as much Fe as Mo–Fe protein. These data suggest that V-nitrogenase is similar to Mo-nitrogenase with the exception that V replaces Mo and the resultant V–Fe protein is less stable than the corresponding Mo–Fe protein.

Comparison of the kinetics of V-nitrogenase and Mo-nitrogenase implicates Mo in substrate and inhibitor binding, electron transfer, and product release. The K_m for N_2 is similar for both Mo- and V-nitrogenase, indicating that Mo is not involved in the initial complexation of N_2; on the other hand, the K_m for C_2H_2 and nitriles are about fivefold larger for V- than for Mo-nitrogenase, indicating an intimate role of Mo in the complexation of these substrates. Carbon monoxide is a potent inhibitor of the reduction of all substrates of nitrogenase except H_3O^+. The K_i of CO for inhibition of N_2, as well as C_2H_2, reduction is about fivefold higher for V-nitrogenase than Mo-nitrogenase. This effect of CO suggests that Mo is involved in complexation with CO and that Mo is involved in N_2 reduction as well as C_2H_2 reduction. Another inhibitor of the reduction of N_2, but not other substrates, is H_2. The K_i of this inhibitor is similar for V- and Mo-nitrogenase. Additional evidence for an intimate role of Mo in N_2 fixation is the increased electron allocation for reduction of H_3O^+ versus exogenous substrates such as N_2, C_2H_2, etc., by V-nitrogenase.

Acrylonitrile is reduced by nitrogenase to propylene and propane. The propylene:propane product ratio is about 5–7 for Mo-nitrogenase and is reduced to 1–3 for V-nitrogenase. This altered product ratio supports a role for Mo in the complete reduction process, including product dissociation.

It has been suggested that a molybdenum-containing component is shared by all known molybdenum enzymes including nitrogenase (Nason *et al.,* 1971). Basis for this proposal is the formation of assimilatory NADPH-nitrate reductase by *in vitro* incubations of extracts of nitrate-induced *Neurospora crassa* mutant *nit-1* with acid-treated extracts of *Clostridium, Azotobacter,* or *G. max* bacteroid nitrogenase or Mo–Fe protein. It is of possible concern that the least effective source of the proposed molybdenum component was the highest-purity Mo–Fe protein. Successful attempts to replace the Mo–Fe protein of nitrogenase with an acidified extract of a non-nitrogenase molybdenum-containing protein have not been reported.

The mechanism of nitrogenase reaction will be briefly discussed. Various nitrogenase mechanisms and sites have been described (Hardy *et al.,* 1973). All prosposals include Fe and/or Mo as key components. The combined available biological (Sections V and VI) and model data (Section VII) support a dinuclear site containing Mo and a second metal which may be Fe. One such possible site involving Mo and Fe bridged by inorganic sulfur is shown in Fig. 3 (Hardy *et al.,* 1971). This site is drawn to accomodate the various analogue substrates and inhibitor data. Complexation of N_2 or H_2 to the Fe site and of the other substrates and CO to the Mo site is proposed. Subsequent interaction of the N_2 with reduced Mo produces a dinuclear diazene intermediate. Subsequent two electron additions give hydrazine and ammonia and regenerate the site for another reduction cycle.

VII. ABIOLOGICAL DINITROGEN FIXATION BY IRON AND SULFUR

Several abiological N_2 fixation reactions involving iron as well as molybdenum are now known. These reactions are of four basic types: (1) Haber-Bosch, (2) Vol'pin and Shur, (3) transition metal—N_2 complex, and (4) transition metal complex and reductants in protonic media.

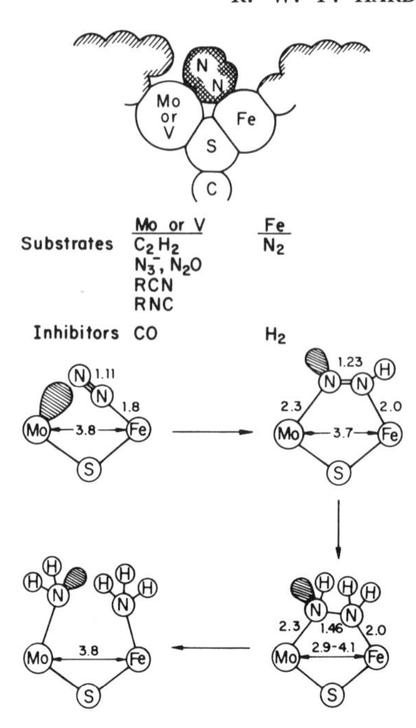

Fig. 3. Proposed model of nitrogenase site and steps in reduction of N_2 (Hardy *et al.,* 1971). Substrates and inhibitors are segregated on the basis of the metal with which they are proposed to complex. Suggested distances are indicated in Ångstrom.

These general types, as well as selected specific examples, will be described since they are relevant to the possible role of Fe and Mo in biological fixation. The more relevant of these Fe or Mo systems contain S as a ligand.

The most important abiological N_2-fixation system is the commercial Haber-Bosch process (Nielsen, 1968; Bridger, 1970). Iron is the basic component of the catalyst in almost all applications with small amounts of Al_2O_3 and K_2O added for stabilization and improved chemisorption of N_2. Molybdenum-based catalysts for N_2 fixation by the Haber process are also active, functioning at lower temperatures than Fe, but these are of shorter useful life. In this heterogenous catalysis, a 3:1 mixture of highly purified H_2 and N_2 is circulated over the catalyst at 450°C and 200–300 atm to obtain a quantitative yield of NH_3. Over 30×10^6 tons of N_2 are fixed annually by this process, which has made possible the high yields of modern agricultural crops in the developed world. Extreme conditions of temperature and pressure as well as the requirement for an anhydrous environment and heterogenous form of the catalyst differentiate this system from biological N_2 fixation.

Vol'pin and Shur (1966, 1969) described the first example of fixation of N_2 under mild conditions. Transition metals including those located on the left side of the transition period, e.g., Ti, Mn, Cr, Mo, W, Fe, and V, and strong reducing agents such as organometallic compounds, metal hydrides, and metals, e.g., RMgX, RLi, R_3Al, $LiAlH_4$, Li, Mg, K, Al, and Li and Na naphthalide were shown to reduce N_2 to hydrazides and nitrides. The latter can be released as NH_3 by hydrolysis. In some cases, N_2 fixation has been demonstrated at 1 atm of pressure and ambient temperature. The instability of the strong reductants in aqueous systems limits the above reactions to aprotonic media versus protonic media for biological N_2 fixation.

Four examples of the Vol'pin and Shur type of N_2-fixing systems will be described. They show similarities to nitrogenase with respect to inhibitors or reduction of alternate substrates of nitrogenase. Evidence obtained with one also suggests the initial formation of a metal·N_2 complex. In the first example (Bell and Brintzinger, 1970), a very strong reducing agent, naphthalene dianion, in combination with $FeCl_3$ reduces N_2 to NH_3 with a maximum yield of one NH_3 per Fe. Stronger reductants appear to be essential for N_2 reduction by Fe in this system. In another combination (Broitman *et al.*, 1971; Borodko *et al.*, 1971), $FeCl_3$ and isopropyl magnesium chloride in the presence of N_2 and Ph_3P form a binuclear iron·N_2 complex identified by its infrared absorption at 1761 and 1700 cm^{-1} for $^{14}N_2$ and $^{15}N_2$, respectively. The N_2 in the complex is replaceable by CO. Addition of HCl at —50°C yields partial reduction to a hydrazine product whose analysis corresponds to $[(Ph_3P)_2Fe(iPr)]_2N_2H$, while heating at $\geqq$90°C produces nitrides which yield ammonia after hydrolysis. In another example (van Tamelen *et al.*, 1971) of this type, a combination of $FeCl_3$ and Mg reduces N_2 to NH_3 and also reduces other substrates of nitrogenase. An alkene was formed from an alkyne, and CH_4, C_2H_4, and C_2H_6 were formed from an isonitrile. A related combination (van Tamelen *et al.*, 1971) of $Fe(acac)_3$ and Na naphthalide reduced another nitrogenase substrate KCN to CH_4, although it failed to reduce N_2.

During the last six years, inorganic chemists have demonstrated the almost ubiquitous ability of transition metals to form stable dinitrogen complexes. Isolable complexes of iron are mainly mononuclear with a single N_2 ligand per metal, e.g., $(Ph_3P)_3Fe(N_2)H_2$ (Sacco and Aresta, 1968), $[(CN)_5Fe(N_2)]^{2-}$ (Feltham *et al.*, 1969), $(EtPh_2P)_2Fe(N_2)H_2$ (Sacco and Aresta, 1968; Campbell *et al.*, 1968; Bancroft *et al.*, 1969), and $[trans\text{-}(Et_2P\cdot CH_2CH_2\cdot PEt_2)_2Fe(N_2)H]^+$ (Bancroft *et al.*, 1970), although examples of binuclear complexes have recently been described, e.g., $\{[\pi\text{-}C_5H_5Fe(Me_2PC_2H_2CH_2PMe_2)]_2N_2\}^{2+}$ (Silverthorn, 1971). Numerous Mo complexes of N_2 have also been isolated, e.g., $[(\pi\text{-}C_6H_6)(Ph_3P)_2Mo]_2N_2$ (Silverthorn, 1971; Green and Silverthorn, 1971) and *trans*-$[(Ph_2PCH_2CH_2PPh_2)_2Mo(N_2)_2]$ (Chatt *et al.*, 1972). Although the γ-(NN) in the transition metal·N_2 complexes is reduced substantially from that of molecular N_2, the complexes including those of either Mo or Fe have shown little indication of activation of the N_2 toward reduction. A recent exception is the reduction of *trans*-bis-dinitrogen molybdenum or tungsten complexes to complexed diazene (Chatt *et al.*, 1972). Some characteristics of the complexes may be of interest with respect to the initial reaction of N_2 with nitrogenase. For example, the ligand affinity sequence of the complexes in which $CO >> H_2 \sim N_2 > NH_3, H_2O$ parallels that of nitrogenase (Hardy *et al.*, 1971, 1973). The Mössbauer spectrum of Fe·N_2 and Fe·CO complexes have been compared (Bancroft *et al.*, 1970). The smaller isomer shift and lower quadrupole splitting in the N_2 complex suggests that N_2 is a less effective δ-donor and/or π-acceptor and that there is less back donation by Fe to N_2 than to CO. The reduction of γ-(NN) by reaction of a $Re(N_2)$ complex with Mo has provided the model basis for a suggested Fe–N_2–Mo intermediate in nitrogenase-catalyzed fixation (Chatt *et al.*, 1969). However, the $Re(N_2)Mo$ complex has shown no tendency toward reduction.

Several examples of N_2 fixation by iron or molybdenum complexes and reductants in protonic media have been recently described. These examples are most relevant to biological dinitrogen fixation. The most exciting of these examples demonstrate that iron and/or molybdenum possess the chemical capabilities of fixing dinitrogen under biological conditions.

Some specific examples of N_2 fixation in protonic media will be described proceeding from what appears to be the least relevant to the most relevant with respect to nitrogenase. Iron and molybdenum complexes with S, N, or P ligands and $NaBH_4$ or $Na_2S_2O_4$ as reductants show substrate reductions parallel to those of nitrogenase (Newton *et al.*, 1971). Azide is reduced to ammonia, cyanide and nitriles to hydrocarbons and ammonia, and alkynes almost exclusively to alkenes. A low level of reduction of N_2 to NH_3 was suggested, but lack of experimental results as well as confirmation with $^{15}N_2$, is in sharp contrast to the following examples. Systems containing Mo and utilizing strong reducing agents such as Cr^{2+} or Ti^{3+} convert N_2 at ambient temperature and pressure to N_2H_4 as the major product as well as some NH_3 (Shilov *et al.*, 1971). Mo is claimed to be the catalyst. It is of interest that V can replace Mo, as it also does in the biological reaction. An oxidation state of Mo of 3+ and V of 2+ is suggested. The V^{2+} does not require

Ti. Activation of these systems by Mg^{2+} is analogous to the requirement for M^{2+} for nitrogenase activity. Further similarities with the nitrogenase system include: H_2 evolution in competition with N_2 reduction, C_2H_2 reduction to C_2H_4, and competitive inhibition by CO.

Thiol–molybdo–iron systems reduce N_2 to NH_3 in protonic media with $NaBH_4$ as reductant, a pN_2 as low as 1 atm and thiols such as cysteine and 1-thioglycerol (Hill and Richards, 1971; Schrauzer and Schlessinger, 1970; Schrauzer and Doemeny, 1971; Schrauzer *et al.*, 1971). This system also catalyzes the evolution of H_2 from $S_2O_4^{2-}$ or BH_4^-, and the evolution is stimulated by ATP somewhat analogous to the ATP-dependent H_2 evolution catalyzed by nitrogenase. Although molybdenum is the dominant metal with a Mo/Fe molar ratio of 50–60 in these systems, iron is absolutely essential for activity of this system. With the best system, yield of ammonia is 1.5 times that of Mo and turnover number (molecules of N_2 reduced/minute/molecule of Mo) of the model is 2×10^{-5} times that of nitrogenase. The reaction is inhibited by a pCO of 0.5 atm as is biological N_2 fixation.

A direct role of S in N_2 fixation has been suggested by two reports of sulfur mediation of abiological N_2 fixation (Owsley and Helmkamp, 1967; Ellermann *et al.*, 1969). One utilized $C[CH_2P(RNa)]_4$ and CS_2 to form an organic complex containing molecular N_2; the other utilized benzenesulfenium ion, $C_6H_5S^+$, to react with N_2. Neither system has been confirmed, and it is concluded that a direct role of S versus an indirect role (as a ligand) in biological N_2 fixation is not indicated by chemical models.

It can be concluded that interesting abiological models of nitrogenase have been found and that some of these models involve molybdenum and iron with sulfur ligands. These abiological systems may be the first models possessing the biological activity of the iron–sulfur enzyme which they are attempting to mimic. The familial relationship of dinitrogen fixation and iron–sulfur proteins continues.

REFERENCES

Allen, A. D. (1971). *Advan. Chem.* **100,** 79.

Allen, A. D., and Bottomley, F. (1968). *Accts. Chem. Res.* **1,** 360.

Bancroft, G. M., Mays, M. J., and Prater, B. E. (1969). *Chem. Commun.* 585.

Bancroft, G. M., Mays, M. J., Prater, B. E., and Stefanini, F. P. (1970). *J. Chem. Soc. A* 2146.

Bell, L. G., and Brintzinger, H. H. (1970). *J. Amer. Chem. Soc.* **92,** 4464.

Benemann, J. R., Yoch, D. C., Valentine, R. C., and Arnon, D. I. (1969). *Proc. Nat. Acad. Sci. U.S.* **64,** 1079.

Benemann, J. R., Yoch, D. C., Valentine, R. C., and Arnon, D. I. (1971). *Biochim. Biophys. Acta* **226**, 205.
Bergersen, F. J. (1969). *Proc. Royal Soc. B.* **172**, 401.
Bergersen, F. J. (1970). *Aust. J. Biol. Sci.* **23**, 1015.
Bergersen, F. J. (1971). *Annu. Rev. Plant Physiol.* **22**, 121.
Bergersen, F. J., and Turner, G. L. (1968). *J. Gen. Microbiol.* **53**, 205.
Bergersen, F. J., and Turner, G. L. (1970). *Biochim. Biophys. Acta* **214**, 28.
Biggins, D. R., and Kelly, M. (1970). *Biochim. Biophys. Acta* **205**, 288.
Biggins, D. R., and Postgate, J. R. (1969). *J. Gen. Microbiol.* **56**, 181.
Biggins, D. R., and Postgate, J. R. (1971). *Eur. J. Biochem.* **19**, 408.
Biggins, D. R., Kelly, M., and Postgate, J. R. (1971). *Eur. J. Biochem.* **20**, 140.
Borodko, Yu. G., and Shilov, A. E. (1969). *Russ. Chem. Rev.* **38**, 355.
Borodko, Yu. G., Broitman, M. O., Kachapina, L. M., Shilov, A. E., and Ukhin, L. Yu. (1971). *Chem. Commun.* 1185.
Bothe, H. (1970). *Ber. Deutsch. Bot. Ges.* **83**, 421.
Bothe, H. (1971). *Eur. Biophys. Congr. Proc., 1st* 109.
Bridger, G. W. (1970). *In* "Catalyst Handbook," p. 126. Springer-Verlag, New York.
Broitman, M. O., Denisov, N. T., Shuvalova, N. I., and Shilov, A. E. (1971). *Kinet. Katal.* **12**, 504.
Bui, P. T., and Mortenson, L. E. (1968). *Proc. Nat. Acad. Sci. U.S.* **61**, 1021.
Bui, P. T., and Mortenson, L. E. (1969). *Biochemistry* **8**, 2462.
Bulen, W. A., and LeComte, J. R. (1966). *Proc. Nat. Acad. Sci. U.S.* **56**, 979.
Bulen, W. A., and LeComte, J. R. (1972). *In* "Methods in Enzymology, Photosynthesis and Nitrogen Fixation" (A. San Pietro, ed.), Vol. **24**, p. 423. Academic Press, New York.
Bulen, W. A., Burns, R. C., and LeComte, J. R. (1964). *Biochem. Biophys. Res. Commun.* **17**, 265.
Bulen, W. A., Burns, R. C., and LeComte, J. R. (1965a). *Proc. Nat. Acad. Sci. U.S.* **53**, 532.
Bulen, W. A., LeComte, J. R., Burns, R. C. and Hinkson, J. (1965b). *In* "A Symposium on Non-heme Iron Proteins: Role in Energy Conversions" (A. San Pietro, ed.), p. 261. Antioch Press, Yellow Springs, Ohio.
Burns, R. C. (1965). *In* "A Symposium on Non-heme Iron Proteins: Role in Energy Conversion" (A. San Pietro, ed.), p. 289. Antioch Press, Yellow Springs, Ohio.
Burns, R. C., and Bulen, W. A. (1966). *Arch. Biochem. Biophys.* **113**, 461.
Burns, R. C., and Hardy, R. W. F. (1971). *Fed. Proc. Fed. Amer. Soc. Exp. Biol.* **30**, 1291.
Burns, R. C., and Hardy, R. W. F. (1972a). *In* "Methods in Enzymology, Photosynthesis and Nitrogen Fixation" (A San Pietro, ed.), Vol. **24**, p. 480. Academic Press, New York.
Burns, R. C., and Hardy, R. W. F. (1973). "Nitrogen Fixation in Bacteria and Higher Plants." Springer-Verlag, New York (in press).
Burns, R. C., and Hardy, R. W. F. (1972b). *164th Meeting Amer. Chem. Soc., New York,* Abstr. 201.
Burns, R. C., Holsten, R. D., and Hardy, R. W. F. (1970). *Biochem. Biophys. Res. Commun.* **39**, 90.
Burns, R. C., Fuchsman, W. H., and Hardy, R. W. F. (1971). *Biochem. Biophys. Res. Commun.* **42**, 353.
Burris, R. H. (1969). *Proc. Royal Soc. B.* **172**, 339.

Burris, R. H. (1971). *In* "The Chemistry and Biochemistry of Nitrogen Fixation" (J. Postgate, ed.), p. 105. Plenum Press, New York.

Campbell, C. H., Dias, A. R., Green, M. L. H., Saito, T., and Swanwick, M. G. (1968). *J. Organometal. Chem.* **14,** 349.

Carnahan, J. E., and Castle, J. E. (1958). *J. Bacteriol.* **75,** 121.

Carnahan, J. E., Mortenson, L. E., Mower, H. F., and Castle, J. E. (1960). *Biochim. Biophys. Acta* **44,** 520.

Chatt, J. (1969). *Proc. Royal Soc. B.* **172,** 327.

Chatt, J. (1970). *Pure Appl. Chem.* **24,** 425.

Chatt, J., Heath, G. A., and Richards, R. L. (1972). *Chem. Commun.* p. 1010.

Chatt, J., and Richards, R. L. (1971). *In* "The Chemistry and Biochemistry of Nitrogen Fixation" (J. Postgate, ed.), p. 57. Plenum Press, New York.

Chatt, J., Dilworth, J. R., Richards, R. L., and Sanders, J. R. (1969). *Nature (London)* **224,** 1201.

Commoner, B. (1971). "Closing the Circle," A. Knopf, New York.

Daesch, G., and Mortenson, L. E. (1968). *J. Bacteriol.* **96,** 346.

Dalton, H., and Mortenson, L. E. (1970). *Bacteriol. Proc.* p. 148.

Dalton, H., and Postgate, J. R. (1969). *J. Gen. Microbiol.* **56,** 307.

Dalton, H., Morris, J. A., Ward, M. A., and Mortenson, L. E. (1971). *Biochemistry* **10,** 2066.

Davis, L. C., Shah, V. K., Brill, W. J., and Orme-Johnson, W. H. (1972). *Biochim. Biophys. Acta* **256,** 512.

Detroy, R. W., Witz, D. F., Parejko, R. A., and Wilson, P. W. (1968). *Proc. Nat. Acad. Sci. U.S.* **61,** 537.

D'Eustachio, A. J., and Hardy, R. W. F. (1964). *Biochem. Biophys. Res. Commun.* **15,** 319.

Dixon, R. A., and Postgate, J. R. (1971). *Nature (London)* **234,** 47.

Dixon, R. A., and Postgate, J. R. (1972). *Nature (London)* **237,** 102.

Drozd, J., and Postgate, J. R. (1970). *J. Gen. Microbiol.* **63,** 63.

Dua, R. D., and Burris, R. H. (1963). *Proc. Nat. Acad. Sci. U.S.* **50,** 169.

Eady, R R., Smith, B. E., Cook, K. A., and Postgate, J. R. (1972). *Biochem. J.* **128,** 655.

Ellermann, J., Poersch, F., Kunstmann, R., and Kramolowsky, R. (1969). *Angew. Chem. Int. Ed.* **8,** 203.

Evans, H. J. (1969). *In* "How Crops Grow—A Century Later" (J. G. Horsfall, ed.), *Conn. Agr. Exp. Sta. Bull.* **708,** 110.

Evans, H. J., and Russell, S. A. (1971). *In* "The Chemistry and Biochemistry of Nitrogen Fixation" (J. Postgate, ed.), p. 191. Plenum Press, New York.

Evans, H. J., Koch, B., and Klucas, R. (1972). *In* "Methods in Enzymology, Photosynthesis and N_2 Fixation" (A. San Pietro, ed.), Vol. **24,** p. 470. Academic Press, New York.

Evans, M. C. W., and Smith, R. V. (1971). *J. Gen. Microbiol.* **65,** 95.

Evans, M. C. W., Telfer, A., Cammack, R., and Smith, R. V. (1971). *FEBS Lett.* **15,** 317.

Feltham, R. D., Metzger, H. G., and Singler, R. (1969). *Proc. XII I.C.C.C., Sydney* p. 225.

Ferguson, J. E., and Love, J. L. (1970). *Rev. Pure Appl. Chem.* **20,** 33.

Fisher, R. J., and Brill, W. J. (1969). *Biochim. Biophys. Acta* **184,** 99.

Fisher, R. J., and Wilson, P. W. (1970). *Biochem. J.* **117,** 1023.

Frolov, E. N., Likhtenshtein, G. I., and Syrtsova, L. A. (1971). *Dokl. Akad. Nauk. SSSR* **196,** 1149.
Fuchsman, W. H., and Hardy, R. W. F. (1972). *Bioinorg. Chem.* **1,** 195.
Gallon, J. R., LaRue, T. A., and Kurz, W. G. W. (1972). *Can. J. Microbiol.* **18,** 327.
Ganelin, V. L., L'vov, N. P., Kirshteine, B. E., Lyubimov, V. I., and Kretovich, V. L. (1969). *Dokl. Akad. Nauk. SSSR* **185,** 1169.
Grau, F. H., and Wilson, P. W. (1963). *J. Bacteriol.* **85,** 446.
Green, M. L. H. and Silverthorn, W. E. (1971). *Chem. Commun.* 557.
Gvozdev, R. I., Yakovlev, V. A., Linde, V. R., Vorob'ev, L. V., and Alfimova, E. Ya. (1969). *Izv. Akad. Nauk. SSSR Ser. Biol.* 215.
Gvozdev, R. I., Sadkov, A. P., Sevchenko, L. A., Kulikov, A. V., and Vorob'ev, L. V. (1971). *Izv. Akad. Nauk. SSSR Ser. Biol.* 246.
Hardy, R. W. F., and Burns, R. C. (1968). *Ann. Rev. Biochem.* **37,** 331.
Hardy, R. W. F., and D'Eustachio, A. J. (1964). *Biochem. Biophys. Res. Commun.* **15,** 314.
Hardy, R. W. F., and Knight, E., Jr. (1966). *Biochim. Biophys. Acta* **122,** 520.
Hardy, R. W. F., and Knight, E., Jr. (1968). *Progr. Phytochem.* **1,** 407.
Hardy, R. W. F., Knight, E., Jr., and D'Eustachio, A. J. (1965). *Biochem. Biophys. Res. Commun.* **20,** 539.
Hardy, R. W. F., Holsten, R. D., Jackson, E. K., and Burns, R. C. (1968). *Plant Physiol.* **43,** 1185.
Hardy, R. W. F., Burns, R. C., and Parshall, G. W. (1971). *Advan. Chem.* **100,** 219.
Hardy, R. W. F., Burns, R. C., and Parshall, G. W. (1973). *In* "Inorganic Biochemistry" (G. Eichhorn, ed.), Vol. 2, p. 745. Elsevier, Amsterdam (in press).
Haystead, A., and Stewart, W. D. P. (1972). *Arch. Mikrobiol.* **82,** 325.
Haystead, A., Robinson, R., and Stewart, W. D. P. (1970). *Arch. Mikrobiol.* **74,** 235.
Henrici-Olivé, G., and Olivé, S. (1969). *Angew. Chem. Int. Ed.* **8,** 650.
Hill, R. E. E., and Richards, R. L. (1971). *Nature (London)* **233,** 114.
Huang, T. C., Zumft, W. G., and Mortenson, L. E. (1973). *J. Bacteriol.* **113,** 884.
Israel, D. W., Howard, R. H., Evans, H. J., and Russell, S. (1972). *Plant Physiol.* Suppl. **49,** 50.
Ivanov, I. D., Moshkovskii, Yu. Sh., Stukan, R. H., Matkhanov, G. I., Mardanyan, S. S., and Belov, Yu. M. (1968). *Mikrobiologya* **37,** 407.
Ivleva, I. N., Medzhidov, A. A., Likhtenshtein, G. I., Sadkov, A. P., and Yakovlev, V. A. (1969a). *Biofizika* **14,** 639.
Ivleva, I. N., Likhtenshtein, G. I., and Sadkov, A. P. (1969b). *Biofizika* **14,** 779.
Jackson, E. K., Parshall, G. W., and Hardy, R. W. F. (1968). *J. Biol. Chem.* **243,** 4952.
Jeng, D. Y., and Mortenson, L. E. (1969). *158th Meeting Amer. Chem. Soc. New York,* Abstr. 227.
Jeng, D. Y., Devanathan, T., and Mortenson, L. E. (1969a). *Biochem. Biophys. Res. Commun.* **35,** 625.
Jeng, D. Y., Devanathan, T., Moustafa, E., and Mortenson, L. E. (1969b). *Bacteriol. Proc.* p. 119.
Jeng, D. Y., Morris, J. A., and Mortenson, L. E. (1970). *J. Biol. Chem.* **245,** 2809.
Kajiyama, S., Matsuki, T., and Nosoh, Y. (1969). *Biochem. Biophys. Res. Commun.* **37,** 711.
Kelly, M. (1966). *Int. Congr. Microbiol. 9th, Moscow* p. 277.

Kelly, M. (1969a). *Biochim. Biophys. Acta* **171**, 9.
Kelly, M. (1969b). *Biochim. Biophys. Acta* **191**, 527.
Kelly, M., and Lang, G. (1970). *Biochim. Biophys. Acta* **223**, 86.
Kelly, M., Klucas, R. V., and Burris, R. H. (1967). *Biochem. J.* **105**, 3C.
Kennedy, I. R. (1970a). *Aust. Biochem. Soc. Proc.* **3**, 11.
Kennedy, I. R. (1970b). *Biochim. Biophys. Acta* **222**, 135.
Klucas, R. V., and Evans, H. J. (1968). *Plant Physiol.* **43**, 1458.
Klucas, R. V., Koch, B., Russell, S., and Evans, H. J. (1968). *Plant Physiol.* **43**, 1906.
Koch, B., Evans, H. J., and Russell, S. (1967a). *Plant Physiol.* **42**, 466.
Koch, B., Evans, H. J., and Russell, S. (1967b). *Proc. Nat. Acad. Sci. U.S.* **58**, 1343.
Koch, B., Wong, P., Russell, S. A., Howard, R., and Evans, H. J. (1970). *Biochem. J.* **118**, 773.
Knight, E., Jr., and Hardy, R. W. F. (1966). *J. Biol. Chem.* **241**, 2752.
Knight, E., Jr., and Hardy, R. W. F. (1967). *J. Biol. Chem.* **242**, 1370.
Knight, E., Jr., D'Eustachio, A. J., and Hardy, R. W. F. (1966). *Biochim. Biophys. Acta* **113**, 626.
Kretovich, V. L., Evstigneeva, Z. G., Aseeva, K. B., Zargaryan, O. N., and Mochalkina, N. A. (1970). *Dokl. Akad. Nauk. SSSR* **190**, 1235.
Kuchynka, K. (1969). *Catal. Rev.* **3**, 111.
Leigh, G. J. (1971). *In* "The Chemistry and Biochemistry of Nitrogen Fixation" (J. Postgate, ed.), p. 19. Plenum Press, New York.
Ljones, T., Burris, R. H., and Wilson, P. W. (1972). *Fed. Proc. Fed. Amer. Soc. Exp. Biol.* **31**, 478.
L'vov, N. P., Sergeev, N. S., Vermova, M. K., Shaposhnikov, G. L., and Kretovich, V. L. (1971). *Dokl. Akad. Nauk. SSSR* **201**, 1493.
Mahl, M. C., and Wilson, P. W. (1968). *Can. J. Microbiol.* **14**, 33.
Manorik, V. A., Starchenkov, Yu. P., Datsenko, V. K., and Yakovleva, N. S. (1970). *Dop. Akad. Nauk. Ukr. RSR, Ser. B* **32**, 177.
Manorik, V. A., Starchenkov, Yu. P., and Datsenko, V. K. (1971). *Dop. Akad. Nauk. Ukr. RSR, Ser. B* **33**, 363.
Matkhanov, G. I., Ivanov, I. D., Vanin, A. F., and Belov, Yu. M. (1969). *Biofizika* **14**, 1124.
McKenna, C. E., Benemann, J. R., and Traylor, T. G. (1970). *Biochem. Biophys. Res. Commun.* **41**, 1501.
Millbank, J. W. (1972). *New Phytol.* **71**, 1.
Mortenson, L. E. (1962). *In* "The Bacteria" (I. C. Gunsalus and R. Y. Stanier, eds.), Vol. 3, p. 718. Academic Press, New York.
Mortenson, L. E. (1964a). *Biochim. Biophys. Acta* **81**, 473.
Mortensen, L. E. (1964b). *Proc. Nat. Acad. Sci. U.S.* **52**, 272.
Mortenson, L. E. (1965). *In* "A Symposium on Non-heme Iron Proteins: Role in Energy Conversion" (A. San Pietro, ed.), p. 243. Antioch Press, Yellow Springs, Ohio.
Mortenson, L. E. (1966). *Biochim. Biophys. Acta* **127**, 18.
Mortenson, L. E. (1968). *Surv. Progr. Chem.* **4**, 127.
Mortenson, L. E. (1972). *In* "Methods in Enzymology, Photosynthesis and Nitrogen Fixation." (A. San Pietro, ed.), Vol. **24**, p. 446. Academic Press, New York.
Mortenson, L. E., Valentine, R. C., and Carnahan, J. E. (1962). *Biochem. Biophys. Res. Commun.* **7**, 448.

Mortenson, L. E., Valentine, R. C., and Carnahan, J. E. (1963). *J. Biol. Chem.* **238,** 794.

Mortenson, L. E., Morris, J. A., and Jeng, D. Y. (1967). *Biochim. Biophys. Acta* **141,** 516.

Mortenson, L. E., Zumft, W. G., and Palmer, G. (1973). *Biochim. Biophys. Acta* **292,** 422.

Moshkovskii, Yu. Sh. and Ivanov, I. D., Stukan, R. A., Matkhanov, G. I., Mardanyan, S. S., Belov, Yu. M., and Goldanskii, V. I. (1967). *Dokl. Akad. Nauk. SSSR* **174,** 215.

Moustafa, E. (1970). *Biochim. Biophys. Acta* **206,** 178.

Moustafa, E., and Mortenson, L. E. (1968). *Anal. Biochem.* **24,** 226.

Moustafa, E., and Mortenson, L. E. (1969). *Biochim. Biophys. Acta* **172,** 106.

Munson, T. O., and Burris, R. H. (1969). *J. Bacteriol.* **97,** 1093.

Murphy, P. M., and Koch, B. L. (1971). *Biochim. Biophys. Acta* **253,** 295.

Murray, R., and Smith, D. C. (1968). *Coord. Chem. Rev.* **3,** 429.

Nakos, G., and Mortenson, L. E. (1971a). *Biochim. Biophys. Acta* **229,** 431.

Nakos, G., and Mortenson, L. E. (1971b). *Biochemistry* **10,** 455.

Nason, A., Lee, K.-Y., Pan, S., Ketchum, P. A., Lamberti, A. and Devries, J. (1971). *Proc. Nat. Acad. Sci. U.S.* **68,** 3242.

Newton, W. E., Corbin, J. L., Schneider, P. W., and Bulen, W. A. (1971). *J. Amer. Chem. Soc.* **93,** 268.

Nielsen, A. (1968). "An Investigation of Promoted Iron Catalysts for the Synthesis of Ammonia," Jul. Gjellerup Forlag, Copenhagen.

Nimek, M. W., and Wilson, P. W. (1963). *Nature (London)* **200,** 709.

Novikov, G. V., Syrtsova, L. A., Likhtenshtein, G. I., Trukhtanov, V. A., Rachek, V. F., and Goldanskii, V. I. (1968). *Dokl. Akad. Nauk. SSSR* **181,** 1170.

Oppenheim, J., and Marcus, L. (1970). *J. Bacteriol.* **101,** 286.

Oppenheim, J., Fisher, R. J., Wilson, P. W., and Marcus, L. (1970). *J. Bacteriol.* **101,** 292.

Orme-Johnson, W. H., Stombaugh, N. A., and Burris, R. H. (1972). *Fed. Proc., Fed. Amer. Soc. Exp. Biol.* **31,** 448.

Owsley, D. C., and Helmkamp, G. K. (1967). *J. Amer. Chem. Soc.* **89,** 4558.

Parejko, R. A., and Wilson, P. W. (1971). *Proc. Nat. Acad. Sci. U.S.* **68,** 2016.

Parshall, G. W. (1967). *J. Amer. Chem. Soc.* **89,** 1822.

Parshall, G. W., and Hardy, R. W. F. (1973). Unpublished results.

Peive, J., Yagodin, B. A., Savich, M. S., Ovcharenko, G. A., Yuferova, S. G., and Malinovskii, A. V. (1971). *Dokl. Akad. Nauk SSSR* **197,** 721.

Postgate, J. R. (1970). *Nature (London)* **226,** 25.

Postgate, J. R. (1971a). *Symp. Soc. Gen. Microbiol.* **21,** 287.

Postgate, J. R. (1971b). *In* "The Chemistry and Biochemistry of Nitrogen Fixation" (J. Postgate, ed.), p. 161. Plenum Press, New York.

Postgate, J. R. (1971c). *In* "Biological Nitrogen Fixation in Natural and Agricultural Habitats" (T. A. Lie and E. G. Mulder, eds.), *Plant Soil,* Special Volume, p. 551.

Sacco, A. and Aresta, M. (1968). *Chem. Commun.* 1223.

Schrauzer, G. N., and Doemeny, P. A. (1971). *J. Amer. Chem. Soc.* **93,** 1608.

Schrauzer, G. N., and Schlessinger, G. (1970). *J. Amer. Chem. Soc.* **92,** 1808.

Schrauzer, G. N., Schlessinger, G., and Doemeny, P. A. (1971). *J. Amer. Chem. Soc.* **93,** 1803.

Shethna, Y. I., Stombaugh, N. A., and Burris, R. H. (1971). *Biochem. Biophys. Res. Commun.* **42,** 1108.
Shilov, A. E., and Likhtenshtein, G. I. (1971). *Izv. Akad. Nauk. USSR, Ser. Biol. No.* *4*, 518.
Shilov, A., Denisov, N., Efimov, O., Shuvalov, N., Shuvalova, N., and Shilova, A. (1971). *Nature (London)* **231,** 460.
Silver, W. S. (1971). *In* "The Chemistry and Biochemistry of Nitrogen Fixation" (J. Postgate, ed.), p. 245. Plenum Press, New York.
Silverstein, R., and Bulen, W. A. (1970). *Biochemistry* **9,** 3809.
Silverthorn, W. E. (1971). *Chem. Commun.* 1310.
Simon, M. A., and Brill, W. J. (1971). *J. Bacteriol.* **105,** 65.
Slepko, G. I., Uzenskaya, A. M., Linde, V. R., and Levchenko, L. A. (1971). *Izv. Akad. Nauk. SSSR Ser. Biol.* 86.
Smith, M. H. (1966). *Nature (London)* **210,** 341.
Smith, R. V., and Evans, M. C. W. (1970). *Nature (London)* **225,** 1253.
Smith, R. V., and Evans, M. C. W. (1971). *J. Bacteriol.* **105,** 913.
Smith, R. V., Noy, R. J., and Evans, M. C. W. (1971a). *Biochim. Biophys. Acta* **253,** 104.
Smith, R. V., Telfer, A., and Evans, M. C. W. (1971b). *J. Bacteriol.* **107,** 574.
Sorger, G. (1971). *Biochem. J.* **122,** 305.
Sorger, G. J., and Trofimenkoff, D. (1970). *Proc. Nat. Acad. Sci. U.S.* **65,** 74.
Stasny, J. T., Burns, R. C., and Hardy, R. W. F. (1971). *Bacteriol. Proc.* p. 139.
Stasny, J. T., Burns, R. C., and Hardy, R. W. F. (1973). Unpublished results.
Starchenkov, E. P., Yakovleva, N. S., and Datsenko, V. K. (1971). *Fiz. Biokhim. Kul't. Rast.* **3,** 33.
Stewart, W. D. P., and Lex, M. (1970). *Arch. Mikrobiol.* **73,** 250.
Streicher, S., Gurney, E., and Valentine, R. C. (1971). *Proc. Nat. Acad. Sci. U.S.* **68,** 1174.
Syrtsova, L. A., Likhtenshtein, G. I., Pisarskaya, T. N., Ganelin, V. L., Frolov, E. N., and Rachek, V. F. (1969). *Mol. Biol.* **3,** 651.
Syrtsova, L. A., Levchenko, L. A., Frolov, E. N., Likhtenshtein, G. I., Pisarskaya, T. N., Vorob'ev, L. V., and Gromoglasova, V. A. (1971). *Mol. Biol.* **5,** 726.
Taylor, K. B. (1969). *J. Biol. Chem.* **244,** 171.
Tsibris, J. C. M., and Woody, R. W. (1970). *Coordin. Chem. Rev.* **5,** 417.
Turner, G. L., and Bergersen, F. J. (1969). *Biochem. J.* **115,** 529.
Updike, J. (1966). "The Couples," A. Knopf, New York.
Vandecasteele, J. P., and Burris, R. H. (1970). *J. Bacteriol.* **101,** 794.
van Lin, B., and Bothe, H. (1972). *Arch. Mikrobiol.* **82,** 155.
van Tamelen, E. E. (1970). *Accts. Chem. Res.* **3,** 361.
van Tamelen, E. E. (1971). *Advan. Chem.* **100,** 95.
van Tamelen, E. E., Rudler, H.. and Bjorklund, C. (1971). *J. Amer. Chem. Soc.* **93,** 3526.
Vol'pin, M. E., and Shur, V. B. (1966). *Nature (London)* **209,** 1236.
Vol'pin, M. E., and Shur, V. B. (1969). *Dokl. Akad. Nauk SSSR* **156,** 1102.
Ward, M. A., Dalton, H., and Mortenson, L. E. (1971). *Bacteriol. Proc.* p. 139.
Weiher, J. F., Burns, R. C., and Hardy, R. W. F. (1973). Unpublished results.
Wilson, P. W. (1968). *In* "Encyclopedia of Plant Physiology" (W. Ruhland, ed.), Vol. 8, p. 9. Springer-Verlag, Berlin.
Winter, H. C., and Arnon, D. I. (1970). *Biochim. Biophys. Acta* **197,** 170.
Winter, H. C., and Burris, R. H. (1968). *J. Biol. Chem.* **243,** 940.

Winter, H. C., and Ober, J. A. (1971). *Fed. Proc., Fed. Amer. Soc. Exp. Biol.* **30,** 1292.
Witz, D. F., and Wilson, P. W. (1967). *Bacteriol. Proc.* p. 112.
Witz, D. F., Detroy, R. W., and Wilson, P. W. (1967). *Arch. Mikrobiol.* **55,** 369.
Wong, P. P., Evans, H. J., Klucas, R., and Russell, S. (1971). *In* "Nitrogen Fixation in Natural and Agricultural Habitats" (T. Lie and E. G. Mulder, eds.), *Plant Soil,* Special Volume, p. 525.
Yakovlev, V. A., Vorob'ev, L. V., Levchenko, L. A., Linde, V. R., Slepko, G. I., and Syrtsova, L. A. (1965). *Biokhimiya* **30,** 1167.
Yates, M. G. (1970). *FEBS Lett.* **8,** 281.
Yates, M. G. (1971a). *In* "The Chemistry and Biochemistry of Nitrogen Fixation" (J. Postgate, ed.), p. 283. Plenum Press, New York.
Yates, M. G. (1971b). *Eur. J. Biochem.* **24,** 347.
Yates, M. G., and Daniel, R. M. (1970). *Biochim. Biophys. Acta* **197,** 161.
Yoch, D. C., and Arnon, D. I. (1970). *Biochim. Biophys. Acta* **197,** 180.
Yoch, D. C., and Arnon, D. I. (1972). *Bacteriol. Proc.* p. 155.
Yoch, D. C., Benemann, J. R., Valentine, R. C., and Arnon, D. I. (1969). *Proc. Nat. Acad. Sci. U.S.* **64,** 1404.
Yoch, D. C., Benemann, J. R., Arnon, D. I., Valentine, R. C., and Russell, S. A. (1970). *Biochem. Biophys. Res. Commun.* **38,** 838.
Zumft, W. G., Palmer, G., and Mortenson, L. E. (1973). *Biochim. Biophys. Acta* **292,** 413.

CHAPTER 4

Iron-Sulfur Proteins in Photosynthesis

CHARLES F. YOCUM, JAMES N. SIEDOW, and
ANTHONY SAN PIETRO

The intimate involvement and mechanism of action of iron–sulfur proteins (ferredoxins) in photosynthesis has been the focal point of intensive research during the past decade. As depicted in Fig. 1, ferredoxin occupies a central role in light-mediated electron transport reactions by plant chloroplasts. Once ferredoxin is reduced, it serves for the reduction of a variety of different compounds (Fig. 1). This chapter, however, is concerned primarily with recent information relating to the photoreduction of ferredoxin, the mechanism of NADPH formation, and the possible role of ferredoxin in cyclic photophosphorylation. A discussion of the role of

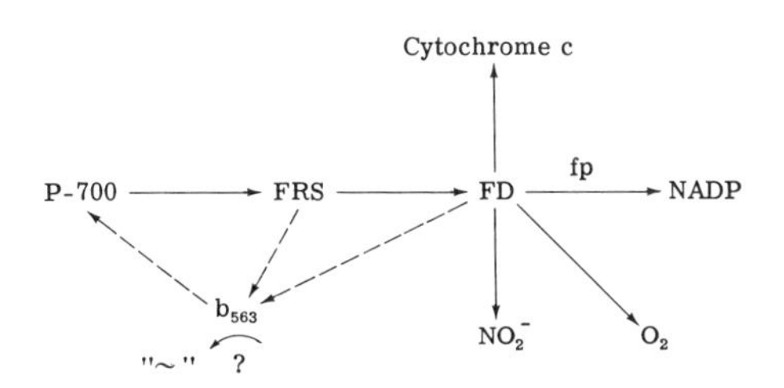

Fig. 1. Ferredoxin-mediated reactions. The Fd-NADP reductase is denoted as fp.

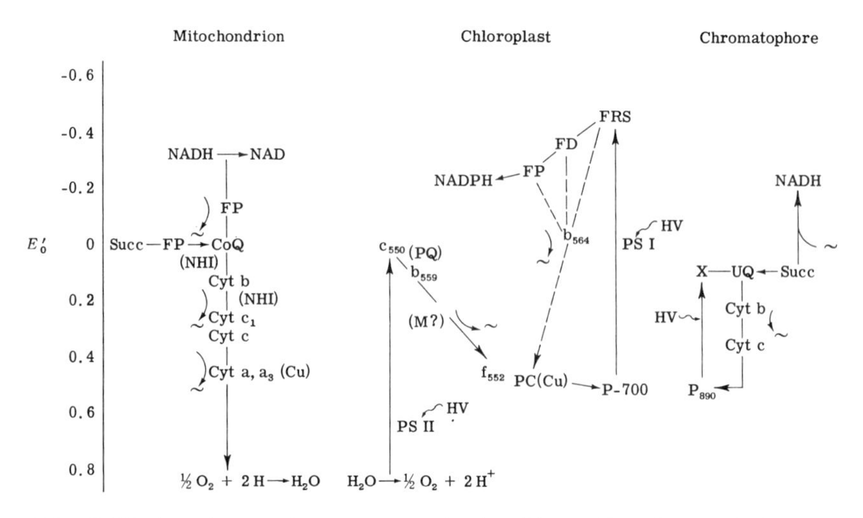

Fig. 2. Electron transport sequences: mitochondrion, chloroplast, and chromatophore.

ferredoxin in the other reactions shown in Fig. 1 was presented earlier (San Pietro, 1968) and will not be treated here.

Inasmuch as iron–sulfur proteins appear to be involved in other energy transducing systems, e.g., mitochondria and chromatophores, a brief comparison of these systems is provided in Fig. 2 and Table III.

I. FERREDOXIN REDUCTION

Continuing investigations on the nature of the photochemical apparatus have led to the conclusion that the first reduced species formed by illuminated chlorophyll is a substance other than reduced ferredoxin. Chance *et al.* (1965) demonstrated convincingly that the rate of the photooxidation of ferrocytochrome f by chloroplasts was uninfluenced by the presence or absence of ferredoxin. That is, the kinetics of the photooxidation process both in intact chloroplasts and in chloroplast fragments from which the ferredoxin had been removed were essentially the same. It was, therefore, concluded that some compound associated with photosystem I, other than ferredoxin, serves to accept the electron provided in the primary photooxidation. Thermodynamically, this was not an unusual finding, since the theoretical change in characteristic potential generated in photosystem I is approximately 1.8 eV. We may assume one terminus to be cytochrome f ($E_0' = +0.34$ V) or the copper protein, plastocyanin ($E_0' = +0.37$ V) either of which may serve as the electron donor to the photosystem I reaction center. If ferredoxin ($E_0' = -0.43$ V) is the reaction partner in this process, the stabilized change in characteristic potential is at most 0.8 V; that is, less than 50% of the thermodynamically theoretical change in characteristic potential available from photosystem I.

The availability of dipyridyl salts (viologen dyes) possessing oxidation–reduction potentials as reducing as —0.72 V permitted experimentation that provided additional circumstantial evidence in support of the conclusions of Chance *et al.* (1965). Zweig and Avron (1965), Kok *et al.* (1965), and Black (1966) observed that isolated chloroplasts under strict anaerobic conditions were capable of reducing these viologen dyes of quite negative oxidation–reduction potential. While the results (Table I) admittedly were variable, the inescapable conclusion was that chloroplasts do indeed have the capacity to reduce compounds with oxidation–reduction potentials more negative than that of ferredoxin. In order to estimate the oxidation–reduction potential of the "unknown compound," it is necessary to assume (1) that it is a cofactor which can exist in an oxidized and reduced form and (2) that the ratio of the oxidized/reduced form is 1 at equilibrium. Under these conditions, the potential of the unknown substance should be approximately —0.55 V, that is, some 0.12 V more negative than ferredoxin.

The relevance of the polarographically determined viologen potentials to biologically significant reactions was clearly demonstrated by Shin and Arnon (1965). They showed that the ferredoxin-mediated reduction of NADP did not proceed at a significant rate when hydrogen gas (plus hydrogenase) was used as the reductant. If, however, either benzyl ($E_0' = -0.35$ V) or methyl ($E_0' = -0.45$ V) viologen was added to the

TABLE I
REDUCTION OF VIOLOGENS BV CHLOROPLASTS

E_0' of dipyridyl (mV)	Percent photoreduction		
	a	*b*	*c*
−342	100	100	—
−426	100	100	84
−521	50	100	30, 20
−636	7	26	—
−656	3	4	—

[a] Black (1966).
[b] Kok *et al.* (1965).
[c] Zweig and Avron (1965).

reaction mixture, good rates were observed, and the rate with the more negative potential dye was highest. The viologen dyes used served as a shuttle for electrons generated by the enzyme system transferring them to ferredoxin, which is not reduced at an appreciable rate by the enzyme alone.

Vernon *et al.* (1965) studied the reduction of NADP by chlorophyllin a, a chemically modified form of chlorophyll a. They showed that when a suitable donor, such as ascorbate, was provided, NADP photoreduction proceeded at the maximal rate with only the flavoprotein, ferredoxin-NADP reductase. Ferredoxin was actually inhibitory. Later experiments by Brune and San Pietro (1970) showed that chlorophyllin a generated a change in characteristic potential, as measured by viologen reduction, equal to that of chlorophyll a. Thus, there appear to be no thermodynamic restrictions on ferredoxin reduction by this system. The difference between the two systems, chloroplasts and chlorophyllin a, is the necessity for structural integrity in the former only as evidenced by the absolute need for ferredoxin for NADP photoreduction.

All the foregoing discussion would remain speculative were it not for a number of reports from several laboratories concerning the isolation of chloroplast-bound factors which undergo photosystem I reduction and which bear a resemblance to the "unknown compound" described above.

Fujita and Myers (1966, 1967) and later Fujita and Murano (1967), reported the isolation of a "cytochrome reducing substance (CRS)" from algae and also from spinach chloroplasts. It is interesting to note that CRS and methyl viologen both restored low rates of ferredoxin-mediated NADP reduction in chloroplast fragments from which CRS had previously been removed.

Yocum *et al.* (1969) reported the isolation of a "lipoidal factor" from sonicated chloroplasts which stimulated NADP reduction by chloroplast fragments devoid of the factor. Yocum and San Pietro (1969, 1970) extended these studies and isolated a compound called "ferredoxin reducing substance (FRS)." In addition to stimulating NADP reduction, FRS is capable of restoring viologen reduction in depleted chloroplast fragments and, upon anaerobic photoreduction, of reducing ferredoxin in a subsequent dark phase. Preliminary experiments by Brune (1970) indicate that FRS has an oxidation–reduction potential of about —0.48 V, considerably below that of ferredoxin.

Regitz *et al.* (1970) showed that rabbits challenged with specially treated chloroplasts developed antibodies which inhibited photosystem I reactions, including NADP and anthroquinone ($E_0' = -0.25$ V) reduction. They showed further that an antigen could be extracted from ether-treated chloroplasts which reversed the antibody-induced inhibition of

TABLE II
Comparison of Substances which Resemble FRS

Substance	Absorbance maxima	Molecular weight	Remarks	Reference
FRS	262	3000–6000	NADP reduction; methyl viologen reduction; dark reduction of ferredoxin and NADP	Yocum and San Pietro (1969, 1970)
CRS	265	5000	Pseudocyclic photophosphorylation; photooxidation of reduced DPIP and plastocyanin; methyl viologen reduction; cytochrome c reduction	Fujita and Myers (1966, 1967) Fujita and Murano (1967)
Phosphodoxin	270, 330	—	Pseudocyclic photophosphorylation; NADP reduction	Black *et al.* (1963)
Protein Factor	260, 315	15,000	Pseudocyclic photophosphorylation; ferricyanide reduction	Saltman and Gee (1966) Gee *et al.* (1970)
D-2	280, 310	5000	NADP reduction; pseudocyclic photophosphorylation	Wu and Myers (1969) Wu *et al.* (1970)
$S_{l\text{-}eth}$	262, 315	—	Cytochrome c reduction; anthraquinone reduction; pseudocyclic photophosphorylation	Regitz *et al.* (1970)
LAF	280, 315	4000–8000	Activates RUDP carboxylase	Wildner and Criddle (1969)
ORS	—	—	Oxygen reduction (*Anabaena*)	Honeycutt and Krogmann (1970)
P-430	430	—	P-700 reaction partner	Hiyama and Ke (1971)
Bound NHI	—	—	EPR signal, $g = 1.94$ (25°K). Chemically reduced by dithionite; not by Cleland's reagent	Malkin and Bearden (1971)
DPS	259, 310	3000–6000	Photoreduction of cytochrome c; dark reduction of cytochrome c and DPIP; inhibits nitrite reductase	Siedow and San Pietro (1971)

anthroquinone reduction. Collaborative studies between Dr. A. Trebst and ourselves have shown that FRS and the chloroplast antigen ($S_{l\text{-}eth}$ in their terminology) have a common identity.

A similar study by Honeycutt and Krogmann (1970) in which the Trebst antibody preparation was tested against an "oxygen reducing substance (ORS)" isolated from blue-green algae gave comparable results. The substance from algae reversed the antibody inhibition of anthroquinone reduction, and it would thus appear that FRS, the $S_{l\text{-}eth}$ antigen, and ORS all exhibit a common reactivity with the antibody against anthroquinone reduction.

In addition to the factors discussed above, other substances have been isolated from (or detected in) photosynthetic tissue whose role is less clearly defined. A summary of a number of these factors, together with those discussed above, is given in Table II. The isolated factors have in common the drastic nature of the procedures required for their liberation from the chloroplast. These procedures range from vigorous, prolonged sonication through chemical extraction (ether, acetone) to extraction at 100°C (phosphodoxin). Second, those factors which catalyze pseudocyclic photophosphorylation most probably are capable of at least reducing oxygen and as such, may have a common identity. Finally, similarities in the spectra of these compounds suggest that they are of a similar nature. More research is required to sort out the various activities catalyzed by these substances. The drastic treatments described above rarely leave residual chloroplast fragments suitable for subsequent assay. Future research in this area should be directed at this problem. It is hoped that further knowledge about the chemical nature of these substances will make the solution of this problem more readily attainable.

II. FERREDOXIN CATALYTIC ACTIVITY

A. NADP Reduction

The mechanism of NADP reduction as summarized by San Pietro (1968) involved (1) photoreduction of ferredoxin (Fd); (2) oxidation of the reduced Fd by the flavoprotein, Fd-NADP reductase (FAD) giving the reduced flavoprotein and oxidized Fd; and (3) oxidation of the reduced flavoprotein by NADP. The present picture of NADP reduction differs only to the extent that a few more of the details are now available. For example, although earlier work had established that Fd was a one-electron carrier (Fry *et al.*, 1963), only recently did Forti *et al.* (1970) demonstrate that the Fd-NADP reductase is also a one-electron carrier *in vivo;* that is, it is reduced only to the level of the semiquinone during photosynthetic electron transport. To demonstrate this, they followed the reduction of the Fd-NADP reductase by spinach chloroplasts by measuring the increase in absorbance at 550 or 540 nm (semiquinone formation) and the decrease in absorbance at 456 nm (loss of the fully oxidized flavoprotein). The decrease in absorbance at 456 nm was completely accounted for by the formation of the semiquinone.

Another aspect of the role of Fd in NADP reduction which has received some attention recently involves the formation of a complex between Fd and Fd-NADP reductase and its possible role in NADP reduction. Some years ago, Lazzarini and San Pietro (1962) obtained data suggestive of an interaction between Fd and the Fd-NADP reductase. They observed NADPH:cytochrome c oxidoreductase activity when both proteins were present; there was essentially no activity with either protein alone. Since then, more direct evidence for the *in vitro* formation of the complex has been reported.

Specifically, perturbations in the visible spectrum of the reductase were observed when spinach Fd and Fd-NADP reductase were mixed. Shin and San Pietro (1968) studied the spectral properties of the complex using the split-compartment cuvette technique and observed two absorbance maxima in the difference spectrum, at 395 and 465 nm. They were then able to utilize this difference spectrum to measure the stoichiometry of complex formation and show that the Fd and Fd-NADP reductase react in an equimolar ratio. In a later paper, Foust *et al.* (1969) not only confirmed the results of Shin and San Pietro but also looked at the thermodynamic properties of the complex and the sensitivity of the complex to ionic strength. They concluded from these studies that the interactions between the Fd and the Fd-NADP reductase were mainly hydrophilic in nature.

While there is no evidence for complex formation *in vivo*, Hiyama *et al.* (1970) have speculated that a light-induced absorption change which they observed around 470 nm at low temperatures in whole cells of *Chlamydomonas reinhardi* might be attributed to the reduction of the complex.

Another interesting property of Fd both with respect to the mechanism of NADP reduction and the evolution of the protein is the functional interchangeability of Fd's, irrespective of source, in catalyzing NADP photoreduction by spinach chloroplasts. This might well be expected among the Fd's of higher plants and algae since all those studied to date, with two exceptions, contain 2 gm atoms each of nonheme iron and acid labile sulfur and have similar molecular weights (approximately 12,000). In addition, they all possess identical EPR signals ($g = 1.94$

below 40°K) and contain similar though rarely identical amino acid compositions. In most studies, where a higher plant or algal Fd has been used in place of spinach Fd in the photoreduction of NADP (by spinach chloroplasts), the resulting rate was approximately that of the control.

With regard to bacterial Fd's, this functional interchangeability comes as more of a surprise. While the bacterial Fd's show redox potentials similar to those of higher plants Fd's ($E_0' = -420$ mV), they are generally smaller in size (MW = 6000–10,000) and usually contain a greater amount of nonheme iron (4–7 Fe/mole). Nonetheless, they are quite capable of replacing spinach Fd in NADP reduction by spinach chloroplasts. The rates varied from 24% of the control with Fd from the sulfate-reducing bacterium, *Desulfovibrio gigas*, to 70% with Fd from *Chromatium*, a purple photosynthetic bacterium (Lee *et al.*, 1970). Further, they made unique use of these differential reactivities to support a proposed pattern of evolutionary relationships between bacteria and higher plants. The resulting evolutionary sequence corresponds rather closely with schemes proposed by other workers using different criteria (Buchanan *et al.*, 1969).

The suggestion has been made by Foust *et al.* (1969) that the differences in activities of the various Fd's reside in their differential ability to transfer electrons to the spinach Fd-NADP reductase. In this regard, the observations of Davenport (1965) are very noteworthy. It was shown by Arnon *et al.* (1958) that the rate of NADP photoreduction by spinach chloroplasts in the presence of spinach ferredoxin was stimulated by inclusion of the phosphorylation cofactors (ADP, P_i, and Mg^{2+}) in the reaction mixture. This observation was confirmed by Davenport (1965) using pea chloroplasts and pea ferredoxin. However, no such stimulation was observed when clostridial ferredoxin was used in place of pea ferredoxin. This means that even in the presence of an excess of clostridial ferredoxin and the phosphorylation cofactors, a reaction other than the formation of ATP was still rate limiting. This must be the transfer of electrons from reduced ferredoxin to the Fd-NADP reductase. Thus, while there is functional interchangeability of ferredoxins, it is clear that the plant Fd-NADP reductase prefers its own ferredoxin.

To look at this question at an even more fundamental level, it is quite possible that the differential NADP photoreduction activities displayed by these various Fd's are manifestations of structural variations in the different Fd's. In any case, this system would appear to represent an area where those students of structural aspects and those of functional aspects of proteins might meet on a middle ground and together shed light on the presently little understood mechanism of electron transfer between Fd and Fd-NADP reductase.

While considering structural differences, some mention should be made of the two Fd's alluded to earlier whose Fe/S ratio deviates from unity. Both of these Fd's were isolated from algae: the first by Yamanaka *et al.* (1969) from the blue-green alga, *Anacystis nidulans,* and the second by Böger (1970) from the alga, *Bumilleriopsis filiformis* Vischer. Both these Fd's show marked similarities to other plant and algal Fd's with respect to absorption spectra (oxidized and reduced), molecular weights, and amino acid composition. However, they both differ from other known plant Fd's in that while each contains two atoms of nonheme iron, there is only one atom of labile sulfur. In terms of reactivity, both are capable of replacing spinach Fd in catalyzing the photoreduction of NADP by spinach chloroplasts; however, in the case of the *Bumilleriopsis* Fd at least, the rates are lower than those observed with spinach Fd (Böger, 1969). This also represents a rather unique property among higher plant and algal Fd's. Böger (1970) also reported that upon dialysis against EDTA part of the iron but none of the labile sulfur was removed, a finding that differs from spinach Fd, where both the iron and the sulfur are removed simultaneously (Fry and San Pietro, 1962). As a result, he postulated that the active center of *Bumilleriopsis* Fd may only consist of one atom each of nonheme iron and labile sulfur.

B. Cyclic Photophosphorylation

Ever since Arnon *et al.* (1954) and Frenkel (1954) independently discovered cyclic photophosphorylation, there has been continued question as to its physiological importance. The concept originally served to explain how noncyclic electron flow, which produced ATP and NADPH in equimolar amounts, could drive the reductive pentose cycle wherein the stoichiometry is three ATP and two NADPH per CO_2 fixed. Utilizing cyclic phosphorylation, the extra ATP could be conveniently supplied as needed by a proper balance between cyclic and noncyclic electron flow. Recently, however, evidence has been presented (Izawa and Good, 1968; Neumann *et al.,* 1971) which suggests that the true stoichiometry of noncyclic electron flow may be two ATP per NADP reduced. If this is so, the problem of an ATP deficit is turned around and now is one of ATP excess. As a consequence, the need for a cyclic form of ATP synthesis no longer exists. However, the last word on ATP formation by chloroplasts is not in. Even if the stoichiometry of noncyclic electron transport proposed by Izawa and Good (1968) is correct, it may simply mean that cyclic photophosphorylation is not necessary for photosynthesis. This does not necessarily mean that it cannot exist possibly as an evolutionary vestige of some earlier, less efficient, system as a sort of nonspecific ATP generating system.

Early attempts to show cyclic photophosphorylation in chloroplasts were hampered by the fact that it could only be demonstrated when a catalytic amount of some electron transfer cofactor was added, and the best of these cofactors is the nonphysiological dye, phenazine methosulfate. Nonetheless, Forti and Parisi (1963) submitted evidence for an *in vivo* cyclic photophosphorylation when they demonstrated a light-stimulated increase in the ATP level in spinach leaves in the presence of a concentration of *p*-chlorophenyldimethylurea (CMU) that completely inhibited O_2 evolution and CO_2 fixation. This was confirmed in later studies by Tanner *et al.* (1968) and Teichler (1967), who were also able to show that the turnover rate of the cyclic phosphorylation pathway was very slow. Assuming that a cyclic path of phosphorylation does exist, the central problem in recent years has been to determine which of the components of the photosynthetic electron transport chain is (or are) the true endogenous mediator(s) of cyclic electron flow. Although the final answer to this problem has yet to be provided, several possibilities have been suggested.

Arnon (1969) has reported that Fd can mediate a cyclic phosphorylation in spinach chloroplasts. The rather large concentrations of Fd needed in these experiments, however, has cast some doubt upon the "catalytic" role which Fd is said to have played in these reactions. As it now stands, the general acceptance of Fd as the true mediator of cyclic phosphorylation must await further, more conclusive evidence.

The Fd-NADP reductase is another component which has received some support as the mediator of cyclic phosphorylation. This possibility was suggested by Zanetti and Forti (1966) in an attempt to explain the presence of NADPH:cytochrome f oxidoreductase activity which they found to be associated with the crystalline Fd-NADP reductase. Evidence in favor of this hypothesis was presented by Fewson *et al.* (1963), who earlier demonstrated that Fd-supported cyclic phosphorylation could be suppressed by an antibody to the Fd-NADP reductase. The main drawback to this hypothesis lies in the fact that no one has yet been able to demonstrate directly Fd-NADP reductase-mediated cyclic phosphorylation.

In addition to the above, recent work by Hiyama *et al.* (1970) with the pale-green mutant of *Chlamydomonas reinhardi* has led to an even more complicated story. They looked at light- and O_2-induced changes in cytochromes f, b_{563}, and b_{559} utilizing a combination of inhibitors and lowered temperature. Their results led to the conclusion that *Chlamydomonas reinhardi* contains two pathways of cyclic electron flow operating in parallel. One feeds off somewhere between P-700 and Fd and goes back into the electron transport chain in the vicinity of cytochrome f, by way of cytochrome b_{563}. The second cycle comes off of the Fd-Fd-NADP reductase complex (exactly which of the two components donates the electron is not known) and reduces cytochrome f directly. They did not investigate the site of phosphorylation. Thus, which of the two cyclic schemes, or both, is coupled to ATP synthesis remains unknown.

One final aspect which should be considered is the postulated role of cyclic photophosphorylation in those plants containing the C_4 dicarboxylic acid cycle of CO_2 fixation (Hatch and Slack plants). There are two main reasons why cyclic phosphorylation is postulated to be rather active in these plants (Lee *et al.*, 1970). The first and probably most compelling argument involves the theoretical stoichiometry of photosynthesis in C_4 cycle plants first postulated by Chen *et al.* (1969). They note that two moles extra of ATP are required by C_4 plants in order to regenerate phosphoenolpyruvate (PEP) thus giving a requirement of five ATP and two NADPH per CO_2 fixed as compared to three ATP and two NADPH per CO_2 fixed in Calvin cycle plants. If this stoichiometry is real, and even if the noncyclic schemes of Izawa and Good (1968) and Neumann *et al.* (1971) (referred to earlier) are correct, there is still the need for one ATP more than the noncyclic system can supply. We thus find ourselves back in the same dilemma which originally led to the idea of cyclic phosphorylation. The second reason for postulating a role for cyclic phosphorylation in C_4 plants involves the light saturation curves for the two processes. Unlike Calvin cycle plants which saturate near 2000–3000 ft-c, photosynthesis in the C_4 plants does not saturate until intensities near 8000–10,000 ft-c are reached (Black *et al.*, 1969). In studies carried out on various chloroplast reactions, only cyclic phosphorylation required such high light intensities to reach saturation (Chen *et al.*, 1969). Taken together, these two observations tend to argue strongly in favor of a role for cyclic phosphorylation in supplying the extra ATP.

In order to test this hypothesis, attempts have been made to see if C_4 plants contain larger amounts of various photosystem I components, since cyclic electron flow is considered to be a photosystem I reaction. As a result, Black and Mayne (1970) found that, on a chlorophyll basis, there is about 50% more P-700 in C_4 cycle plants than in those plants which contain only the Calvin cycle. Similarly, Lee *et al.* (1970) found that C_4 plants also contained 50% more Fd, again on a chlorophyll basis, than Calvin cycle plants. If the cyclic hypothesis is true, then these data might argue that Fd plays some role in the cyclic process, although it

TABLE III
COMPARISON OF OXIDATIVE AND PHOTOSYNTHETIC ENERGY TRANSDUCTION

Parameter	Organelle		
	Mitochondrion	Chloroplast	Chromatophore
Energy source	Glycolysis, TCA cycle (chemical)	Light (400–700 nm) (physical)	Light (>750 nm) (physical)
G^0 donor acceptor	52 kcal/2 e^- (NADH → H_2O)	60 kcal/2 e^- (H_2O → FRS)	20 kcal/2 e^- (Bchl → X)
Substrates			
Donor	NADH, succinate	H_2O	Succinate
Terminal acceptor	O_2	NADP	NAD
Other	ADP, P_i	ADP, P_i	ADP, P_i
Products	H_2O, ATP, fumarate, NAD	O_2, ATP, NADPH	ATP, NADH, fumarate
ATP formation			
Coupling factors	Yes	Yes	Yes
Proton pumping	In → out	Out → in	Out → in
Ion transfer	Yes	Yes	? Yes
Uncouplers	DNP, CCCP, arsenate	NH_4^+, DNP, CCCP, nigericin (K^+)	Nigericin (K^+)
Effect on electron flow	Stimulates	Stimulates	Inhibits
Inhibitors	Oligomycin	DIO-9, phloridzin, DCCD, synthalin	
Coupling sites	3	2 (3?)	1 (2?)
Inhibitors of electron flow	Antimycin A, CN^-, N^{3-}, Amytal, rotenone, BAL	DCMU	Antimycin A

Regulation	ADP/P_i ratio	(NADPH/NADP ratio)	Energy charge
Components			
Heme Fe^{2+}	Cytochrome b, cyt c_1, cyt c, $(a + a_3)$	Cytochrome b_{559}, cyt b_{564}, c_{552}	Cytochrome c_2, b
Cu^{2+} protein	Cytochrome oxidase	Plastocyanin	—
Nonheme Fe^{2+}	3 sites (?)	Ferredoxin (M?)	—
Flavoproteins	1 FAD (succinate $\rightarrow$ CoQ) 1 FMN (NADH $\rightarrow$ CoQ)	1 FAD (Fd-NADP reductase)	—
Mg^{2+} porphyrins	—	Chl a	Bchl
Quinones	Yes	Yes	Yes
Site of electron transport	Membrane bound	Membrane bound	Membrane bound
Fragmentation			
Methods	Detergent and sonication	Detergent and sonication	Sonication
Number of resulting complexes	(4) NADH $\rightarrow$ CoQ Succinate $\rightarrow$ CoQ $Co_Q \rightarrow$ Cyt c Cyt c $\rightarrow O_2$	(2) $H_2O \rightarrow Fe(CN)_6^{3-}$ Asc, DPIP / NADP	(1) Succinate $\rightarrow$ NAD

does not necessarily have to represent the site at which electrons leave the NADP reduction pathway.

In view of the earlier discussion on the functional interchangeability of various Fd's, it should be noted that Lee *et al.* (1970) made a detailed study of the Fd from nutsedge (a C_4 plant) and found it was very similar to spinach Fd with respect to molecular weight, absorption spectrum, iron and sulfur content, and amino acid composition; it did differ from spinach Fd in its amino acid sequence and isoelectric point. Not surprisingly, the nutsedge Fd was just as efficient as spinach Fd in supporting NADP photoreduction by spinach chloroplasts.

In summary, it should be noted that cyclic photophosphorylation is far from being totally characterized, and the possible role which Fd might play in it, if any, is likewise still open to debate. The best evidence to date has Fd on the path of cyclic phosphorylation, but whether it is the point of departure from the noncyclic chain of flow has yet to be shown conclusively.

III. PHOTOSYNTHETIC AND OXIDATIVE ENERGY TRANSDUCTION

Although not entirely within the realm of this chapter, a brief comparison of the electron transport sequences in mitochondria, chloroplasts, and chromatophores is included (Fig. 2 and Table III) because of its possible usefulness to graduate students and investigators new to this area. An excellent and more detailed account of these systems is available (Racker, 1970).

The major problem unique to both photosynthesis (chloroplasts and chromatophores) and respiration (mitochondria) is the nature of those events, both chemical and physical, involved in the transduction of electron transfer energy into phosphate bond energy. Each system is comprised of chains of electron transport carriers and enzymes fixed in membranes which function to transfer electrons and protons from the hydrogen donor to the acceptor. Part of the energy generated thereby is conserved generally in the form of ATP; other energy forms such as ion gradients or membrane conformational changes are possible.

In these systems, the electron transport sequence involves flavoproteins and cytochromes which function to transfer electrons (and protons) in a thermodynamic stepwise fashion leading ultimately to the reduction of the final acceptor (Fig. 2 and Table III). The tentative sites of "coupling" of electron transport to energy conservation are indicated in Fig. 2.

Although the electron donors in mitochondria are indicated to be

NADH and succinate (Fig. 2 and Table III), the following quote from Racker (1970) is particularly noteworthy.

> In both mitochondria and chloroplasts the major hydrogen donor is water, but the final acceptor is different. In mitochondria, it is oxygen, in chloroplasts it is TPN. The importance of water as a hydrogen donor in mitochondria oxidations is not frequently emphasized. During the metabolic acrobatics of pyruvate oxidation in the Krebs cycle, water is incorporated into intermediates at three steps: in the hydration of fumarate to malate, and in the course of the utilization of acetyl CoA and succinyl CoA. Thus, for each molecule of pyruvate which donates four hydrogens, water contributes six additional hydrogens which are channeled into the mitochondrial oxidation chain. Mitochondria, therefore, like chloroplasts, cleave water, albeit by a different mechanism. In chloroplasts the oxygen of water is liberated as molecular oxygen; in mitochondria it is incorporated into the CO_2 liberated during the oxidation of the Krebs cycle intermediates.

While the possible involvement of nonheme iron (NHI) in the energy transduction process is clearly not established, there appears to be a good correlation between the presence of NHI and "coupling" sites, especially in mitochondria. It is possible that NHI may well be involved also in chloroplasts and chromatophores.

The excellent studies of Levine and Goodenough (1970) with chloroplasts from mutants of *Chlamydomonas reinhardi* indicate the presence of a cofactor (M) of unknown nature which functions in the electron transport sequence interconnecting the two photosystems. It is tempting to speculate that the cofactor M, by analogy to mitochondria, is a nonheme iron protein and perhaps even similar or identical to the "bound NHI" in chloroplasts discovered recently by Malkin and Bearden (1971). There is clearly no evidence to date to support this view. Perhaps when additional evidence is available, it will be shown that the present tentative correlation between NHI and "coupling" sites should really be viewed rather as a "functional" correlation.

NOTE ADDED IN PROOF. Since the completion of this chapter, much additional evidence relating specifically to the role of "bound" iron–sulfur proteins in the primary photochemical reactions of photosynthesis has appeared. The plant system is discussed in the recent reviews of Siedow *et al.* (1973) and Ke (1973); the bacterial system is the subject of a recent paper by Dutton *et al.* (1973).

ACKNOWLEDGMENTS

Some of the research described in this chapter was supported by the National Institutes of Health, Grant GM 16314 (to A.S.P.). The authors wish to thank Jamie San Pietro for the illustrations.

REFERENCES

Arnon, D. I. (1969). *In* "Progress in Photosynthesis Research" (H. Metzner, ed.), Vol. III, p. 1444. Tübingen.

Arnon, D. I., Allen, M. B., and Whatley, F. R. (1954). *Nature* **174**, 394.

Arnon, D. I., Whatley, F. R., and Allen, M. B. (1958). *Science* **127**, 1026.

Black, C. C. (1966). *Biochim. Biophys. Acta* **120**, 332.

Black, C. C., Chen, T. M.. and Brown, H. R., (1969). *Weed Sci.* **17**, 338.

Black, C. C., and Mayne, B. C. (1970). *Plant Physiol.* **45**, 738.

Black, C. C., San Pietro, A., Limbach, D., and Norris, G. (1963). *Proc. Nat. Acad. Sci. U.S.* **50**, 37.

Böger, P. F. (1969). *Pflanzenphysiologie* **61**, 447.

Böger, P. (1970). *Planta* **92**, 105.

Brune, D. (1970). Personal communication.

Brune, D., and San Pietro, A. (1970). *Arch. Biochem. Biophys.* **141**, 371.

Buchanan, B. B., Matsubara, H., and Evans, M. C. W. (1969). *Biochim. Biophys. Acta* **189**, 46.

Chance, B., San Pietro, A., Avron, M., and Hildreth, W. W. (1965). *In* "Non-Heme Iron Proteins" (A. San Pietro, ed.), p. 225. Antioch Press, Yellow Springs, Ohio.

Chen, T. M., Brown, H. R., and Black, C. C. (1969). *Plant Physiol.* **44**, 649.

Davenport, H. E. (1965). *In* "Non-Heme Iron Proteins" (A. San Pietro, ed.), p. 115. Antioch Press, Yellow Springs, Ohio.

Dutton, P. L., Lcigh, J. S., and Reed, D. W. (1973). *Biochim. Biophys. Acta* **292**, 654.

Fewson, C. A., Black, C. C., and Gibbs, M. (1963). *Plant Physiol.* **38**, 680.

Forti, G., and Parisi, B. (1963). *Biochim. Biophys. Acta* **71**, 1.

Forti, G., Melandri, B. A., San Pietro, A., and Ke, B. (1970). *Arch. Biochem. Biophys.* **140**, 107.

Foust, G. P., Mayhew, S. G., and Massey, V. (1969). *J. Biol. Chem.* **244**, 964.

Frenkel, A. W. (1954). *J. Amer. Chem. Soc.* **76**, 5568.

Fry, K. T., and San Pietro, A. (1962). *Biochem. Biophys. Res. Commun.* **9**, 218.

Fry, K. T., Lazzarini, R. A., and San Pietro, A. (1963). *Proc. Nat. Acad. Sci. U.S.* **50**, 652.

Fujita, Y., and Murano, F. (1967). *Plant Cell Phys.* **8**, 269.

Fujita, Y., and Myers, J. (1966). *Plant Cell Phys.* **7**, 599.

Fujita, Y., and Myers, J. (1967). *Arch. Biochem. Biophys.* **119**, 8.

Gee, R., Kylin, A., and Saltman, P. (1970). *Biochem. Biophys. Res. Commum.* **40**, 642.

Honeycutt, R., and Krogmann, D. W. (1970). *Fed. Proc.* **29**, 536.

Hiyama, T., and Ke, B. (1971). *Proc. Nat. Acad. Sci. U.S.* **68**, 1010.

Hiyama, T., Nishimura, M., and Chance, B. (1970). *Plant Physiol.* **64**, 163.

Izawa, S., and Good, N. E. (1968). *Biochim. Biophys. Acta* **162**, 380.

Ke, B. (1973). *Biochim. Biophys. Acta* **301**, 1.

Kok, B. H., Rurainski, J., and Owens, O. V. H. (1965). *Biochim. Biophys. Acta* **109**, 347.

Lazzarini, R. A., and San Pietro, A. (1962). *Biochim. Biophys. Acta* **62**, 417.

Lee, S. S., Travis, J., and Black, C. C. (1970). *Arch. Biochem. Biophys.* **141**, 676.

Levine, R. P., and Goodenough, U. W. (1970). *Amer. Rev. Genet.* **4**, 397.

Malkin, R., and Bearden, A. J. (1971). *Proc. Nat. Acad. Sci. U.S.* **68,** 16.
Neumann, J., Arnutzen, C. J., and Dilley, R. A. (1971). *Biochemistry* **10,** 866.
Racker, E. (1970). *In* "Membranes of Mitochondria and Chloroplasts" (E. Racker, ed.), p. 127. Van Nostrand Reinhold, Princeton, New Jersey.
Regitz, G., Berzborn, R., and Trebst, A. (1970). *Planta* **91,** 8.
Saltman, P., and Gee, R. (1966). *Symp. Use Isotopes Plant Nutr. Physiol., Vienna.* Int. At. Energy Agency SM-77125.
San Pietro, A. (1968). *In* "Biological Oxidations" (T. P. Singer, ed.), p. 515. Wiley (Interscience), New York.
Shin, M., and Arnon, D. I. (1965). *J. Biol. Chem.* **240,** 1405.
Shin, M., and San Pietro, A. (1968). *Biochem. Biophys. Res. Commun.* **33,** 38.
Siedow, J. N., and San Pietro, A. (1971). Unpublished results.
Siedow, J. N., Yocum, C. F., and San Pietro, A. (1973). *In* "Current Topics in Bioenergetics" (D. R. Sanadi, ed.), Vol. V, p. 107. Academic Press, New York.
Tanner, W., Loffler, M., and Kandler, O. (1968). *Plant Physiol.* **44,** 422.
Teichler, D. (1967). *Arch. Biochem. Biophys.* **120,** 227.
Vernon, L. P., San Pietro, A., and Limbach, D. A. (1965). *Arch. Biochem. Biophys.* **109,** 92.
Wildner, G. F., and Criddle, R. S. (1969). *Biochem. Biophys. Res. Commun.* **37,** 952.
Wu, M., and Myers, J. (1969). *Arch. Biochem. Biophys.* **132,** 430.
Wu, M., Myers, J., and Forrest, H. S. (1970). *Arch. Biochem. Biophys.* **140,** 391.
Yamanaka, T., Takenami, S., Wada, K., and Okunuki, K. (1969). *Biochim. Biophys. Acta* **180,** 196.
Yocum, C. F., and San Pietro, A. (1969). *Biochem. Biophys. Res. Commun.* **36,** 614.
Yocum, C. F., and San Pietro, A. (1970). *Arch. Biochem. Biophys.* **140,** 152.
Yocum, C. F., Kyle, J. L., and Gross, J. A. (1969). *In* "Progress in Photosynthesis Research" (H. Metzner, ed.), Vol. I, p. 122. Tübingen.
Zanetti, G., and Forti, G. (1966). *J. Biol. Chem.* **241,** 279.
Zweig, G., and Avron, M. (1965). *Biochem. Biophys. Res. Commun.* **19,** 347.

CHAPTER 5

Ferredoxin and Carbon Assimilation

BOB B. BUCHANAN

I. INTRODUCTION

More than 25 years ago, Lipmann (1946) discussed, on thermodynamic grounds, the feasibility of a reductive carboxylation of what is now known as acetyl coenzyme A. Lipmann envisaged that the formation of pyruvate in this manner would offer to a photosynthetic or chemosynthetic cell an ideal mechanism of incorporating carbon dioxide into a key metabolic intermediate. However, early attempts to achieve a net synthesis of pyruvate from carbon dioxide and a two-carbon unit met with no success. Several investigators (Wilson *et al.*, 1948; Wolfe and O'Kane, 1955) were able to show a rapid exchange between [^{14}C]bicarbonate and the carboxyl group of pyruvate in cell-free extracts of certain nonphotosynthetic bacteria but could not demonstrate an incorporation into pyruvate of acetyl coenzyme A [or a precursor, acetyl phosphate, which, in the presence of phosphotransacetylase (Stadtman and Barker, 1950), is converted to acetyl coenzyme A]. Not until 1959 was a synthesis of pyruvate from a two-carbon unit and carbon dioxide achieved in an enzymic reaction. In that year, Mortlock and Wolfe (1959) reported that the nonphysiological reductant, sodium dithionite, could drive at a low rate the synthesis of pyruvate from acetyl phosphate and carbon dioxide in cell-free extracts of the fermentative bacterium, *Clostridium butylicum*. But there was no indication that dithionite could be replaced by a physiological reductant or that the reaction was of physiological significance.

Thus, for years there was general agreement that pyruvate synthesis at best was of questionable importance in carbon dioxide assimilation. An appraisal in 1963 of the evidence for reversibility of the phosphoroclastic and other α-decarboxylation reactions led Wood and Stjernholm (1963) to conclude ". . . that the utilization of CO_2 by the reversal of α-decarboxylation is of little practical significance in the heterotrophic assimilation of CO_2, at least in the organisms so far studied."

Further progress on pyruvate synthesis was not made until ferredoxin had been isolated and certain key properties determined (Mortenson *et al.*, 1962; Tagawa and Arnon, 1962). Accordingly, when Tagawa and Arnon (1962, 1968) reported that the redox potential of ferredoxin was 100 mV more reducing than that of nicotinamide adenine dinucleotides, the possibility arose that the strong reducing power of ferredoxin might be used in pyruvate synthesis. However, there was at the time no experimental evidence that ferredoxin could participate directly as a reductant in this or any other enzymic reaction concerned with carbon assimilation. [The indirect participation of ferredoxin in carbon dioxide assimilation by way of nicotinamide adenine dinucleotides—with an attendant drop of 100 mV in reducing potential—was not in doubt because ferredoxins were then known to act as electron carriers in the reduction of adenine dinucleotides by illuminated chloroplasts (Shin and Arnon, 1965; Shin *et al.*, 1963) and by cell-free bacterial extracts (Valentine *et al.*, 1962; Buchanan and Bachofen, 1968; Buchanan and Evans, 1969).]

Two years later, Bachofen *et al.* (1964) obtained the first evidence that reduced ferredoxin can drive the reductive synthesis of pyruvate from acetyl coenzyme A and carbon dioxide in cell-free extracts of the fermentative (heterotrophic) bacterium *Clostridium pasteurianum* [Eq. (1)]:

$$\text{Acetyl CoA} + CO_2 + \text{Ferredoxin}_{red} \rightarrow \text{Pyruvate} + \text{CoA} + \text{Ferredoxin}_{ox} \quad (1)$$

Named pyruvate synthetase, reaction (1) was the first demonstration of ferredoxin as a direct reductant in carbon dioxide assimilation. Pyruvate synthase with reduced ferredoxin achieves a reversal of the "phosphoroclastic" splitting of pyruvate [known in fermentative organisms (Koepsell *et al.*, 1944; Wolfe and O'Kane, 1953) since 1944 and found by Mortenson *et al.* (1962) to depend on ferredoxin] and thus represents the reductive carboxylation that Lipmann (1946) had proposed and Wood and Stjernholm (1963) had concluded to be of doubtful significance.

Soon after its discovery in 1964 in *Clostridium pasteurianum,* the pyruvate synthetase system was found in the photosynthetic bacterium, *Chromatium* (Buchanan *et al.,* 1964)—a finding which indicated that ferredoxin-dependent carboxylation is not unique to fermentative organisms. Ferredoxin-dependent carbon dioxide fixation has since been demonstrated in numerous anaerobic bacteria, both photosynthetic and nonphotosynthetic, but not in photosynthetic organisms which evolve oxygen (i.e., algae and higher plants) or in aerobic nonphotosynthetic cells. A common feature of all organisms showing ferredoxin-linked carboxylation is a requirement for an anaerobic environment for growth.

The finding of a second ferredoxin-linked carboxylation—the α-ketoglutarate synthetase reaction (Buchanan and Evans, 1965) [Eq. (2)]—showed that ferredoxin can promote the synthesis of α-keto acids other than pyruvate.

$$\text{Succinyl CoA} + CO_2 + \text{Ferredoxin}_{red} \rightarrow \alpha\text{-Ketoglutarate} + \text{CoA} + \text{Ferredoxin}_{ox} \quad (2)$$

Later work demonstrated ferredoxin-dependent reductive carboxylations that lead to the synthesis of α-ketoisovalerate (Allison and Peel, 1968), α-ketobutyrate (Buchanan, 1969), and phenyl pyruvate (Gehring and Arnon 1971) (Fig. 1). The ferredoxin-linked carboxylations

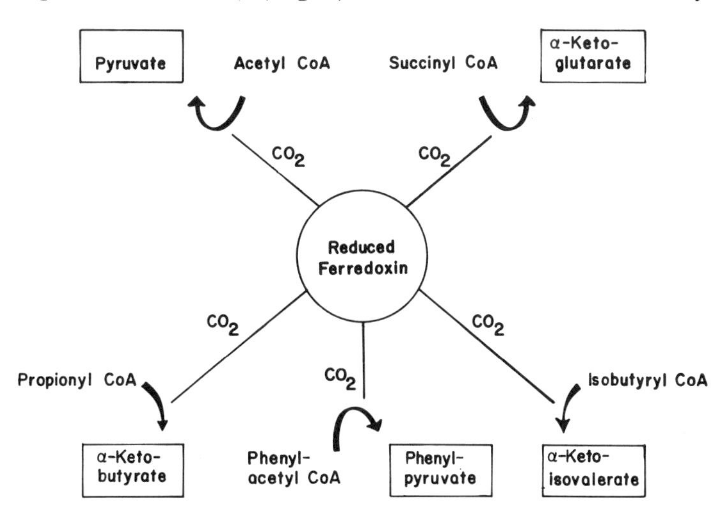

Fig. 1. Ferredoxin-dependent carboxylation reactions in photosynthetic and fermentative (heterotrophic) bacteria.

described so far involve the reductive carboxylation of an acyl coenzyme A derivative to an α-keto acid [Eq. (3)]

$$\text{Acyl CoA} + CO_2 + \text{Ferredoxin}_{red} \rightarrow \alpha\text{-Keto Acid} + \text{CoA} + \text{Ferredoxin}_{ox} \quad (3)$$

All α-keto acids shown to be synthesized via a ferredoxin-linked carboxylation are important intermediates in the biosynthesis of amino acids. However, the role of α-keto acids formed in ferredoxin-dependent carboxylation reactions is not limited to amino acid biosynthesis. There is now considerable evidence that carboxylation reactions driven by reduced ferredoxin are essential for the operation of two new cyclic mechanisms for the assimilation of carbon dioxide: the reductive carboxylic acid cycle of photosynthetic bacteria (Evans *et al.,* 1966; Buchanan *et al.,* 1967) and the reductive monocarboxylic acid cycle of fermentative bacteria (Thauer *et al.,* 1970).

This chapter summarizes current evidence for the role of ferredoxin in carbon assimilation with particular reference to the new carbon cycles. Evidence pertaining to the physiological significance of the new pathways is also described.

II. REDUCTIVE CARBOXYLIC ACID CYCLE OF BACTERIAL PHOTOSYNTHESIS

The ferredoxin-dependent syntheses of pyruvate [Eq. (1)] and α-ketoglutarate [Eq. (2)] form the basis for the reductive carboxylic acid cycle—a new pathway of carbon dioxide assimilation in bacterial photosynthesis (Evans *et al.,* 1966; Buchanan *et al.,* 1967). The new cycle is independent of the reductive pentose phosphate cycle (Bassham and Calvin, 1962) and provides another cyclic mechanism for carbon dioxide assimilation that continuously regenerates an acceptor for carbon dioxide. One complete turn of the reductive carboxylic acid cycle (Fig. 2) incorporates four molecules of carbon dioxide and results in the net synthesis of oxalacetate, which is itself an intermediate in the cycle. Thus, beginning with one molecule of oxalacetate, one complete turn of the reductive carboxylic acid cycle will regenerate it and yield, in addition, a second molecule of oxalacetate formed by the reductive fixation of four molecules of carbon dioxide.

The carboxylations of the reductive carboxylic acid cycle include, apart from the pyruvate and α-ketoglutarate synthetase reactions, isocitrate dehydrogenase (Ochoa and Weisz-Tabori, 1954; Moyle, 1956) which catalyzes reversibly the carboxylation of α-ketoglutarate to isocitrate [Eq. (4)] and phosphoenolpyruvate carboxylase (Bandurski and Greiner,

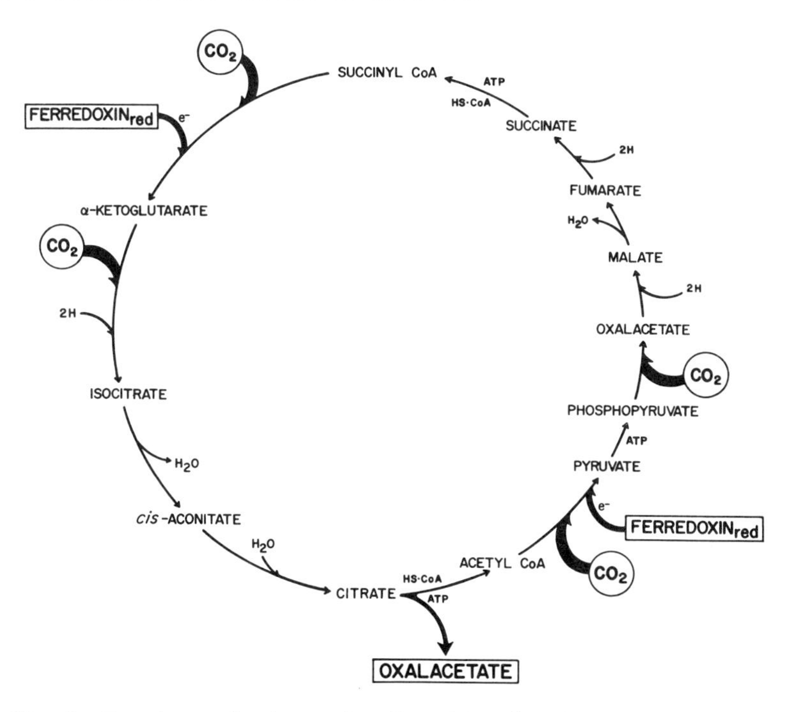

Fig. 2. Complete reductive carboxylic acid cycle of bacterial photosynthesis. Reversibility of the reactions is not indicated.

1953) which catalyzes the carboxylation of phosphoenolpyruvate to oxalacetate [Eq. (5)].

$$\alpha\text{-Ketoglutarate} + CO_2 + NADPH_2 \rightleftharpoons \text{Isocitrate} + NADP \quad (4)$$

$$\text{Phosphoenolpyruvate} + CO_2 \rightarrow \text{Oxalacetate} + P_i \quad (5)$$

A variant of the complete reductive carboxylic acid cycle is the "short" reductive carboxylic acid cycle (Fig. 3) which, in one turn, incorporates two molecules of carbon dioxide and yields one molecule of acetate. The complete cycle (Fig. 2) and the short cycle (Fig. 3) have the same sequence of reactions from oxalacetate to citrate. Thus, beginning again with oxalacetate, a complete turn of the short reductive carboxylic acid cycle would result in the regeneration of the oxalacetate and the synthesis of acetyl coenzyme A from two molecules of carbon dioxide.

In its overall effect, the short reductive carboxylic acid cycle (Fig. 3) which generates acetyl coenzyme A from two molecules of carbon dioxide is a reversal of the Krebs citric acid cycle, which degrades acetyl coenzye A to two molecules of carbon dioxide (Krebs, 1953; Krebs and

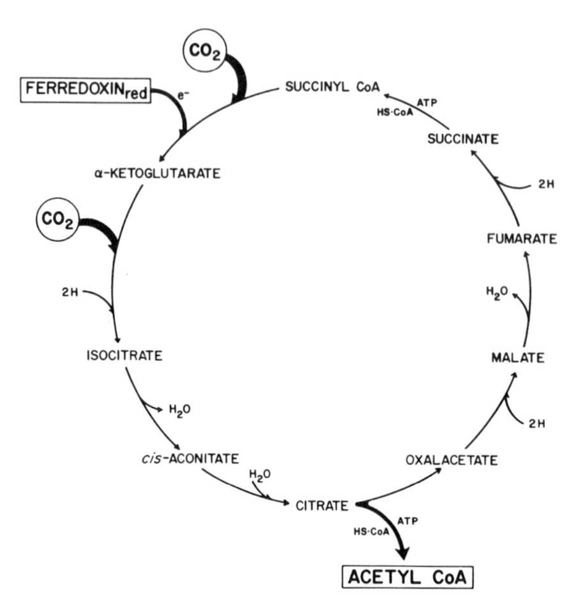

Fig. 3. Short reductive carboxylic acid cycle of bacterial photosynthesis. Reversibility of the reactions is not indicated.

Lowenstein, 1960). A basic distinction between the two cycles is that the reductive carboxylic acid cycle is endergonic in nature and hence must be linked with energy-yielding reactions which in this instance are the photoreduction of ferredoxin and photophosphorylation (Arnon, 1967). Moreover, although several reversible enzyme reactions of the citric acid cycle function also in the reductive carboxylic cycle, only the reductive cycle has the pyruvate and α-ketoglutarate synthetases which, by reversing two steps that are irreversible in the citric acid cycle, permit the reductive carboxylic acid cycle to function as a pathway for carbon dioxide assimilation.

The operation of the reductive carboxylic acid cycle at the expense of radiant energy involves bacterial photophosphorylation and photoreduction of ferredoxin. The evidence for the latter in subcellular preparations from photosynthetic bacteria is not nearly so extensive as in chloroplasts. Evans and Buchanan (1965), using chlorophyll-containing particles from *Chloribium thiosulfatophilum,* were able to show the formation of reduced ferredoxin that was strictly dependent on light and on an added electron donor, such as sodium sulfide. Furthermore, Buchanan and Evans (1969) have shown that, in cell-free preparations of *C. thiosulfatophilum,* photoreduced ferredoxin can serve as an electron donor for the reduction of NAD and thus provide a supply of $NADH_2$ (or

$NADPH_2$, which is formed less effectively) for the operation of the reductive carboxylic acid cycle. The reduced ferredoxin (and nicotinamide adenine dinucleotides) required for operation of the cycle may also be supplied, independently of light, by hydrogen gas (Buchanan *et al.*, 1964; Weaver *et al.*, 1965; Buchanan and Bachofen, 1968; Buchanan and Evans, 1969) and hydrogenase—an enzyme native to photosynthetic bacteria.

The operation of the reductive carboxylic acid cycle (Evans *et al.*, 1966; Buchanan *et al.*, 1967) in bacterial photosynthesis rests on the identification in cell-free extracts of *C. thiosulfatophilum* and *R. rubrum* of the enzymes, listed in Table I, that are required to catalyze the sequence of reactions shown in Fig. 2. The activities of the individual enzymes (in μmole/mg protein/hr) ranged from 0.012 for α-ketoglutarate synthetase to 159 for malate dehydrogenase. Since these measurements were made to establish the presence of these enzymes in cell-free extracts, without a systematic search for optimal experimental conditions, they give no definitive information about the relative activities of these enzymes *in vivo*.

Other evidence includes phosphoenolpyruvate synthetase, an enzyme that catalyzes the synthesis of phosphoenol pyruvate from ATP and

TABLE I

ACTIVITIES OF ENZYMES OF THE REDUCTIVE CARBOXYLIC ACID CYCLE IN EXTRACTS OF *C. thiosulfatophilum* AND *R. rubrum*[a]

Enzyme	Enzyme activity (μmole/mg protein/hour)	
	C. thiosulfatophilum	*R. rubrum*
Acetyl CoA synthetase	0.8	24
Pyruvate synthetase	0.2	0.06
Phosphoenolpyruvate synthetase	2.3	0.7
Phosphoenolpyruvate carboxylase	4.8	6.0
Malate dehydrogenase	37	159
Fumarate hydratase	118	128
Succinate dehydrogenase	0.85	1.2
Succinyl CoA synthetase	1.6	3.3
α-Ketoglutarate synthetase	0.4	0.012
Isocitrate dehydrogenase	102	70
Aconitate hydratase	3.1	7.3
Citrate lyase	0.15	0.17

[a] From Buchanan *et al.* (1967) and Evans *et al.* (1966).

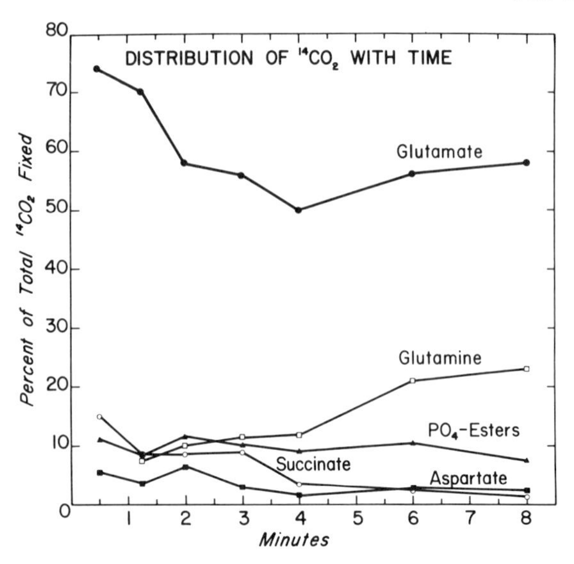

Fig. 4. Products of short-term photosynthesis by *Chlorobium thiosulfatophilum.*

pyruvate [Eq. (6)] in *Escherichia coli,* as reported by Cooper and Kornberg (1965).

$$\text{Pyruvate} + \text{ATP} \rightarrow \text{Phosphoenol pyruvate} + \text{P}_i + \text{AMP} \quad (6)$$

The presence of phosphoenolpyruvate synthetase in *C. thiosulfatophilum* and *R. rubrum* (and *Chromatium*) was demonstrated by Buchanan and Evans (1966). The equilibrium of the phosphoenolpyruvate synthetase reaction lies far on the side of phosphoenol pyruvate formation and would thus favor the operation of the reductive carboxylic acid cycle. A similar effect would also result from the irreversibility of phosphoenolpyruvate carboxylase (Bandurski and Greiner, 1953) [Eq. (5)].

The reductive carboxylic acid cycle appears to function as a biosynthetic pathway that is particularly well suited to provide carbon skeletons for the amino acids that are the main products of photosynthesis in *C. thiosulfatophilum* (Fig. 4) (Hoare and Gibson, 1964; Evans *et al.,* 1966) and other bacteria (Losada *et al.,* 1960; Fuller *et al.,* 1961; Hoare, 1963). Thus the reactions of the new cycle supply α-ketoglutarate for the synthesis of glutamate, oxalacetate for aspartate, and pyruvate for alanine. A principal product of the short cycle is acetyl coenzyme A which would be used for biosynthetic reactions, particularly the synthesis of fatty acids.

The new reductive carboxylic acid cycle invites comparison with the reductive pentose phosphate cycle (Bassham and Calvin, 1962) which hitherto has been regarded as the sole cyclic mechanism for carbon dioxide assimilation and which has been reported in *C. thiosulfatophilum* (Smillie *et al.*, 1962) and *R. rubrum* (Anderson and Fuller, 1967) [and *Chromatium* (Fuller *et al.*, 1961; Latzko and Gibbs, 1969)]. Until recently, it was not possible to assess the importance of the new carboxylic acid cycle in relation to the pentose cycle. However, the recent demonstration of the inhibition of cellular photosynthesis by low levels of the inhibitor fluoroacetate led Sirevåg and Ormerod (1970a,b) to conclude that in *C. thiosulfatophilum* carbon dioxide assimilation occurs largely via the reductive carboxylic acid cycle, while the reductive pentose phosphate cycle, if functional, is of minor significance. Such a quantitative assessment of the relative importance of the new cycle in *R. rubrum* cannot yet be made; but pertinent to this point is the demonstration by Shigesada *et al.* (1966) that whole cells of *R. rubrum* converted anaerobically in the light $^{14}CO_2$ and [^{14}C]succinate into glutamate by a condensation that can now be explained by the operation of α-ketoglutarate synthetase. Similar short-exposure experiments with $^{14}CO_2$ by Yoch and Lindstrom (1967) are also consistent with the primary operation of α-ketoglutarate synthetase in carbon dioxide assimilation by a related purple photosynthetic bacterium, *Rhodopseudomonas palustris*.

III. REDUCTIVE MONOCARBOXYLIC ACID CYCLE OF FERMENTATIVE METABOLISM

The fermentative bacterium, *Clostridium kluyveri*, can convert carbon dioxide directly to formate in C_1 metabolism (Jungermann *et al.*, 1968). Until the recent work of Thauer *et al.* (1970), no mechanism was known to account for this conversion in *C. kluyveri*.

The new evidence shows that carbon dioxide can be converted to formate via a cyclic mechanism of carbon dioxide fixation (Fig. 5). Desig-

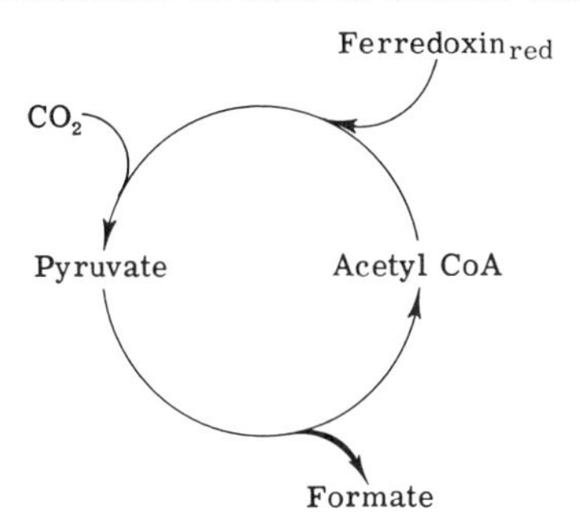

Fig. 5. Reductive monocarboxylic acid cycle of fermentative metabolism. Reversibility of the reactions is not indicated.

nated the reductive monocarboxylic acid cycle, the new cycle uses the power of reduced ferredoxin for the reduction of carbon dioxide to formate and leads to a regeneration of the carbon dioxide acceptor, acetyl coenzyme A. The formate produced by the cycle may be used [following conversion to the formyltetrahydrofolate derivative (Thauer *et al.*, 1970)] in a variety of processes—such as thymine biosynthesis—that depend on a C_1 unit.

Evidence for the reductive monocarboxylic acid cycle in *C. kluyveri* consist of (a) demonstration in cell-free extracts of the two component enzymes, pyruvate synthetase [Eq. (1)] and pyruvate formate lyase (Chase and Rabinowitz, 1968; Knappe *et al.*, 1969) [Eq. (7)]

$$\text{Pyruvate} + \text{CoA} \rightarrow \text{Acetyl CoA} + \text{Formate} \tag{7}$$

and (b) demonstration (in a cell-free preparation) of a production of formate from carbon dioxide and reduced ferredoxin in the presence of acetyl coenzyme A. Such a conversion would be predicted from the cycle shown in Fig. 5.

A feature of the reductive monocarboxylic acid cycle of fermentative bacteria—which distinguishes it from the reductive carboxylic acid cycle of bacterial photosynthesis—is the lack of a requirement for ATP. The lack of an ATP requirement shows that the reducing power of ferredoxin suffices to drive the monocarboxylic acid cycle in the direction of formate synthesis. Reduced ferredoxin could be supplied by either hydrogen gas or $NADH_2$ (Gottschalk and Chowdhury, 1969; Thauer *et al.*, 1969). (The effectiveness of $NADH_2$ as a reductant for ferredoxin in the synthesis of pyruvate or in the evolution of hydrogen has been observed for preparations of *C. kluyveri* but not for preparations from other organisms.)

The monocarboxylic acid cycle has been described only for the fermentative bacterium, *C. kluyveri*. It seems likely that other anaerobes—including photosynthetic bacteria—could use this mechanism to convert carbon dioxide to a formate derivate functional in C_1 metabolism.

IV. ENZYMES CATALYZING FERREDOXIN-DEPENDENT CARBOXYLATION REACTIONS

A. Pyruvate Synthetase

Pyruvate synthase (Bachofen *et al.*, 1964; Buchanan *et al.*, 1964) [Eq. (1)] is widely distributed in anaerobic organisms (Table II); it has been found in all photosynthetic bacteria examined (Buchanan *et al.*, 1964,

TABLE II

DISTRIBUTION OF FERREDOXIN-LINKED CARBOXYLATION ENZYMES IN DIFFERENT ORGANISMS

	Photosynthetic bacteria			Fermentative bacteria		
Carboxylation reaction	Green sulfur	Purple sulfur	Purple nonsulfur	Clostridia	Rumen	Sulfur
Pyruvate synthetase	*Chlorobium thiosulfatophilum* *Chloropseudomonas ethylicum*	*Chromatium*	*Rhodospirillum rubrum*	*Clostridium pasteurianum* *Clostridium kluyveri* *Clostridium acidi-urici* *Clostridium butyricum*		*Desulfovibrio desulfuricans*
α-Ketoglutarate synthetase	*Chlorobium thiosulfatophilum* *Chloropseudomonas ethylicum*	—	*Rhodospirillum rubrum*	—	*Bacteroides ruminicola*	—
α-Ketobutyrate synthetase	—	*Chromatium*	—	*Clostridium pasteurianum*	—	*Desulfovibrio desulfuricans*
α-Ketoisovalerate synthetase	—	—		—	*Peptostrepto-coccus elsdenii*	—
Phenylpyruvate synthetase	*Chlorobium thiosulfatophilum* *Chloropseudomonas ethylicum*	*Chromatium*	—	—	—	—

1967; Evans, 1968) and in several types of fermentative bacteria (Andrew and Morris, 1965; Raeburn and Rabinowitz, 1965a; Stern, 1965; Heer and Bachofen, 1966; Bothe, 1969; Keller, in preparation 1973). The enzyme has been partly purified from two photosynthetic bacteria [*C. thiosulfatophilum* (Buchanan and Arnon, 1969; Buchanan *et al.*, 1965) and *Chromatium* (Buchanan *et al.*, 1965)] and, following specific treatment to release cofactor from the enzyme, was shown to require thiamine pyrophosphate. An apparently similar enzyme (named "pyruvate-ferredoxin oxidoreductase"), purified to homogeneity from the fermentative bacterium *Clostridium acidi-urici,* contains an iron–sulfur chromophore (Raeburn and Rabinowitz, 1965b; Uyeda and Rabinowitz, 1967) which may couple directly to ferredoxin in the synthesis (or breakdown) of pyruvate.

In addition to its role in the reductive carboxylic acid cycle of photosynthetic bacteria (Fig. 1) and reductive monocarboxylic acid cycle of fermentative bacteria (Fig. 3), pyruvate synthetase appears to be important in the assimilation of exogenous acetate (and carbon dioxide) in both groups of organisms (Cutinelli *et al.*, 1951; Tomlinson, 1954; Hoare and Gibson, 1964). [External acetate is activated to the coenzyme A thioester in an ATP-dependent reaction prior to assimilation (Buchanan and Arnon, 1969).] The main products formed from acetate and carbon dioxide assimilated via pyruvate synthetase are amino acids—particularly alanine which is formed directly by transamination. Other amino acids, such as aspartate, may be derived from pyruvate following additional carboxylation steps (Evans *et al.*, 1966).

As shown below, the ferredoxin-dependent synthesis of alanine from acetate and carbon dioxide by way of pyruvate is independent of phosphoenolpyruvate—the compound previously considered the main precursor of pyruvate in both photosynthetic and nonphotosynthetic cells (Meister, 1965).

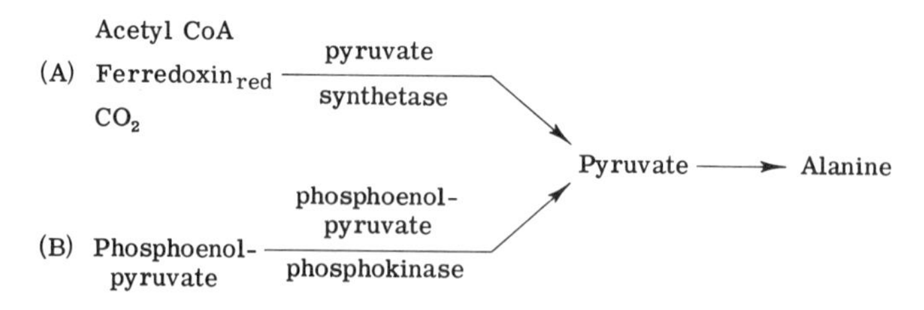

B. α-Ketoglutarate Synthetase

α-Ketoglutarate synthetase [Eq. (2)]—a key enzyme of the reductive carboxylic acid cycle of bacterial photosynthesis (Fig. 1)—was discovered in the green photosynthetic bacterium, *C. thiosulfatophilum*, and for several years was not known to occur in nonphotosynthetic bacteria. The important finding by Allison and Robinson (1970) of α-ketoglutarate synthetase in fermentative bacteria of the rumen (Table II) shows that this enzyme, like pyruvate synthetase, occurs also in nonphotosynthetic anaerobes.

Apart from its role in the reductive carboxylic acid cycle, α-ketoglutarate synthetase provides a key mechanism for assimilation of externally supplied succinate and carbon dioxide by both photosynthetic (Shigesada *et al.*, 1966) and nonphotosynthetic (Allison and Robinson, 1970) anaerobes. Precursors of succinate such as propionate may also be assimilated by this mechanism. [Like acetate, succinate is activated to the coenzyme A thioester in an ATP-dependent reaction prior to assimilation (Buchanan and Arnon, 1969).] The principal products formed from succinate and carbon dioxide are amino acids—especially glutamate which is derived directly from α-ketoglutarate by transamination.

The biosynthesis of glutamate by the α-ketoglutarate synthetase reaction (A) is independent of isocitrate (B) derived from the citric acid cycle (Krebs, 1953; Krebs and Lowenstein, 1960).

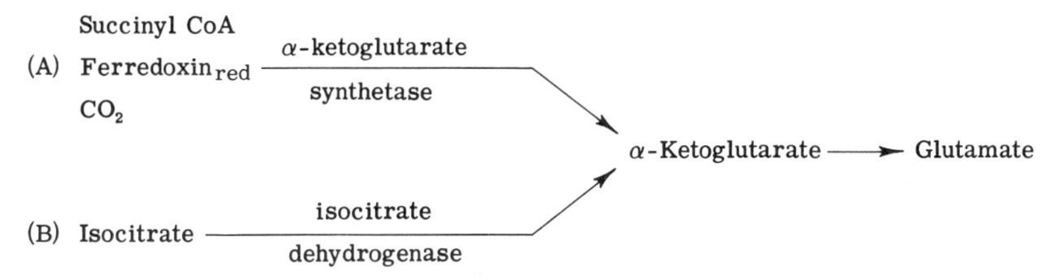

α-Ketoglutarate synthetase has been partly purified from the photosynthetic bacterium *C. thiosulfatophilum* and shown to catalyze the breakdown as well as the synthesis of α-ketoglutarate (Gehring and Arnon, 1972). The enzyme, like pyruvate synthetase, shows a requirement for thiamine pyrophosphate (Buchanan and Arnon, 1969).

C. α-Ketobutyrate Synthetase

α-Ketobutyrate synthetase (Buchanan, 1969) catalyzes a ferredoxin-dependent reductive carboxylation of propionyl coenzyme A to α-ketobutyrate [Eq. (8)].

$$\text{Propionyl CoA} + \text{Ferredoxin}_{\text{red}} + CO_2 \rightarrow \alpha\text{-Ketobutyrate} + \text{CoA} + \text{Ferredoxin}_{\text{ox}} \quad (8)$$

The enzyme has been found in both photosynthetic and fermentative bacteria (Buchanan, 1969) (Table II) but has not been purified.

α-Ketobutyrate synthetase appears to be important in a new pathway for the biosynthesis of α-aminobutyrate and isoleucine (Buchanan, 1969). The new pathway is independent of threonine and threonine deaminase—the key enzyme in the mechanism previously considered to account for the biosynthesis of α-aminobutyrate and isoleucine (Meister, 1965). In the new pathway (A), propionate and carbon dioxide replace threonine (B) as carbon source for formation of the α-ketobutyrate required in the synthesis of α-aminobutyrate or isoleucine.

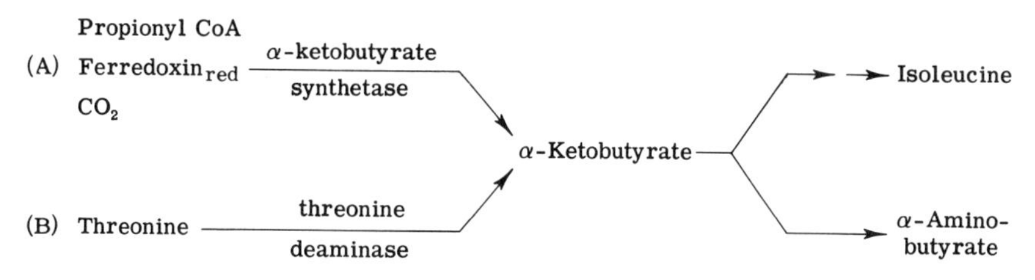

Aside from occurrence of the key enzyme α-ketobutyrate synthetase in several organisms (Table II), evidence for the new path of isoleucine synthesis has been obtained mainly for the photosynthetic bacterium, *Chromatium.* Growth experiments (Buchanan, 1969) showed that *Chromatium* cells assimilated [^{14}C]propionate into a variety of nonvolatile compounds, particularly amino acids (Fig. 6). Of the compounds labeled, isoleucine was the most prominent and accounted for 16% of the [^{14}C]propionate assimilated (Fig. 7). Threonine, a precursor of isoleucine in the known pathway, showed less than one-fifth the activity of isoleucine, and aspartate, a known precursor of threonine in other organisms (Meister, 1965), was only slightly labeled. The high labeling in isoleucine, relative to threonine, supports the conclusion that in *Chromatium* a significant part of the α-ketobutyrate needed for the synthesis of isoleucine (and α-aminobutyrate) is formed via α-ketobutyrate synthase.

D. α-Ketoisovalerate Synthetase

A ferredoxin-dependent reductive carboxylation of isobutyryl coenzyme A to α-ketoisovalerate [herein referred to as the α-ketoisovalerate synthetase reaction, Eq. (9)] was found by Allison and Peel (1968, 1971) in cell-free extracts from the fermentative bacterium, *Peptostreptococcus elsdenii.* α-Ketoisovalerate formed in this reaction

$$\text{Isobutyryl CoA} + \text{Ferredoxin}_{\text{red}} + CO_2 \rightarrow \alpha\text{-Ketoisovalerate} + \text{CoA} + \text{Ferredoxin}_{\text{ox}} \quad (9)$$

is converted by transamination to valine.

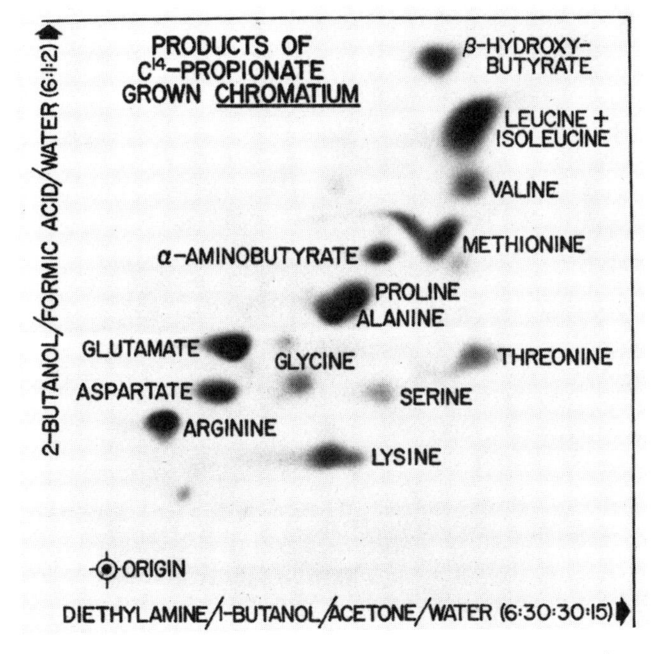

Fig. 6. Products of [^{14}C]propionate assimilation by *Chromatium* cells.

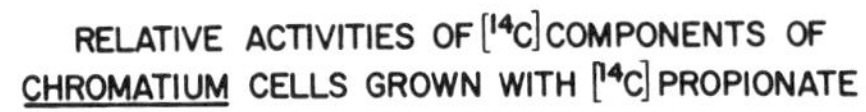

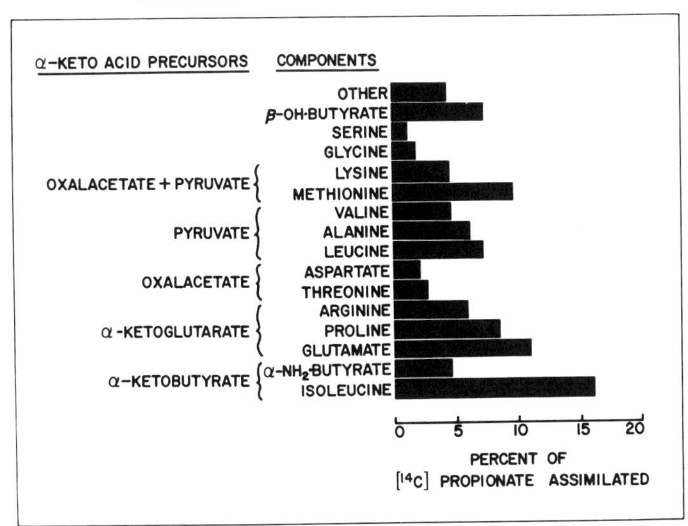

Fig. 7. Extent of labeling in products of [^{14}C]propionate assimilation by *Chromatium* cells.

As with other ferredoxin-linked carboxylations which lead to amino acids, the α-ketoisovalerate mechanism for valine biosynthesis does not involve steps of the mechanism previously established. The carbon for α-ketoisovalerate is derived from isobutyrate and carbon dioxide in the new mechanism (A) rather than from α,β-dihydroxyisovalerate (which is produced from α-acetolacetate) in the previously established path (B) (Meister, 1965).

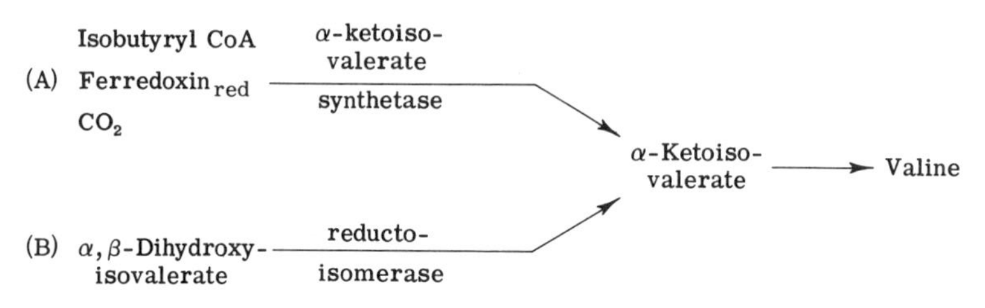

α-Ketoisovalerate synthetase has so far been found only in preparations from the rumen bacterium, *P. elsdenii*. Whether the ferredoxin-dependent route of valine biosynthesis is used by photosynthetic or other fermentative bacteria is an open question.

E. Phenylpyruvate Synthetase

The phenylpyruvate synthetase reaction was first proposed by Allison and Robinson (1967) on the basis of experiments with whole cells of *Chromatium* and *R. rubrum*. These authors described ^{14}C-labeling data which were consistent with a synthesis of phenylalanine via phenyl pyruvate from a condensation of phenyl acetate and carbon dioxide.

Direct evidence for phenylpyruvate synthetase in a cell-free system and its dependence on ferredoxin, however, has been obtained only recently. Gehring and Arnon (1971) have obtained evidence for a ferredoxin-dependent reductive carboxylation of phenylacetyl coenzyme A to phenyl pyruvate [Eq. (10)].

$$\text{Phenylacetyl CoA} + \text{Ferredoxin}_{red} + CO_2 \rightarrow \text{Phenyl pyruvate} + \text{CoA} + \text{Ferredoxin}_{ox} \quad (10)$$

The phenyl pyruvate so formed may be converted to phenylalanine by transamination.

The biosynthesis of phenylalanine by the phenylpyruvate synthetase mechanism (A) extends to aromatic amino acids a ferredoxin route of biosynthesis; as for the ferredoxin-mediated pathways of amino acid synthesis described above, the new phenylalanine pathway (A) is independent of the mechanism—the shikimate pathway (Meister, 1965) —previously established (B).

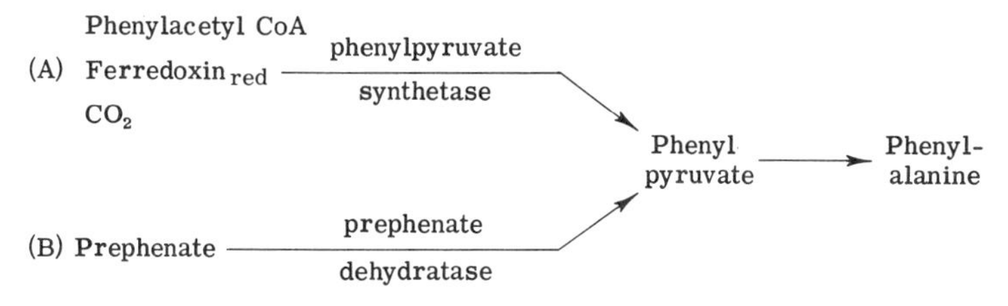

Although phenylpyruvate synthetase has been found only in preparations from photosynthetic sulfur bacteria (*Chromatium, Chlorobium thiosulfatophilum,* and *Chloropseudomonas ethylicum,* Table II), the ^{14}C experiments of Allison and associates with whole cells would indicate that the enzyme is present in rumen bacteria (Allison, 1965) as well as in the nonsulfur photosynthetic bacterium *R. rubrum* (Allison and Robinson, 1967).

V. CONCLUDING REMARKS

The discovery of a role for ferredoxin in the reductive synthesis of pyruvate in both photosynthetic and nonphotosynthetic anaerobic bacteria has led to demonstration of other ferredoxin-linked carboxylations in these organisms. Each of these reactions involves a reductive carboxylation of an acyl coenzyme A derivative to an α-keto acid. Present evidence indicates that each reaction is catalyzed by a specific enzyme.

The new ferredoxin-dependent reactions function in the assimilation by bacterial cells of carbon dioxide and organic acids (such as acetate or succinate). The α-keto acids formed are key metabolic intermediates, particularly important in the synthesis of amino acids. The synthesis of amino acids by the ferredoxin-dependent reactions involves, in each case, a new pathway which differs from that previously established.

The contribution of the new ferredoxin-dependent reactions is not restricted to amino acid biosynthesis. The pyruvate and α-ketoglutarate synthetase reactions form the bases for two new cycles of carbon dioxide assimilation: the reductive carboxylic acid cycle of bacterial photosynthesis and the reductive monocarboxylic acid cycle of fermentative metabolism. The operation of both of these cycles is dependent on the strong reducing potential of ferredoxin.

The ability of both fermentative and photosynthetic bacteria (but not plants) to use ferredoxin as a reductant in carbon dioxide assimilation

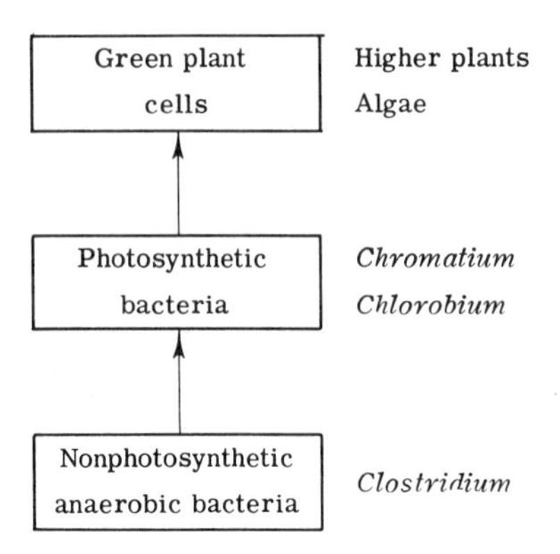

Fig. 8. A current view of the evolutionary development of photosynthesis.

is of particular interest from the standpoint of evolution. A hypothesis favored by this laboratory holds that photosynthesis manifested itself first in photosynthetic anaerobes (such as *Chlorobium*) which, in turn, evolved from fermentative organisms of the *Clostridium* type (Fig. 8). Bacterial photosynthesis would then have been followed by algal and higher plant photosynthesis which added oxygen to the earth's atmosphere. This view—advanced by Arnon *et al.* (1958, 1961) on the basis of a comparative analysis of certain metabolic features of the pertinent organisms—is supported by the findings on ferredoxin-linked carbon assimilation summarized here. This hypothesis has also received important support from recent comparative studies on amino acid composition and sequence of ferredoxins (Tanaka *et al.*, 1966; Matsubara *et al.*, 1967; Keresztes-Nagy *et al.*, 1969; Benson and Yasonobu, 1969; Buchanan *et al.*, 1969; Buchanan and Arnon, 1970).

Prior to the accumulation of the earth's atmosphere of oxygen derived from plant photosynthesis, the metabolic energy available to anaerobic cells was limited to the relatively low amount released by fermentation. The addition of oxygen to the earth's atmosphere made possible the development of respiration and the complete oxidation of organic substrates to carbon dioxide and water via the citric acid cycle (Krebs, 1953; Krebs and Lowenstein, 1960)—a development that increased the yield of ATP nineteenfold over that obtained anaerobically in glucose breakdown by glycolysis alone.

In such a course of biochemical evolution, the reductive carboxylic acid cycle would represent a primitive biosynthetic pathway that has survived to this day in photosynthetic bacteria. The striking resemblance of certain features of the anaerobic "short" cycle (for formation of acetyl coenzyme A from carbon dioxide) to the aerobic citric acid cycle (for oxidation of acetyl coenzyme A to carbon dioxide) may therefore be of evolutionary significance. It seems possible that, with the advent of atmospheric oxygen, the earlier anaerobic reductive carboxylic acid cycle (used as a mechanism for reductive carbon dioxide assimilation) was converted to the citric acid cycle which serves as a mechanism for the oxidation of acetyl coenzyme A to carbon dioxide and water. In this view, the present use of reactions of the citric acid cycle for biosynthesis (Krebs, 1953; Krebs and Lowenstein, 1960) would be analogous to an earlier stage in evolution when these reactions were part of the reductive carboxylic acid cycle that functioned solely for biosynthesis.

NOTE ADDED IN PROOF. Recent evidence indicates that a green photosynthetic bacterium (*Chlorobium thiosulfatophilum*) lacks a functional reductive pentose phosphate cycle—the photosynthetic carbon reduction cycle characteristic of green plants—and exclusively uses the reductive carboxylic acid cycle (and its associated reactions) not only, as previously recognized, in the photosynthetic conversion of CO_2 to amino acids and organic acids but also to carbohydrates (Buchanan *et al.*, 1972). The results show that CO_2 concentration influences the type of products formed by the reductive carboxylic acid cycle, with a low level of CO_2 favoring the synthesis of carbohydrates. There is also now strong evidence that a key enzyme of the cycle, pyruvate synthetase ("pyruvate-ferredoxin oxidoreductase") from the fermentative bacterium *Clostridium acidi-urici* is an iron–sulfur protein with an absorption maximum at 400 nm (Uyeda and Rabinowitz, 1971a, 1971b). The enzyme was reduced (bleached) by reduced ferredoxin in the synthesis of pyruvate or by pyruvate plus CoA in the breakdown of pyruvate. This finding raises the possibility that an iron–sulfur group may be a general feature of the ferredoxin-linked carboxylation enzymes.

REFERENCES

Allison, M. J. (1965). *Biochem. Biophys. Res. Commun.* **18**, 30.
Allison, M. J., and Peel, J. L. (1968). *Bacteriol. Proc.* p. 142.
Allison, M. J., and Peel, J. L. (1971). *Biochem. J.* **121**, 431.
Allison, M. J., and Robinson, I. M. (1967). *J. Bacteriol.* **93**, 1269.
Allison, M. J., and Robinson, I. M. (1970). *J. Bacteriol.* **104**, 63.
Anderson, L., and Fuller, R. C., (1967). *Plant Physiol.* **42**, 497.
Andrew, I. G., and Morris, J. G. (1965). *Biochim. Biophys. Acta* **97**, 176.
Arnon, D. I. (1967). *Physiol. Rev.* **47**, 317.
Arnon, D. I., Whatley, F. R., and Allen, M. B. (1958). *Science* **127**, 1026.
Arnon, D. I., Losada, M., Nozaki, M., and Tagawa, K. (1961). *Nature (London)* **190**, 601.

Bachofen, R., Buchanan, B. B., and Arnon, D. I. (1964). *Proc. Nat. Acad. Sci. U.S.* **51**, 690.

Bandurski, R. S., and Greiner, C. M. (1953). *J. Biol. Chem.* **204**, 781.

Bassham, J. A., and Calvin, M. (1962). "The Photosynthesis of Carbon Compounds." Benjamin, New York.

Benson, A. M., and Yasonobu, K. T. (1969). *J. Biol. Chem.* **244**, 955.

Bothe, H. (1969). In "Progress in Photosynthesis Research" (H. Metzner, ed.), Vol. 3, p. 1483. Laupp, Tübingen.

Buchanan, B. B. (1969). *J. Biol. Chem.* **244**, 4218.

Buchanan, B. B., and Arnon, D. I. (1969). In "Methods in Enzymology" (S. P. Colowick and N. O. Kaplan, eds.), Vol. 13, p. 170. Academic Press, New York.

Buchanan, B. B., and Arnon, D. I. (1970). *Advan. Enzymol.* **33**, 119.

Buchanan, B. B., and Bachofen, R. (1968). *Biochim. Biophys. Acta* **162**, 607.

Buchanan, B. B., and Evans, M. C. W. (1965). *Proc. Nat. Acad. Sci. U.S.* **54**, 1212.

Buchanan, B. B., and Evans. M. C. W. (1966). *Biochem. Biophys. Res. Commun.* **22**, 484.

Buchanan, B. B., and Evans, M. C. W. (1969). *Biochim. Biophys. Acta* **180**, 123.

Buchanan, B. B., Bachofen, R., and Arnon, D. I. (1964). *Proc. Nat. Acad. Sci. U.S.* **52**, 839.

Buchanan, B. B., Evans, M. C. W., and Arnon, D. I. (1965). In "Non-Heme Iron Proteins: Role in Energy Conversion" (A. San Pietro, ed.), p. 175. Antioch Press, Yellow Springs, Ohio.

Buchanan, B. B., Evans, M. C. W., and Arnon, D. I. (1967). *Arch. Mikrobiol.* **59**, 32.

Buchanan, B. B., Matsubara, H., and Evans, M. C. W., (1969). *Biochim. Biophys. Acta* **189**, 46.

Buchanan, B. B., Schürmann, P., and Shanmugam, K. T. (1972). *Biochim. Biophys. Acta* **283**, 136.

Chase, T. C., and Rabinowitz, J. C. (1968). *J. Bacteriol.* **96**, 1065.

Cooper, R. A., and Kornberg, H. L. (1965). *Biochim. Biophys. Acta* **104**, 618.

Cutinelli, C., Ehrensvärd, G., Reio, L., Saluste, E., and Stjernholm, R. (1951). *Ark. Kemi.* **3**, 315.

Evans, M. C. W. (1968). *Biochem. Biophys. Res. Commun.* **33**, 146.

Evans, M. C. W., and Buchanan, B. B. (1965). *Proc. Nat. Acad. Sci. U.S.* **53**, 1420.

Evans, M. C. W., Buchanan, B. B., and Arnon, D. I. (1966). *Proc. Nat. Acad. Sci. U.S.* **55**, 928.

Fuller, R. C., Smillie, R. M., Sisler, E. C., and Kornberg, H. L. (1961). *J. Biol. Chem.* **236**, 2140.

Gehring, U., and Arnon, D. I. (1971). *J. Biol. Chem.* **246**, 4518.

Gehring, U., and Arnon, D. I. (1972). *J. Biol. Chem.* **247**, 6963.

Gottschalk, G., and Chowdhury, A. A. (1969). *FEBS Lett.* **2**, 342.

Heer, E., and Bachofen, R. (1966). *Arch. Mikrobiol.* **54**, 1.

Hoare, D. S. (1963). *Biochem. J.* **87**, 284.

Hoare, D. S., and Gibson, J. (1964). *Biochem. J.* **91**, 546.

Jungermann, K., Thauer, R. K., and Decker, K. (1968). *Eur. J. Biochem.* **3**, 351.

Keresztes-Nagy, S., Perini, F., and Margoliash, E. (1969). *J. Biol. Chem.* **244**, 981.

Knappe, J., Schacht, J., Möckel, W., Höpner, T., Vetter, H., Jr., and Edenharder, R. (1969). *Eur. J. Biochem,* **11,** 316.
Koepsell, H. J., Johnson, M. J., and Meek, J. S. (1944). *J. Biol. Chem.* **154,** 535.
Krebs, H. A. (1953). Les Prix Nobel.
Krebs, H. A., and Lowenstein, J. M. (1960). In "Metabolic Pathways" (D. M. Greenberg, ed.), Vol. 1, p. 129. Academic Press, New York.
Latzko, E., and Gibbs, M. (1969). *Plant Physiol.* **44,** 295.
Lipmann, F. (1946). *Advan. Enzymol.* **6,** 231.
Losada, M., Trebst, A. V., Ogata, S., and Arnon, D. I. (1960). *Nature (London)* **186,** 753.
Matsubara, H., Saski, R. M., and Chain, R. K. (1967). *Proc. Nat. Acad. Sci. U.S.* **57,** 439.
Meister, A. (1965). "Biochemistry of the Amino Acids," Vol. 2. Academic Press, New York.
Mortenson, L. E., Valentine, R. C., and Carnahan, J. E. (1962). *Biochem. Biophys. Res. Commun.* **7,** 448.
Mortlock, R. P., and Wolfe, R. S. (1959). *J. Biol. Chem.* **234,** 1657.
Moyle, J. (1956). *Biochem. J.* **63,** 552.
Ochoa, S., and Weisz-Tabori, E. (1954). *J. Biol. Chem.* **159,** 245.
Raeburn, S., and Rabinowitz, J. C. (1965a). *Biochem. Biophys. Res. Commun.* **18,** 303.
Raeburn, S., and Rabinowitz, J. C. (1965b). In "Non-Heme Iron Proteins: Role in Energy Conversion" (A. San Pietro, ed.), p. 189. Antioch Press, Yellow Springs, Ohio.
Shigesada, K., Hidaka, K., Katsuki, H., and Tanaka, S. (1966). *Biochim. Biophys. Acta* **112,** 182.
Shin, M., and Arnon, D. I. (1965). *J. Biol. Chem.* **240,** 1405.
Shin, M., Tagawa, K., and Arnon, D. I. (1963) *Biochem. Z.* **338,** 84.
Sirevåg, R., and Ormerod, J. G. (1970a). *Science* **169,** 186.
Sirevåg, R., and Ormerod, J. G. (1970b). *Biochem. J.* **120,** 399.
Smillie, R. M., Rigopoulos, N., and Kelly, H. (1962). *Biochim. Biophys. Acta* **56,** 612.
Stadtman, E. R., and Barker, H. A. (1950). *J. Biol. Chem.* **184,** 769.
Stern, J. R. (1965). In "Non-Heme Iron Proteins: Role in Energy Conversion" (A. San Pietro, ed.), p. 199. Antioch Press, Yellow Springs, Ohio.
Tagawa, K., and Arnon, D. I. (1962). *Nature (London)* **195,** 537.
Tagawa, K., and Arnon, D. I. (1968). *Biochim. Biophys. Acta* **153,** 602.
Tanaka, M., Nakashima, T., Benson, A., Mower, H., and Yashunobu, K. T. (1966). *Biochemistry* **5,** 1666.
Thauer, R. K., Jungermann, K., Rupprecht, E., and Decker, K. (1969). *FEBS Lett.* **4,** 108.
Thauer, R. K., Rupprecht, E., and Jungermann, K. (1970). *FEBS Lett.* **8,** 304.
Tomlinson, N. (1954). *J. Biol. Chem.* **209,** 597.
Uyeda, K., and Rabinowitz, J. C. (1967). *Fed. Proc.* **26,** 561.
Uyeda, K., and Rabinowitz, J. C. (1971a). *J. Biol. Chem.* **246,** 3111.
Uyeda, K., and Rabinowitz, J. C. (1971b). *J. Biol. Chem.* **246,** 3120.
Valentine, R. C., Brill, W. J., and Wolfe, R. S. (1962). *Proc. Nat. Acad. Sci. U.S.* **48,** 1856.

Weaver, P., Tinker, K., and Valentine, R. C. (1965). *Biochem. Biophys. Res. Commun.* **21**, 195.
Wilson, J., Krampitz, L. O., and Werkman, C. H. (1948). *Biochem. J.* **42**, 598.
Wolfe, R. S., and O'Kane, D. J. (1953). *J. Biol. Chem.* **205**, 755.
Wolfe, R. S., and O'Kane, D. J. (1955). *J. Biol. Chem.* **215**, 637.
Wood, H. G., and Stjernholm, R. (1963). *In* "The Bacteria" (I. C. Gunsalus and R. Y. Stainer, eds.), Vol. 3, p. 41. Academic Press, New York.
Yoch, D. C., and Lindstrom, E. S. (1967). *Biochem. Biophys. Res. Commun.* **28**, 65.

CHAPTER 6

Structure and Reactions of a Microbial Monoxygenase: The Role of Putidaredoxin

I. C. GUNSALUS and J. D. LIPSCOMB

I. INTRODUCTION

A. Roles and Classes of Monoxygenase-Reactive Iron–Sulfur Proteins

The two electron reduction and cleavage of molecular oxygen with the formation of water and the addition of oxygen to a carbon compound is termed a monoxygenase reaction. Monoxygenases function in some aerobic bacteria to degrade a wide range of compounds as carbon and energy sources (Peterson *et al.*, 1966; Bradshaw *et al.*, 1959). In mammalian tissue, monoxygenases participate in the metabolism of steroid hormones and effect the solubilization or degradation of a variety of foreign compounds including some carcinogens, insecticides, drugs, fatty acids, aromatics, and complex hydrocarbons (Kimura and Suzuki, 1965; Omura *et al.*, 1965; Lu and Coon, 1968). Iron–sulfur proteins function in several, although not all, monoxygenases of both bacterial and mammalian tissue origin. In each of these multicomponent monoxygenases, the iron–sulfur protein links a reduced pyridine nucleotide–flavoprotein primary reductant with a terminal oxidase of varying nature. The iron–sulfur proteins found in these systems are of intermediate potential, i.e., E_0' of —100 to —270 mV, rather than the higher-potential or low-potential groups which range around E_0' + 300 or —400 mV. The latter are associated with ATP coupled reactions, with the accumulation of reduced pyridine nucleotide or with "nitrogenase" proteins. Monoxygenase iron–sulfur proteins vary in physical properties, composition, primary structure, the presence or absence of acid labile sulfur, and the number of iron atoms per molecule. Although several iron–sulfur proteins have been identified and isolated in large quantities and high purity, the total knowledge is still sketchy, particularly with regard to their function or biological role. Perhaps as a result, agreement among the specialists in each field is far from complete. Thus caution—and more experiments with generalized, testable hypotheses—are paramount to the delineation of the unifying functional characteristics.

1. Protein 450 (P-450 Cytochromes)

Among the hydroxylase monoxygenases, several methylene, methyl, and general "detoxification" systems employ b-type cytochromes as the terminal oxidases. These bind the substrate, are reduced, and bind oxygen in a cyclic manner. As they do so, characteristic visible and other resonance spectral changes occur. These cytochromes are termed protein 450 (Omura and Sato, 1964) or simply P-450 to connote the appearance of a strong absorbance near 450 nm when the dithionite reduced protein is exposed to carbon monoxide. These proteins are broadly distributed in nature and are critical to the synthetic and degradative systems in which they participate. The range of their substrate specificity is unclear, perhaps because for most systems completely pure proteins are unavailable. To date the liver P-450's are considered the least selective; those associated with specific tissue, such as the steroid synthesizing enzymes of the adrenyl cortex mitochondria, more specific; and bacterial P-450, most selective. Of the latter class, the P-450_{cam} isolated in high purity from *Pseudomonas putida* with its associated flavoprotein and iron–sulfur protein is the only one whose metabolism can be said to be well defined. The monoterpene camphor and closely related lactones are the only compounds which have been demonstrated to show high substrate affinity.

Hydroxylation occurs specifically at methylene carbon-4 relative to the camphor ketone group to form the *exo* epimer. Many aspects of substrate structure of the cytochrome P-450 and putidaredoxin and of their intermediate states in the catalytic cycles which form product, have been studied in detail and will be described later in this chapter, but the critical questions of oxygen and substrate activation remain largely unclarified.

2. Putidaredoxin

Principally, this chapter will be focused on the two iron, two labile sulfur iron–sulfur protein from the P-450_{cam} hydroxylase, trivially named putidaredoxin following the ferredoxin nomenclature coined for the 8-iron, 8-labile sulfur redox protein of the *Clostridium pasteurianum* nitrogen reduction system (Mortenson *et al.*, 1962). Putidaredoxin has yielded to many biochemical and physical probes which have allowed the measurement of its physical parameters with a precision afforded few other proteins. Its enzymic functions have proven equally fascinating, but far more elusive. The iron–sulfur protein which most closely resembles putidaredoxin is, perhaps, adrenodoxin, which fills a completely analogous position in the P-450-mediated 11β-hydroxylase of the adrenal cortex mitochondria first purified by Kimura and Suzuki (1965). The two proteins are remarkably similar in molecular weight, composition, magnetic resonance spectra, redox potential, and apparent biological role, i.e., a specific hydroxylation of a single methylene group. Putidaredoxin and adrenodoxin are interchangeable with regard to the first electron in reduction of either P-450, but not for the second electron transfer or for the substrate product conversion (Gunsalus *et al.*, 1973; Schleyer *et al.*, 1973). A dual function for putidaredoxin, i.e., enzymic functions beyond electron transfer, is thus implied and will be discussed in greater detail below.

The characterization and evaluation of the putidaredoxin has been a joint effort of a group of geneticists, chemists, and physicists whose individual contributions are frequently interwoven and cannot readily be distinguished, except for the basic educational skill which is easily attributed to those having the experience and the theoretical background in the various areas. This fruitful interaction which has grown both in breadth and depth has led to new precision in research directed towards an operational understanding of monoxygenase reactions on what we hope is a more definitive level. This refers both to the physical and chemical changes and to the time dependence of the various reactions. We hope it can eventually lead to an understanding of the nature of electron transport and the nature of protein–protein interaction in effector as well as oxidation–reduction systems.

B. The Methylene Hydroxylase System of *Pseudomonas putida*

Pseudomonas putida was shown by Bradshaw and co-workers (1959) to harbor an inducible system of enzymes which allows the organism to utilize camphor or one of several closely related monoterpenes as the sole source of carbon and energy. Hedegaard and Gunsalus (1965) demonstrated the first reaction of this approximately eleven-step degradation to the level of isobutyrate to be a hydroxylation of camphor in the 5-*exo*

TABLE I
AMINO ACID COMPOSITION AND PROPERTIES OF METHYLENE HYDROXYLASE COMPONENTS[a]

	5-*exo*-Camphor			11β-Steroid
Amino acid	Cytochrome E_C	Reductase E_A	Putidaredoxin E_B	Adrenodoxin
Asp	27	25	10 (14)	25
Asn	9	15	4	
Thr	19	20	5	10
Ser	21	18	7	7
Glu	42	16	9 (12)	12
Gln	13	24	3	
Pro	27	18	4	1
Gly	26	33	9	8
Ala	34	47	10	7
Val	24	34	15	7
Met	9	6	3	3
Ile	24	24	6	8
Leu	40	42	7	13
Tyr	9	6	3	1
Phe	17	10	2	4
His	12	6	2	3
Lys	13	13	3	5
Trp	1	3	1	0
Arg	24	24	5	4
Cys/2	6	6	6	5
Total	397	393	114	118
Free SH	6	6	4	5
N-terminus	Asx	Ser	Ser	Ser
C-terminus	Val	Ala	Gln	Ala
MW	45,000	43,500	12,500	13,100
Prosthetic group	Ferri-heme	1 FAD CHO	$(FeS)_2$	$(FeS)_2$

[a] From Tsai *et al.* (1971).

position. Cushman and co-workers (1967) first purified the acid-labile iron–sulfur protein from this system and termed it putidaredoxin. Later Katagiri *et al.* (1968) purified the flavoprotein reductase and the b-type cytochrome termed P-450_{cam}. Further purification and characterization by Tsai (1971), Yu (1970), and Dus (1970), each working with Gunsalus and several associates, revealed prosthetic groups, physical parameters of composition, and some features of primary and tertiary structure of the homogeneous proteins as partially summarized in Table I. The pure P-450_{cam} was prepared in crystalline form by Yu and Gunsalus (1970).

The reaction sequence originally proposed by Katagiri (1968), appears in revised form in Fig. 1 (Tyson *et al.*, 1970). The rich resonance, including visible, spectra of each of the enzymes has shown P-450_{cam} to carry the substrate and O_2 reactive sites (Tsai *et al.*, 1970; Tyson *et al.*, 1970). The flavoprotein is reduced specifically by NADH and will reduce putidaredoxin rapidly and stoichiometrically. Substrate-free P-450_{cam} is not reduced by putidaredoxin due to unfavorable electrode potential differences. Substrate binding to P-450_{cam}, however, occurs at least one order of magnitude faster than any other reaction in the sequence (Gunsalus *et al.*, 1971) and causes a shift in the maximum Soret absorbance from 417 to 391 nm. A concomitant increase in electrode potential (G. S. Wilson, unpublished results) allows a rapid one electron

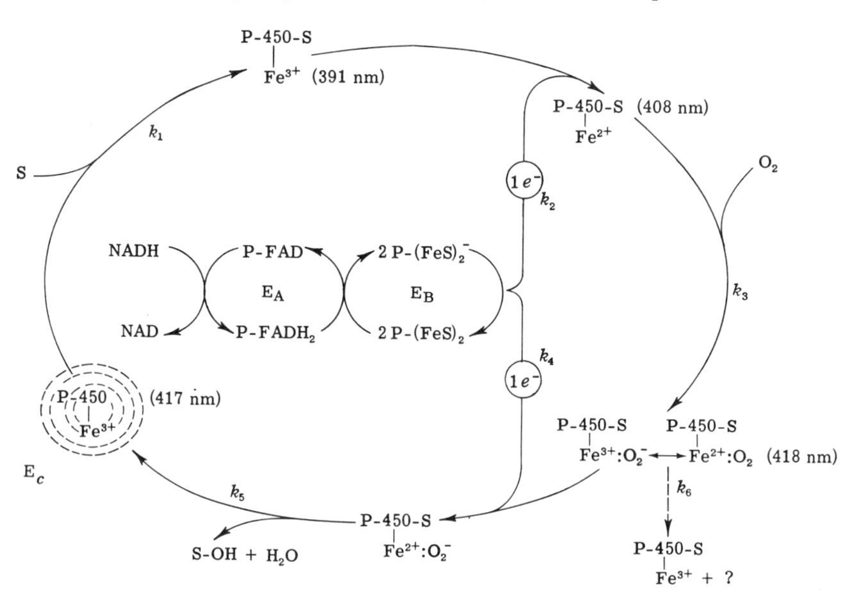

Fig. 1. Cytochrome P-450 reduction–oxygenation cycle.

reduction of the P-450_{cam} substrate complex (P-450–S) by putidaredoxin (k_2 = 35 sec^{-1}). Oxygen binding then occurs rapidly (k_3 = 470 sec^{-1}) to give a species with a Soret band at 418 nm (Tyson *et al.*, 1970; Gunsalus *et al.*, 1970, 1971; Ishimura *et al.*, 1971). This oxy-P-450_{cam} substrate (P-450–S–O_2) complex can react specifically with reduced putidaredoxin to yield hydroxylated product and water in the rate limiting step (k_4 = 17 sec^{-1}).

The genetics of the camphor metabolizing system have been extensively studied, notably by Chakrabarty and Gunsalus (1971) with Rheinwald (1973) as reviewed by Marshall and Gunsalus (1971). The genes which code for the hydroxylase proteins were shown to be located on a small extrachromosomal element or plasmid. The plasmid is transmittable between organisms so that only a few members of the population need carry the genetic information when growing on the more usual substrates. This sort of genetic conservation may account in part for the vast number of substrates amenable to psuedomonads. Several mutants have been isolated which lack activity in each of the hydroxylase enzymes. No mutant has been obtained which is specific for the hydroxylation of D- or L-camphor enantiomer. The approximate equality of D- and L-camphor as substrates, suggests that the hydroxylase is not stereospecific. In each case the product is a 5-*exo*—OH camphor with retention of the substrates stereospecificity.

II. PUTIDAREDOXIN—STRUCTURAL PROPERTIES

A. Chemical and Biological Characterization

Several recent reviews (Gunsalus, 1968; Gunsalus *et al.*, 1971, 1972, 1973; Gunsalus and Lipscomb, 1972) have summarized the biochemical and physical characteristics of P-450_{cam} and, to a lesser extent, putidaredoxin and the flavoprotein reductase. Tsibris and Woody (1970) reviewed the properties of putidaredoxin and related proteins.

The characteristic iron–sulfur absorption spectrum of putidaredoxin is shown in Fig. 2 together with the reduced spectrum produced by adding one-half equivalent of dithionite. Tsibris and associates (1968b) clearly demonstrated the two-iron, two acid-labile sulfur stoichiometry of putidaredoxin through resolution of the prosthetic group by anaerobic 20% TCA treatment and then reconstitution of the holoenyzme with $FeCl_3$ and Na_2S in the presence of 2-mercaptoethanol. Reconstitution with iron and sulfur or with selenium isotopes (Tsibris *et al.*, 1968b), has allowed careful visible (Fig. 2) and other resonance spectroscopic

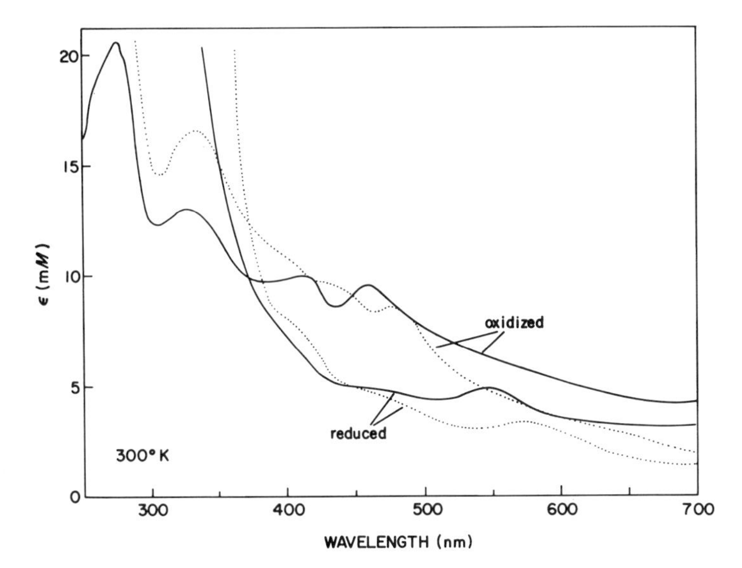

Fig. 2. Optical absorption spectra of puitdaredoxin in oxidized and reduced forms. Tris-Cl buffer pH 7.4, 50 mM; reduction with $Na_2S_2O_4$, twofold excess. Solid lines, native enzyme; dotted lines, enzyme reconstituted with ^{77}Se or ^{80}Se (selenide).

studies of the protein in its redox states. The isoelectric point of the protein was approximated at pH 3.4 by isoelectric focusing (Dus *et al.*, 1971), but precipitation of the protein at this point limited accuracy. The electrode potential was determined to be —239 mV (J. C. M. Tsibris and G. S. Wilson, unpublished data), irrespective of the presence of substrate. The protein exhibits a negative slope potential as a function of above 7.0.

The circular dichroic (CD) spectrum of putidaredoxin is shown in Fig. 3. The selenide analogue shows a red shift of 10 to 30 nm in spectrum and possesses a slightly low potential, but is otherwise similar (Tsibris and Woody, 1970). The red shift is consistent with a ligand to metal charge transfer which is postulated to account for the visible absorption bands. The magnetic CD spectra of putidaredoxin and of adrenodoxin are insensitive to the magnetic field, a property at variance with the two-iron two-sulfide plant ferredoxins (Ulmer, 1971, personal communication, Tsibris and Woody, unpublished data).

Putidaredoxin was partially sequenced by Tsai *et al.* (1971) and shown to have four of the six cysteine residues in two widely spaced sequences of cys–X–Y–cys, unlike the plant redoxin and rubredoxin (Lode and Coon, 1971). These four cysteines presumably form the organic sulfur ligands for the two iron atoms of the single active site.

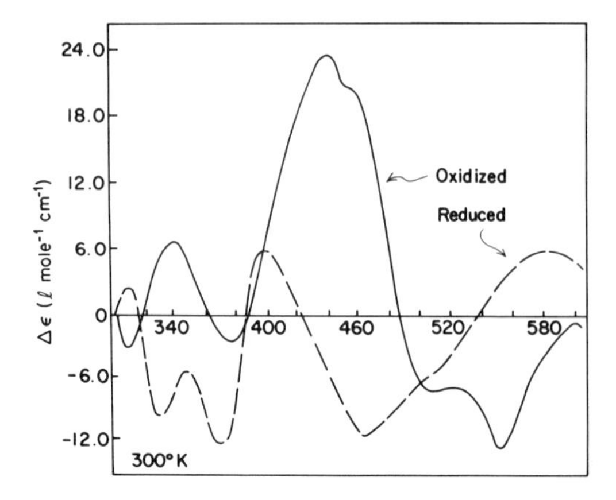

Fig. 3. Circular dichroic spectra of putidaredoxin in oxidized and reduced forms. Tris-Cl buffer pH 7.4, 50 m*M*; redoxin at 36 μ*M*, reduction by NADH, 200 μ*M*, with 0.5 μ*M* NADH:putidaredoxin reductase. From J. C. M. Tsibris (unpublished data).

B. Physical Characterization

The structure of the active site of putidaredoxin has been studied by various physical techniques including optical absorption, ORD, magnetic CD susceptibility, EPR, ENDOR, NMR, and Mössbauer spectroscopy. Many of these measurements take advantage of the fact that the iron and the acid-labile sulfur can be replaced by enriched isotopes.

The oxidized protein is diamagnetic as shown by susceptibility measurements (Moleski *et al.*, 1970) and the absence of EPR signals. Mössbauer spectra (Fig. 4) taken in an externally applied magnetic field corroborate these results. A detailed analysis of the Mössbauer spectra taken with and without a magnetic field shows that the two iron atoms of the oxidized form have similar parameters and are thus not distinguished by this technique (Cooke *et al.*, 1968).

Upon reduction, putidaredoxin exhibits an EPR signal of $S = \frac{1}{2}$ centered around $g_{\perp} \sim 1.94$ and $g_{\parallel} \sim 2.01$ (Tsibris *et al.*, 1968a). Isotopic enrichment with ^{57}Fe demonstrated that the EPR signal originates from iron (Orme-Johnson *et al.*, 1968). Titration showed clearly that this protein undergoes a single-electron reduction. Studies with enzyme from bacteria grown on ^{33}S indicate that both acid-labile sulfur and amino acid sulfur presumably contribute to the signal; the enzyme molecule contains six cysteine and three methionine residues (Table I).

Earlier studies on putidaredoxin have shown that substitution of selenide for the acid-labile sulfur results in a biologically active enzyme (Tsibris *et al.*, 1968b). EPR studies demonstrated that the physical parameters of the active site of the selenium-substituted enzyme, putidaredoxin (Se), must be very similar to those of the native enzyme.

Much valuable information about the structure of the active site has been obtained by Mössbauer and ENDOR studies of reduced putidaredoxin and of its selenium analogue. The Mössbauer spectra observed near room temperature consist of two quadrupole doublets, each of which can be associated with one iron atom (Münck *et al.*, 1972). Based on the isomer shift and the quadrupole splitting, one site clearly can be identified as in the high-spin ferrous state while the parameters for the other site match those of high-spin ferric found for the oxidized protein. ENDOR experiments carried out by Sands and co-workers (Fritz *et al.*, 1971) confirmed the inequivalence of the two iron atoms sites. Using the ENDOR results, it was possible to analyze the very complex low-temperature Mössbauer spectra of reduced putidaredoxin. Figure 5 shows the resolution of a spectrum into overlapping high-spin ferric ($S = \frac{5}{2}$) and ferrous ($S = 2$) components (Münck *et al.*, 1972). Measurements in a strong magnetic field show that the spin residing on the individual iron atoms couples antiferromagnetically to give a total spin of

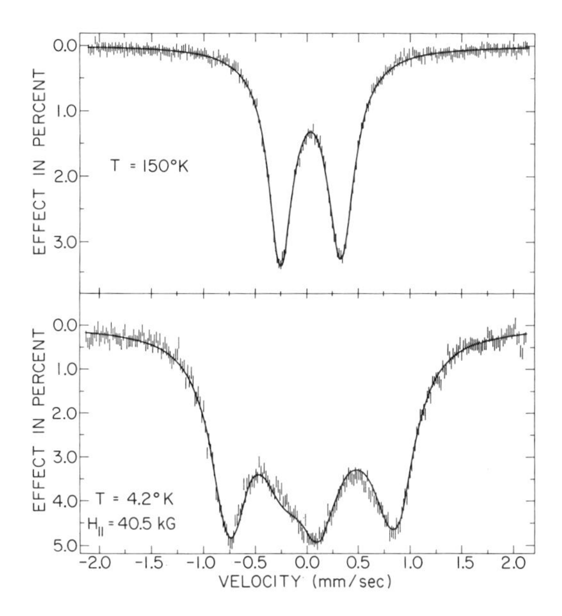

Fig. 4. Mössbauer spectra of oxidized putidaredoxin. Tris buffer pH 7.4 50 m*M*; proteins ~20 mg, at 2 m*M*; velocities relative to ^{57}Co source diffused into copper; magnetic field: upper = 0, lower = 40.5 kG, || to observed γ rays; solid lines, least squares fit to the data, computer generated and Calcomp plotted. Two Lorentzians of equal area fitted to zero field, 150°K data. From Münck *et al.* (1972).

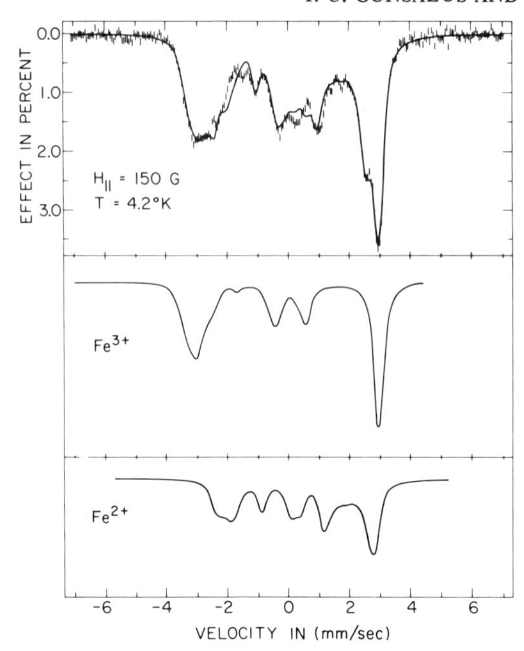

Fig. 5. Mössbauer spectra of reduced putidaredoxin. Conditions as in Fig. 4; magnetic field 150 G, || to observed γ rays. For computer simulation and Calcomp plot, solid line, center part of spectra (± 0.7 m/second) were ignored. Lower curves show decomposition of simulated spectra into Fe^{3+} and Fe^{2+} spectra. From Münck *et al.* (1972).

$S = \frac{1}{2}$. This is in agreement with the model suggested by Gibson and co-workers (Thornley *et al.*, 1966; Gibson *et al.*, 1966). The iron which remains ferric upon the one-electron reduction of putidaredoxin has a small magnetic hyperfine tensor and is highly anisotropic, implying strong convalency and low ligand symmetry. The paramters of the ferrous atom are consistent with a distorted tetrahedral symmetry. The Gibson model is equally well supported by the Mössbauer spectra on the oxidized protein which contains two high-spin ferric iron atoms that couple antiferromagnetically to give a total spin of $S = 0$.

III. PUTIDAREDOXIN–CYTOCHROME P-450$_{cam}$ INTERACTIONS

A. Physical Measurements of Protein–Protein Affinity

The enzymes of the methylene hydroxylase are completely soluble. The purification procedure (Gunsalus *et al.*, 1973) is devoid of sonification and other steps which might disrupt a membrane-bound enzyme, yet only a small carbohydrate moiety is found in P-450 and in the reductase. None is found in putidaredoxin. The nature of electron transfer between enzymes free in solution is not understood. Putidaredoxin reacts very specifically with P-450_{cam} and is required for product formation (Gunsalus *et al.*, 1973; Tyson *et al.*, 1972). The nature of this interaction has been probed by several techniques, each of which suggests that a complex forms between the two molecules.

The CD spectra of a mixture of P-450 and putidaredoxin in the oxidized and reduced states are shown in Fig. 6 (Dus *et al.*, 1971; Hsu *et al.*, 1973). The CD spectrum of P-450_{cam} consists only of an induced component caused by distortion of the inherently symmetrical heme chromophore. It is thus very sensitive to small alterations in the heme environment. Clearly, in both the oxidized and reduced states the mixed CD spectra differ from the sum of the components, strongly suggesting that a variation has occurred in the heme environment of the P-450.

Isoelectric focusing experiments performed by Dus *et al.* (1971) show that a focused band of P-450–S is broken into several overlapping bands by putidaredoxin made to migrate through it, as shown in Fig. 7. Bands containing several ratios of P-450–S to putidaredoxin ranging from 1:1 to 1:6 can be distinguished.

The EPR spectrum of P-450_{cam} at 13°K (Fig. 8) show low-spin ferric character which upon addition of substrate converts partially to high-spin

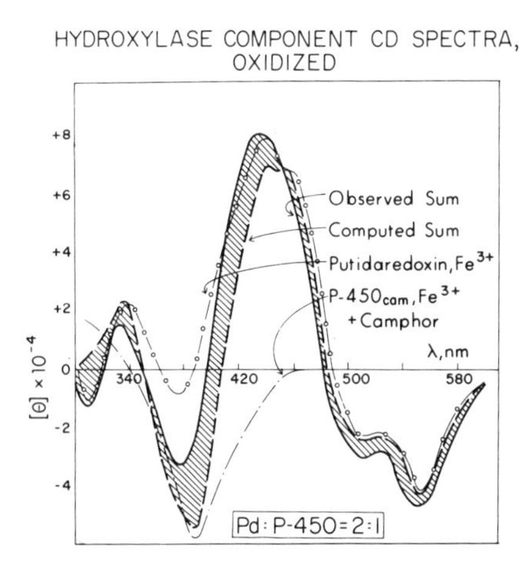

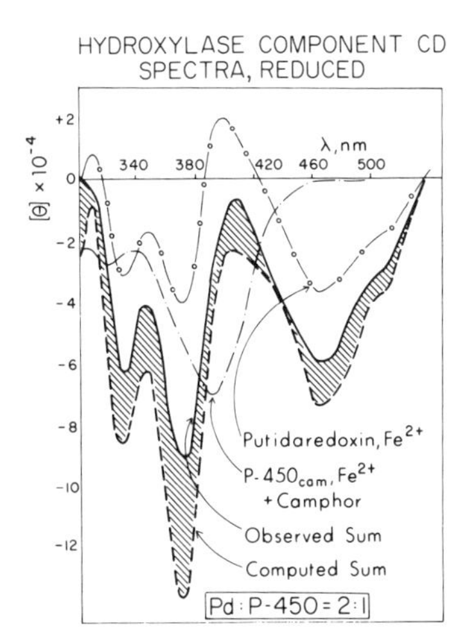

Fig. 6. Circular dichroic spectra of putidaredoxin + cytochrome P-450_{cam}: observed versus sum. Buffer 10 m*M* K^+PO_4 pH 7 + 400 μM camphor. Concentration Pd: P-450 = 37:18 μM; reduced with NADH. From Hsu *et al.* (1973).

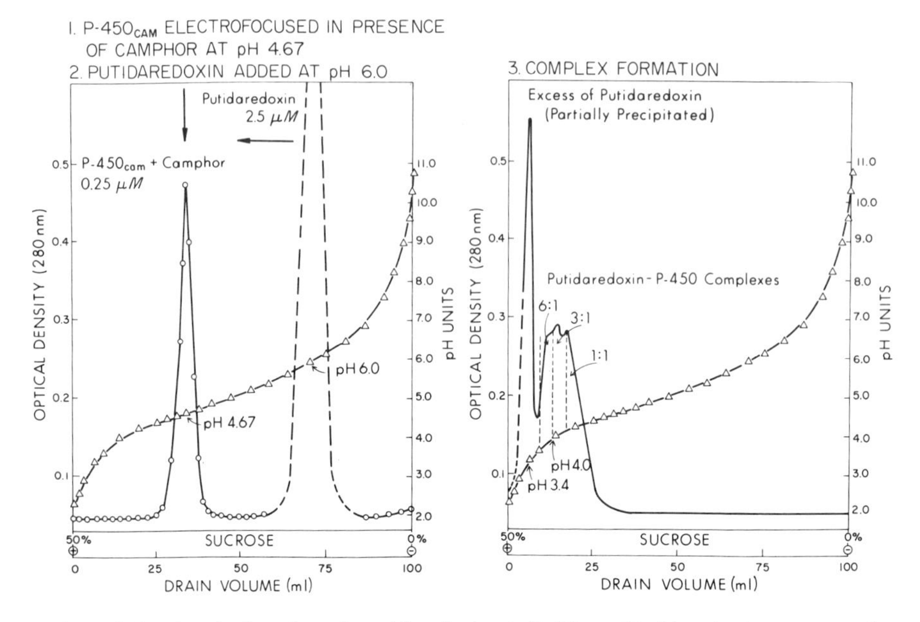

Fig. 7. Isoelectric focusing of putidaredoxin + P-450$_{cam}$. Putidaredoxin was caused to migrate through banded P-450$_{cam}$. Ratios determined by amino acid analysis. From Dus *et al.* (1971).

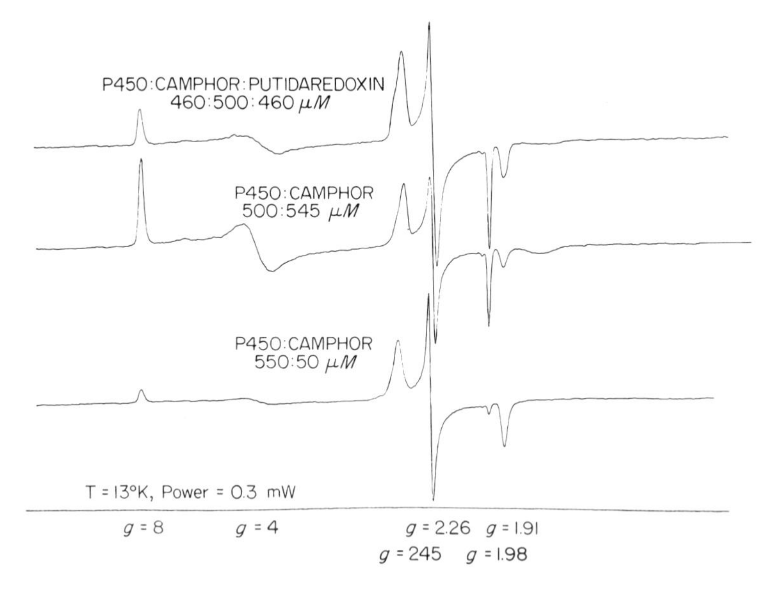

Fig. 8. EPR spectra of putidaredoxin + P-450$_{cam}$ + camphor. Buffer 50 mM PO_4, pH 7. Concentration of protein and cam as on the figure. From H. Beinert, H. W. Orme-Johnson, and J. D. Lipscomb (unpublished data).

ferric heme (Tsai *et al.*, 1970). The EPR bands at $g = 8$, 4, and 1.80 result from the high-spin component; the others from the low-spin form. The band at $g = 1.98$ has not been assigned.

Titration of the P-450–S with putidaredoxin, results in a decrease in the high-spin character and formation of a new species as evidenced by the broadening of the low-spin peaks and the increase in the $g = 1.98$ signal as shown in Fig. 8 (H. Beinert, W. H. Orme-Johnson, J. D. Lipscomb, unpublished data). The effect is maximal at a 1:1 ratio of proteins to the lowest concentration measured, i.e., 80 μM.

The steady-state kinetics of NADH turnover by the hydroxylase system as a function of putidaredoxin concentration was studied by Tyson (Gunsalus *et al.*, 1973, Tyson *et al.*, 1972). As shown in Fig. 9, at low concentrations of P-450_{cam} a twentyfold excess of putidaredoxin is needed to effect saturation (Gunsalus, 1968). Increasing the absolute concentra-

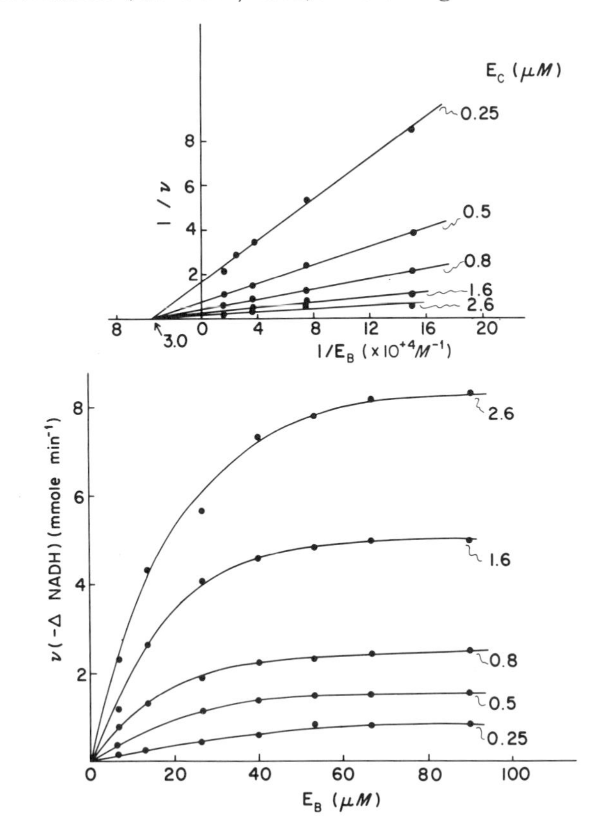

Fig. 9. Putidaredoxin titration of P-450_{cam} versus hydroxylation rate. Buffer 50 mM K^+PO_4 pH 7, 300 μM camphor, 25°C. From Tyson *et al.* (1972).

tion of P-450 decreases the optimum ratio of putidaredoxin to P-450_{cam} to approximately 6:1 at the highest concentrations measured. The linear double reciprocal plots of $1/V$ versus 1/[putidaredoxin], shown in Fig. 9, have a nonzero intercept. The K_m value of about $30 \mu M$ derived from this plot may be representative of a complex which would be consistent with the observed concentration dependence and variation in optimal ratio.

B. Biochemical Studies of Concerted Catalysis

Through careful control of conditions, P-450_{cam} can be stabilized in the oxidized form with and without substrate (P-450, P-450–S) or in the reduced form with substrate and with substrate plus oxygen (P-450^r–S, P-450^r–S–O_2). The products of the reaction of each of these forms with putidaredoxin are shown in Table II. This summary represents measurements by Tyson and co-workers (Tyson *et al.*, 1972, Gunsalus *et al.*, 1973). Stoichiometric product yields based on the total energy supplied are obtained from the reaction of any substrate bound form of P-450_{cam} and reduced putidaredoxin. Oxidized putidaredoxin also gives nearly stoichiometric yields when combined with any reduced substrate bound form of P-450. In the absence of putidaredoxin the reoxidation of the P-450^r–S to P-450–S occurs slowly and without hydroxylated production formation.

Several common chemical and biochemical reducing agents reduce the P-450–S but none of these can adduce product formation. Attempts were made to simulate the implied structural specificity of putidaredoxin by substituting other biochemical macromolecules (Tyson *et al.*, 1972;

TABLE II
PRODUCT YIELDS FROM THE *P. putida* METHYLENE HYDROXYLASE[a,b]

Reaction	Yield: Theory	Yield: Observed
NADH + E_{ABC} + Camphor + O_2	1.0	0.95
E_B^r + E_C^0 + Camphor + O_2	0.5	0.45
E_C^r + Camphor + O_2	0.5	0
E_B^0 + E_C^r + Camphor + O_2	0.5	0.45
E_B^0 + $E_C^rSO_2$ + Camphor + O_2	0.5	0.35
E_B^r + $E_C^rSO_2$ + Camphor + O_2	1.0	0.90

[a] Yield based on 2 e^- per mole of 5-*exo*-OH-camphor 50 mM K^+PO_4 buffer pH 7.0, 25°C.

[b] From Gunsalus and Lipscomb (1972a).

TABLE III
REDOX POTENTIAL AND PROPERTIES OF "HYDROXYLASE COMPONENTS"

Protein	E_0 (mV)	Redox site Fe	Redox site S	MW ($\times 10^3$)	Reference
Ferredoxin					
Spinach	−432	2	2	13	*a*
C. acidi-urici	−420	8	8	6	*a*
E. coli	−360	2	2	12	*b*
P-450$_{cam}$ cytochrome	−270	Heme		43	*c*
Adrenodoxin	−260	2	2	13	*a*
Putidaredoxin	−240	2	2	12.5	*d*
P-450$_{cam}$–S	−170	Heme		43	*e*
Rubredoxin					
C. pasteurianum	− 57	1	0	6	*a*
P. oleovorans	∼− 60	1–2	0	19	*a*
P-450$_{cam}$ · SO_2	0	Heme		45	*f*
Cyt b_5 (rabbit)	+ 30	Heme		14	*a*
Mb (whale)	+ 46	Heme		17	*b*
Cyt b_{562} (*E. coli*)	+130	Heme		12	*a*
Cyt c	+230	Heme		13	*a*
HIPIP (*Chromatium*)	+350	4	4	10	*a*

[a] Mahler, H. R., and Cordes, E. H. (1971). "Biological Chemistry," p. 676. Harper, New York.
[b] Mahler, H. R., and Cordes, E. H. (1971). "Biological Chemistry," p. 668. Harper, New York.
[c] Knappe, J. (unpublished data).
[d] Wilson G. S., in P. Debrunner, *et al.* (eds.) *Mössbauer Spectorscopy in Biol. Syst. Allerton House Meeting Proc.* March, 17–18, 1969.
[e] Wilson, G. S., and Tsibris, J. T. M. (unpublished data).
[f] J. D. Lipscomb (unpublished results).

Lipscomb *et al.*, 1972) the characteristics of which are summarized in Table III. When the possible effector molecules are reduced and reacted with P-450–S + O_2 the yields of product are uniformly very low.

Proteins with structural similarity to putidaredoxin and lower electrode potentials such as adrenodoxin and spinach ferredoxin quite readily reduce P-450–S but produce no product whatsoever. Conversely, one-iron rubredoxin from *C. pasteurianum* or *P. oleovorans* and mammalian liver cytochrome b_5 produce about 10% of the expected yield base on $2e^-$ per mole of product despite structural differences and electrode potentials much higher than P-450–S.

Reactions of the oxidized macromolecules with P-450–S–O_2 give very low product yields with the exception of the rubredoxins and cytochrome b_5 which induce substantial yields of product. With very high concentrations of these effectors product yields are achieved which approach those afforded by putidaredoxin. The reactions and product yields with rubredoxin and cytochrome b_5 are summarized in Table IV.

The electrode potential of P-450_{cam} increases sharply on substrate binding. Camphor binding increases the potential from −270 to −170 mV at pH 7 (G. S. Wilson, unpublished data). Oxygen binding causes a further increase (Lipscomb, unpublished data) such that electrons can be readily transferred to cytochrome b_5 ($E_0' = +30$ mV) but not to rubredoxin ($E_0' = -60$ mv). The effector proteins thus serve as energy sinks when combined with certain forms of P-450_{cam} and energy is dissipated through autoxidation of the effectors rather than in camphor hydroxylation. The lower product yields observed for combination of reduced effectors with P-450–S–O_2 is attributed to similar autoxidative bypass. The energy contributed by the effectors is lost through the slower autoxidation for lack of efficient transfer to P-450–S. The yield of hydroxylated product from the reduced cytochrome b_5:P-450–S–O_2 reaction, increases with cytochrome b_5 concentration. The maximum corresponds to 0.5 mole of product per mole of P-450–S–O_2 supplied (Fig. 10) (Gunsalus and Lipscomb, 1972).

Putidaredoxin at maximum product formation increases the P-450–S–O_2 decay rate a thousandfold and to do so requires a concentration only 2% of that of any other effector protein observed. The similarities between the effector role of putidaredoxin at low concentration and the other effector reactions at high concentration suggests that the effectors interact

TABLE IV
CAMPHOR HYDROXYLATION YIELDS: P-450_{cam} WITH EFFECTORS[a]

Reaction[b]	Theory[c]	$x\dagger$ = PD[d]	RD[e]	Cyt b_5
$Ec^0 + X^r + O_2 + S$	0.5	0.45	0.05	0.05
$Ec^r + X^0 + O_2 + S$	0.5	0.45	0.05	0.05
$Ec^rSO_2 + X^0 + O_2 + S$	0.5	0.35	0.30	0.15
$Ec^rSO_2 + X^r + O_2 + S$	1.0	0.90	0.50	0.45

[a] Zero product with ferredoxin from adrenal, spinach, and *E. coli;* cytochrome c, b_{562} from *E. coli;* myoglobin, pyocyan, *Chromatium* HIPIP, phosphatidyl choline, NADH or NADPH. Gunsalus and Lipscomb (1972).

[b] S, +-camphor.

[c] 2 e^-/mole product.

[d] $x\dagger$, effector for product formation; PD, putidaredoxin.

[e] RD, *P. oleovorans* (octane grown).

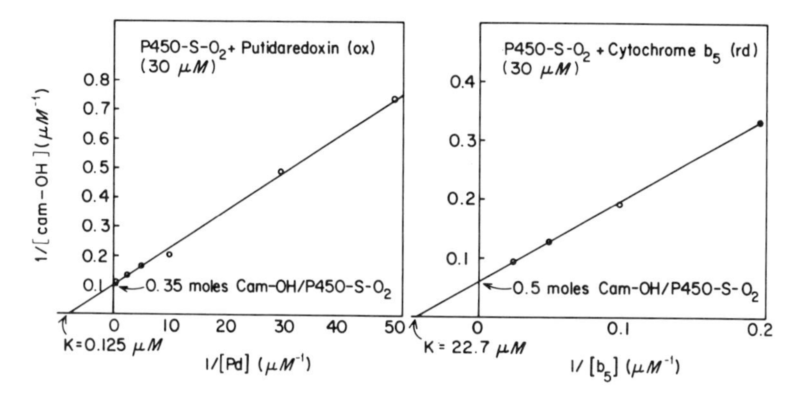

Fig. 10. Reciprocal plots product formed versus effector concentration. Conditions as in Fig. 9. Total hydroxylated product measured after 15 minute incubation.

directly with the P-450–S–O_2 species. The results are consistent with the occurrence of an active P-450–S–O_2:effector complex which breaks down to yield product at a rate commensurate with its stability.

The details of the second electron reaction are not clear. The increased electrode potential of the oxygenated P-450 species may allow transfer from high-potential effectors of specific structure of conformation. Alternatively, one can visualize a two-site model in which this second electron may not be transferred to the P-450 iron center. In either case, the effector proteins, which mediate product formation, but do not reduce P-450–S, clearly delineate the separate and dual functions of putidaredoxin in this monoxygenase.

IV. PROGRESS AND PROBLEMS

A. The Monoxygenase Effector Functions

The active sites of many iron–sulfur proteins have been studied in great detail. Though they show remarkable structural similarity, their redox potentials and functions vary widely. The studies described here suggest that some iron–sulfur proteins may, in fact, be multifunctional, deriving both the unique redox properties and the specific effector roles from primary amino acid sequences, the symmetry or distortion in the active center, and the tertiary structure of the folded peptide chains.

In fact, considerations superior to the composition of the active sites seem to be critical to the specificity of iron sulfide proteins in their functions in monoxygenases. The state of the iron and/or the amino acid composition and configuration near the active site may provide the unifying features for the function of such different proteins, as putidaredoxin, rubredoxin, and cytochrome b_5 in the final step of the P-450_{cam} hydroxylase reaction.

Although it is clear that iron–sulfur proteins can serve an enzymic function in more than one electron transfer reaction in a given monoxygenase reaction cycle, the nature and selectivity especially for the second function remains unclear. Certainly, the activity of putidaredoxin in the cleavage of P-450_{cam}–S–O_2 needs to be resolved in more than highly selective allosteric arguments and must be resolved in some structural terms beyond the binding and conversion of an unstable substrate intermediate. Two of the proteins which will replace putidaredoxin in the P-450_{cam} system have been implicated in other monoxygenase systems. Rubredoxin from *P. oleovarans* is a required element in the three protein hydrocarbon ω-hydroxylase studied by Coon and co-workers (Coon *et al.*, 1973; Lode and Coon, 1971). Reduced rubredoxin can function independently to convert synthetically produced alkyl peroxides to the alkyl alcohol and water as normally yielded by the complete system. Similar activity has not been observed for putidaredoxin, but the peroxides of its substrates are not conveniently prepared. Cytochrome b_5 has been postulated by Estabrook and co-workers (Hildebrandt and Estabrook, 1971) to function as the second electron donor in the liver microsomal P-450 monoxygenase which lacks an iron–sulfur protein requirement. Although much evidence has been cited in support of the cytochrome b_5 role, the data are not definitive. This microsomal system also exhibits an essential requirement for a phospholipid as characterized by Coon and co-workers (Strobel *et al.*, 1970). Strobel and Coon (1972) found that superoxide anion will activate turnover in the microsomal system containing only the phospholipid and P-450. Superoxide dismutase will inhibit the intact system. The P-450_{cam} hydroxylase is not grossly affected by superoxide nor superoxide dismutase—suggesting either a separate mechanism or lack of availability to these reagents. The evidence presented here that several effector proteins are convergent in function in the most homogeneous and selective monoxygenase system available leaves open the possibility that their roles are in actuality similar.

B. Summary of Findings

1. Putidaredoxin is one component of a three-element monoxygenase isolated from *P. putida* cells when grown on camphor as the sole source of carbon and energy. In conjunction with a flavoprotein (reductase) and a specific P-450-type cytochrome, it transports energy from NADH to split O_2 and to produce 5-*exo*-OH camphor and water.

2. Putidaredoxin has been purified to homogeneity and characterized with respect to molecular weight, amino acid composition, and partial sequence, electrode potential, isoelectric point, and visible, EPR, Mössbauer, and CD spectra. The genes which code for putidaredoxin have been shown to reside on a transmittable plasmid.

3. Resonance spectral measurements on putidaredoxin and isotopically substituted reconstituted analogues have adduced the probable structure of the active site composed of two iron atoms with two acid-labile sulfurs and possibly four cysteine sulfurs from in the backbone of the protein.

4. Evidence from isoelectric focusing, CD, EPR, and steady-state kinetics suggest that a complex forms between putidaredoxin and P-450_{cam}.

5. Putidaredoxin is not required for reduction of P-450_{cam} from ferric to ferrous form, but is required for product formation. Two proteins, of a large number tested, were found able to replace putidaredoxin to direct product formation, but none of these will reduce the P-450_{cam} substrate complex nor the P-450–S–O_2 complex. These effectors, two rubredoxins and cytochrome b_5, are 100-fold less active in the rate of product formation than putidaredoxin, but give high yield of product, suggesting that P-450_{cam}–effector complex may be rate controlling.

REFERENCES

Bradshaw, W. H., Conrad, H. E., Corey, E. J., Gunsalus, I. C., and Lednicer, D. (1959). *J. Amer. Chem. Soc.* **81,** 5507.

Chakrabarty, A. M., and Gunsalus, I. C. (1971). *Bacteriol. Proc.* **G137,** 46.

Cooke, R., Tsibris, J. C. M., Debrunner, P. G., Tsai, R. L., Gunsalus, I. C., and Frauenfelder, H. (1968). *Proc. Nat. Acad. Sci. U.S.* **59,** 1045.

Coon, M. J., Astor, A. P., Boyer, R. F., Lode, E. T., and Strobel, H. W. (1973). *In* "Oxidases and Related Redox Systems" (Proc. 2nd Int. Symp.) (T. E. King, H. S. Mason, and M. Morrison, eds.). Univ. Park Press, Baltimore, Maryland (in press).

Cushman, D. W., Tsai, R. L., and Gunsalus, I. C. (1967). *Biochem. Biophys. Res. Commun.* **26,** 577.

Dus, K., Katagiri, M., Yu, C-A., Erbes, B. L., and Gunsalus, I. C. (1970). *Biochem. Biophys. Res. Commun.* **40,** 1423.

Dus, K., Lipscomb, J. D., and Gunsalus, I. C. (1971). 162nd ACS Nat. Meeting, Washington, D.C. Sept. 12, 1971. Abstracts in *Biochemistry* **44.**

Fritz, J., Anderson, R., Fee, J., Petering, D., Palmer, G., Sands, R. H., Tsibris, J. C. M., Gunsalus, I. C., Orme-Johnson, W. H., and Beinert, H. (1971). *Biochim. Biophys. Acta* **253,** 110.

Gibson, J. F., Hall, D. O., Thornley, J. H. M., and Whately, F. R. (1966). *Proc. Nat. Acad. Sci. U.S.* **56,** 987.

Gunsalus, I. C., (1968). *Hoppe-Seylers Z. Physiolog. Chem.* **349,** 1610.

Gunsalus, I. C., Lipscomb, J. D. (1972a). *In* "The Molecular Basis of Electron Transport" (J. Schultz and B. F. Cameron, eds.) (*Miami Winter Symposia* **4,** 179). Academic Press, New York.

Gunsalus, I. C., Tsai, R. L., Tyson, C. A., Hsu, M-C., and Yu, C-A. (1970). *Int. Conf. on Magn. Resonance Biolog. Syst., 4th, Oxford* p. 26.

Gunsalus, I. C., Tyson, C. A., Tsai, R. L., Lipscomb, J. D. (1971). *Chem. Biol. Interactions* **4,** 75.

Gunsalus, I. C., Lipscomb, J. D., Marshall, V., Frauenfelder, H., Greenbaum, E., and Münck, E. (1972). *In* "Biological Hydroxylase Mechanisms" (G. S. Boyd and R. M. S. Smellie, eds.) (*Biochem. Soc. Symposia* **34,** 135). Academic Press, New York.

Gunsalus, I. C., Tyson, C. A., and Lipscomb, J. D. (1973). *In* "Oxidases and Related Redox Systems" (T. E. King, H. S. Mason, and M. Morrison, eds.) (*Proc. 2nd Int. Symposia*) p. 583. Univ. Park Press, Baltimore, Maryland.

Hedegaard, J., and Gunsalus, I. C. (1965). *J. Biol. Chem.* **240,** 4038.

Hildebrandt, A., and Estabrook, R. W. (1971). *Arch. Biochim. Biophys.* **143,** 66.

Hsu, M.-C., Yu, C-A., Tsai, R. L., and Gunsalus, I. C. (1973). Personal communication.

Ishimura, Y., Ullrich, V., and Peterson, J. A. (1971). *Biochem. Biophys. Res. Commun.* **42,** 140.

Katagiri, M., Ganguli, B. N., and Gunsalus, I. C. (1968). *J. Biol. Chem.* **243,** 3543.

Kimura, K., and Suzuki, K. (1965). *Biochem. Biophys. Res. Commun.* **19,** 340.

Lipscomb, J. D., Namtvedt, M. J., and Gunsalus, I. C. (1972). *Fed. Proc. Fed. Amer. Soc. Exp. Biol.* **31,** 448.

Lode, E. T., and Coon, M. J. (1971). *J. Biol. Chem.* **246,** 791.

Lu, A. Y. H., and Coon, M. J. (1968). *J. Biol. Chem.* **243,** 1331.

Marshall, V., and Gunsalus. I. C. (1971). *CRC Critical Rev. Microbiol.* **1,** 291.

Mortenson, L. E., Valentine, R. C., and Carnahan, J. E. (1962). *Biochem. Biophys. Res. Commun.* **7,** 448.

Moleski, C., Moss, T. H., Orme-Johnson, W. H., and Tsibris, J. C. M. (1970). *Biochim. Biophys. Aca.* **214,** 584.

Münck, E., Debrunner, P. G., Tsibris, J. C. M., and Gunsalus, I. C. (1972). *Biochemistry* **11,** 855.

Omura, T., and Sato, R. (1964). *J. Biol. Chem.* **237,** 2370.

Omura, Sato, R., Cooper, D. Y., Rosenthal, O., and Estabrook, R. W. (1965). *Fed. Proc. Fed. Amer. Soc. Exp. Biol.* **24,** 1181.

Orme-Johnson, W. H., Hansen, R. E., Beinert, H., Tsibris, J. C. M., Bartholomaus, R. C., and Gunsalus, I. C. (1968). *Proc. Nat. Acad. Sci. U.S.* **60,** 368.

Peterson, J. A., Basu, D., and Coon, M. J. (1966). *J. Biol. Chem.* **241,** 5162.

Rheinwald, J. G., Chakrabarty, A. M., and Gunsalus, I. C. (1973). *Proc. Nat. Acad. Sci. U.S.* **70,** 885.

Schleyer, H., Cooper, D. Y., and Rosenthal, O. (1973). *In* "Oxidases and Related Redox Systems" (Proc. 2nd Int. Symp.) (T. E. King, H. S. Mason, and M. Morrison, eds.). Univ. Park Press, Baltimore, Maryland (in press).

Strobel, H. W. and Coon, M. J. (1971). *J. Biol. Chem.* **24,** 7826.

Strobel, H. W., Lu, A. Y. H., Heidema, J., and Coon, M. J. (1970). *J. Biol. Chem.* **245,** 4851.

Thornley, J. H. M., Gibson, J. F., Whatley, F. R., and Hall, D. O. (1966). *Biochem. Biophys. Res. Commun.* **24,** 877.

Tsai, R. L., Yu, C-A., Gunsalus, I. C., Peisach, J., Blumberg, W., Orme-Johnson, W. H., and Beinert, H. (1970). *Proc. Nat. Acad. Sci. U.S.* **66,** 1157.

Tsai, R. L., Dus, K., and Gunsalus, I. C. (1971). *Biochem. Biophys. Res. Commun.* **45,** 1300.

Tsibris, J. C. M., Tsai, R. L., Gunsalus, I. C., Orme-Johnson, W. H., Hansen, R. E., and Beinert, H. (1968a). *Proc. Nat. Acad. Sci. U.S.* **59,** 959.

Tsibris, J. C. M., Namtvedt, M. J., and Gunsalus, I. C. (1968b). *Biochem. Biophys. Res. Commun.* **30,** 323.

Tsibris, J. C. M., and Woody, R. W. (1970). *Coord. Chem. Rev.* **5,** 417.

Tyson, C. A., Tsai, R. L., and Gunsalus, I. C. (1970). *J. Amer. Oil Chem. Soc.* **47,** 161.

Tyson, C. A., Lipscomb, J. D., and Gunsalus, I. C. (1972). *J. Biol. Chem.* **247,** 5777.

Yu, C-A., and Gunsalus, I. C. (1970). *Biochem. Biophys. Res. Commun.* **40,** 1431.

CHAPTER 7

Role of Rubredoxin in Fatty Acid and Hydrocarbon Hydroxylation Reactions

EGLIS T. LODE and MINOR J. COON

I. BIOLOGICAL UTILIZATION OF HYDROCARBONS

Hydrocarbons of the alkane series readily undergo biological oxidation despite their chemical unreactivity and poor solubility. A variety of organisms such as bacteria, yeasts, and fungi are known to utilize alkanes

(McKenna and Kallio, 1965; van der Linden and Thijsse, 1965; Fühs, 1961), and animal tissues have also been shown to possess hydrocarbon-oxidizing ability (Stetten, 1943; McCarthy, 1964; Coon *et al.*, 1967; Das *et al.*, 1968; Mitchell and Hübscher, 1968; Ullrich, 1969; Ichihara *et al.*, 1969; Lu *et al.*, 1970). This chapter is concerned primarily with the function of rubredoxin in bacterial enzyme systems which oxidize hydrocarbons and fatty acids. Bacteria which grow on alkanes are of fundamental biochemical interest because their cell constitutents and energy must both be derived from this unusual metabolic pathway. In addition, the potential importance of the microbiological production of foodstuffs from petroleum indicates the value of understanding the mechanism of hydrocarbon oxidation.

A. Mechanism of Initial Oxidative Attack

The involvment of molecular oxygen in hydrocarbon oxidation was first shown by manometric methods and subsequently confirmed by studies with the heavy isotope of oxygen. Stewart *et al.* (1959) found that 75% of the theoretical amount of $^{18}O_2$ was incorporated into hexadecane to form cetyl palmitate in cultures of a gram-negative coccus. More recently, Cardini and Jurtshuk (1970) incubated *n*-octane with $^{18}O_2$ in cell-free extracts of a *Corynebacterium* and recovered *n*-octanol containing the expected amount of heavy oxygen. The initial attack on *n*-alkanes occurs predominantly at the methyl carbon atom, but reactions may occur at other positions in some microorganisms as reviewed by McKenna and Kallio (1965) and van der Linden and Thijsse (1965). While constitutive hydrocarbon-degrading enzymes may be present at a low concentration in many microorganisms, such enzyme systems are generally present at high levels only in cells grown on hydrocarbons or grown on other substrates and then exposed to a hydrocarbon or other suitable inducer (van Eyk and Bartels, 1968).

Two general mechanisms which lead to the formation of *n*-alcohols from alkanes but differ in the nature of the initial attack have been proposed. Senez, Azoulay, and co-workers (Azoulay and Senez, 1958; Senez and Azoulay, 1961; Azoulay *et al.*, 1963) have presented evidence for the anaerobic dehydrogenation of heptane to form the 1-alkene, followed by aerobic formation of the alcohol via the epoxide as a suggested intermediate, as shown at the top of p. 175.

They have reported that an extract of a strain of *Pseudomonas aeruginosa* catalyzes the reversible heptane-dependent reduction of NAD or pyocyanine under anaerobic conditions. The infrared spectrum of a car-

$$CH_3-(CH_2)_4-CH_2-CH_3 \underset{\text{(dehydrogenase)}}{\overset{\text{NAD}}{\rightleftharpoons}} CH_3-(CH_2)_4-CH=CH_2$$

$$\downarrow O_2$$

$$CH_3-(CH_2)_4-CH_2-CH_2OH \quad \xleftarrow[?]{\text{NADH}} \quad CH_3-(CH_2)_4-\overset{H}{C}H-CH_2 \text{ (epoxide, -O-)}$$

bon tetrachloride extract of their reaction mixture indicated the presence of 1-heptene as a free intermediate (Chouteau *et al.*, 1962). Alkenes have also been isolated from microbial extracts or cultures in other laboratories (Wagner *et al.*, 1967; Abbott and Casida, 1968; Iizuka *et al.*, 1969).

The other mechanism, which was proposed by Coon and his associates and is thought to occur more widely, is a hydroxylation reaction which involves an initial attack by molecular oxygen and leads directly to the formation of the alcohol. Baptist and Coon (1959), Baptist *et al.*, (1963), and Gholson *et al.* (1963) prepared a cell-free enzyme system from *Pseudomonas oleovorans* which utilizes oxygen and NADH for the conversion of alkanes to alcohols and fatty acids to their ω-hydroxy derivatives (Kusunose *et al.*, 1964ab). The stoichiometry of the reaction corresponds to that of a monooxygenation or hydroxylation reaction (Peterson *et al.*, 1969; Peterson, 1969; McKenna and Coon, 1970), as follows:

$$RCH_3 + O_2 + NADH + H^+ \rightarrow RCH_2OH + H_2O + NAD^+$$

where R is either an alkyl or ω-carboxyalkyl residue.

Although the stoichiometry does not rule out possible intermediates, only the expected alcohol has been isolated from the reaction mixtures. It may be noted that earlier workers proposed various intermediates such as the alkyl hydroperoxide (Imilik, 1948a–c; Stewart *et al.*, 1959; Leadbetter and Foster, 1960), epoxide (McKenna and Kallio, 1965; van der Linden and Thijsse, 1965), or alkyl free radical (Foster, 1962; Fredricks, 1967). The only detectable product formed when 1-octene was added to the *P. oleovorans* enzyme system was identified as 7-octen-1-ol (Boyer and Coon, 1970). Preferential oxidative attack at the methyl carbon atom of 1-heptene has been observed with resting cells of *P. aeruginosa* (Thijsse and van der Linden, 1963; Huybregtse and van der Linden, 1964) and with cell suspensions of *Micrococcus cerificans* (Stewart *et al.*, 1960). Studies on the oxidation of deuterated ethane in *Pseudomonas methanica* (Leadbetter and Foster, 1960) also indicated that desaturation had not occurred. Therefore, it may be concluded that an alkene is not a free intermediate in the hydroxylation reaction.

B. Products of Hydrocarbon Metabolism

For pseudomonads grown on heptane or hexane, the formation of the corresponding fatty acids via oxidation of the primary alcohols and aldehydes has been established, and it seems likely that this may also be a major pathway for hydrocarbon degradation in other organisms. The NAD-dependent primary alcohol dehydrogenases have been reported in various pseudomonads (Baptist *et al.*, 1963; Azoulay and Heydeman, 1963; Payne, 1963). More recently, the occurrence of inducible alcohol dehydrogenases has been described (van der Linden and Huybregtse, 1969). The aldehyde was shown to be an intermediate by Baptist *et al.* (1963), who trapped the octaldehyde formed from [1-^{14}C]octane and identified the 2,4-dinitrophenylhydrazone derivative. A pyridine nucleotide-dependent aldehyde dehydrogenase has also been studied by Heydeman and Azoulay (1963).

Fatty acids have long been implicated as intermediates in hydrocarbon metabolism (cf. Imilik, 1948a–c; Treccani and Canonica, 1953; Senez and Konovaltschikoff-Mazoyer, 1956). β-Oxidation is probably the major mechanism of fatty acid degradation (Thijsse and van der Linden, 1961, 1963; Heringa *et al.*, 1961), although α-oxidation may also occur to a limited extent (Senez and Konovaltschikoff-Mazoyer, 1956). The acetate (acetyl CoA) formed from β-oxidation is used to synthesize other cellular components, thereby leading to an increase of the biomass of the culture. Some of the fatty acids produced may also be converted via ω-hydroxylation and subsequent dehydrogenation to the α,ω-dicarboxylic acids, which may then be further degraded by β-oxidation. Kester and Foster (1963) and Ali Khan *et al.* (1963) have isolated dioic acids from cultures of a *Corynebacterium* and a *Pseudomonas,* respectively, and Kusunose *et al.* (1964a,b) have shown the ω-hydroxylation of fatty acids by the cell-free enzyme system from *P. oleovorans*. In addition to the attack at methyl carbon atoms, an attack at methylene groups occurs in substrates such as cyclohexane and methylcyclohexane, which form, respectively, cyclohexanol and *trans*-4-methylcyclohexanol (Boyer and Coon, 1970; van der Linden, 1970). However, the major pathway for the conversion of hydrocarbons to biomass by pseudomonads may be envisioned as shown below.

$$
\begin{aligned}
CH_3\text{—}(CH_2)_n\text{—}CH_3 &\xrightarrow[NADH]{O_2} CH_3\text{—}(CH_2)_n\text{—}CH_2OH \\
&\xrightarrow{NAD} CH_3\text{—}(CH_2)_n\text{—}CHO \\
&\xrightarrow{NAD} CH_3\text{—}(CH_2)_n\text{—}COOH \\
&\xrightarrow{\beta\text{-oxidation}} (n+1)\ CH_3COOH \rightarrow \text{Biomass}
\end{aligned}
$$

II. RUBREDOXIN AND OTHER COMPONENTS OF A BACTERIAL ENZYME SYSTEM CATALYZING HYDROCARBON AND FATTY ACID HYDROXYLATION

A. Resolution of *P. oleovorans* Enzyme System into Three Components and Identification of Rubredoxin

The hydrocarbon-oxidizing enzyme system of *P. oleovorans* was resolved by Peterson *et al.* (1966) into three fractions, all of which are required for substrate hydroxylation to occur. One of the fractions had an absorption spectrum similar to that of a red, nonheme iron protein of unknown function previously isolated from *Clostridium pasteurianum* by Lovenberg and Sobel (1965) and named rubredoxin. The other fractions were shown to contain an NAD-rubredoxin reductase and the ω-hydroxylase. The electron transfer scheme shown in Fig. 1 has been proposed for the reconstituted hydroxylation system (Peterson *et al.*, 1967). The involvement of rubredoxin and other protein components in ω-hydroxylation appears to be a general phenomenon in pseudomonads grown on alkanes, as judged by recent reports on *Pseudomonas denitrificans* (Kusunose *et al.*, 1967b, 1968) and on a strain of *P. aeruginosa* (van Eyk and Bartels, 1970). It may be noted that similar schemes involving an iron–sulfur protein, a reductase, and an oxygenase have been proposed for steriod hydroxylation in adrenal mitochondria (Omura *et al.*, 1965; Kimura and Suzuki, 1965), fatty acid desaturation in *Euglena* and spinach preparations (Nagai and Bloch, 1966), and camphor methylene hydroxylation in *Pseudomonas putida* (Cushman *et al.*, 1967). In contrast, these three latter systems all involve iron–sulfur proteins of the ferredoxin type as discussed in other chapters of this book.

B. Properties of NADH-Rubredoxin Reductase

The *P. oleovorans* reductase which catalyzes electron transfer from NADH to rubredoxin has recently been obtained in homogenous form

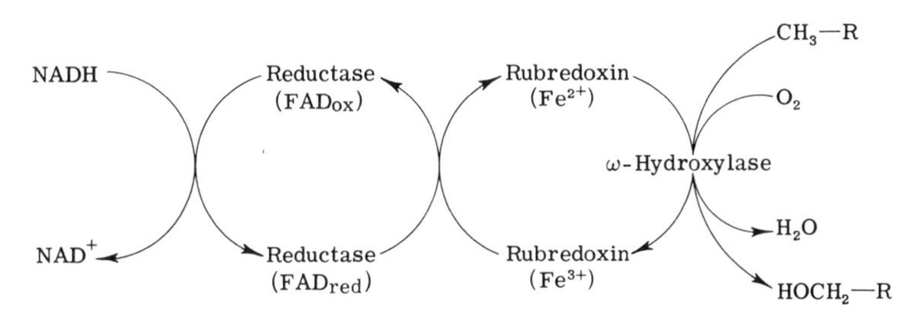

Fig. 1. Role of electron carriers in ω-hydroxylation.

(Ueda *et al.*, 1972). It contains FAD as the prosthetic group and has a molecular weight of about 50,000. NADH is the preferred electron donor for the bacterial rubredoxin reductase, but NADPH will serve to a limited extent. The reductase forms a complex with rubredoxin in a 1:1 ratio, as has also been shown for other reductases and iron–sulfur proteins (Foust *et al.*, 1969). Spinach ferredoxin–NADP reductase can replace the bacterial reductase since it also reduces rubredoxin, but in that case NADPH is used as the electron donor (Peterson *et al.*, 1966; 1967). As a further example of nonspecificity, putidaredoxin–NAD reductase is capable of substituting for the NADH-rubredoxin reductase (Coon and Gunsalus, 1970). A flavin is also apparently involved in the *P. denitrificans* enzyme system, since FAD, FMN, and riboflavin stimulate alkane hydroxylation as well as the reduction of rubredoxin (Kusunose *et al.*, 1967a, 1968). The reductase fraction may be replaced in this system by spinach ferredoxin–NADP reductase or by adrenodoxin–NADP reductase due to their broad specificity.

C. Properties of ω-Hydroxylase

Because of its relative instability and high molecular weight, the ω-hydroxylase has proved to be the most difficult of the three proteins to characterize (McKenna and Coon, 1970). The partially purified enzyme has an apparent molecular weight of 2×10^6 as estimated by gel filtration on a calibrated agarose column. However, upon incubation in sodium dodecyl sulfate and mercaptoethanol, the hydroxylase dissociates into smaller units of molecular weight about 42,000. The partially purified hydroxylase contains nonheme iron and labile sulfide and exhibits an electron paramagnetic resonance (EPR) spectrum with signals at $g = 1.94$ and 2.03 characteristic of ferredoxins. A free radical signal at $g = 2.003$ and a rhombic iron signal at $g = 4.3$ were also found. Further studies are necessary, however, to relate the resonance signals to the function of the iron atoms as catalytic centers in the hydroxylase.

Cyanide is an effective inhibitor in the ω-hydroxylation system, with an apparent K_i of 3.1×10^{-4} M. Cyanide appears to act directly on the ω-hydroxylase but has no significant effect on the transfer of electrons from NADPH to rubredoxin in the presence of the appropriate reductase or from NADPH to cytochrome c with the reductase and rubredoxin serving as intermediate electron carriers. 8-Hydroxyquinoline is also a potent inhibitor, whereas EDTA, Tiron, α,α-dipyridyl, and sodium diethyldithiocarbamate are less effective.

Since only traces of heme are present in the hydroxylase fraction and carbon monoxide neither inhibits hydroxylation nor causes spectral changes at 450 nm after the addition of dithionite, the hydroxylase clearly does not contain hemoprotein P-450 (Peterson *et al.,* 1966; McKenna and Coon, 1970). Spectral studies with cell suspensions of *P. oleovorans* also indicated the absence of P-450 (Peterson, 1970). In contrast, hemoprotein P-450 functions in steroid hydroxylation in adrenal mitochondria (Omura *et al.*, 1965; Kimura and Suzuki, 1965), in camphor methylene hydroxylation in *P. putida* (Katagiri *et al.*, 1968), and in hydrocarbon, fatty acid, and drug hydroxylation in a solubilized enzyme system from liver microsomes (Lu and Coon, 1968; Lu *et al.,* 1969a,b, 1970), as well as in hydrocarbon hydroxylation in a *Corynebacterium* (Cardini and Jurtshuk, 1968; 1970) and in a yeast, *Candida tropicalis* (Lebeault *et al.,* 1971). A requirement for an iron–sulfur protein has not been demonstrated for the latter three enzyme systems.

III. CHARACTERIZATION OF RUBREDOXIN FROM *P. oleovorans*

Pseudomonas oleovorans rubredoxin was purified extensively by Peterson and Coon (1968), who found that many of the properties of the protein resembled those of rubredoxin previously isolated from an anaerobe (*C. pasteurianum*) by Lovenberg and Sobel (1965). Similar red proteins have also been isolated from a variety of other anaerobic bacteria (Stadtman, 1965; Gillard *et al.,* 1965; LeGall and Dragoni, 1966; Lovenberg, 1966; Bachmayer *et al.,* 1967; Newman and Postgate, 1968; Laishley *et al.,* 1969). Rubredoxins from these anaerobes have a molecular weight of about 6000 and contain one atom of iron per molecule but no acid-labile sulfur, in contrast to ferredoxins, which contain iron and labile sulfur in equimolar amount. The chemical and physical properties of rubredoxins from anaerobes have been investigated extensively (see Volume II, Chapter 3). Chemical studies (Lovenberg and Sobel, 1965; Lovenberg and Williams, 1969), spectrophotometric studies (Eaton and Lovenberg, 1970), and X-ray crystallography (Herriot *et al.,* 1970) have shown that the iron atom of *C. pasteurianum* rubredoxin is bonded in nearly tetrahedral coordination by the sulfur atoms of the four cysteine residues.

A. Physical Properties

Rubredoxin from *P. oleovorans* contains a single polypeptide chain of about 19,500 MW as determined by sedimentation studies, amino acid analysis, and polyacrylamide gel electrophoresis in the presence of sodium

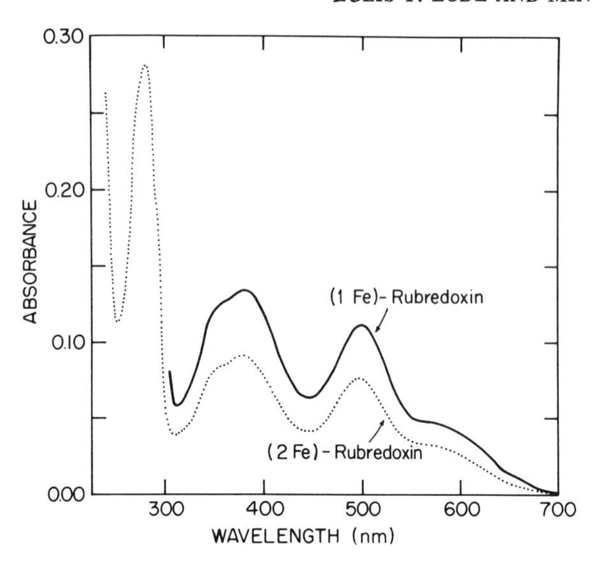

Fig. 2. Absorption spectra of the oxidized forms of (1 Fe)–rubredoxin (0.33 mg per ml) and (2 Fe)–rubredoxin (0.13 mg per ml) in 0.1 *M* tris-chloride buffer pH 7.3.

dodecyl sulfate (Lode and Coon, 1970; 1971). The sedimentation ($s_{20°,w}$) and diffusion ($D_{20°,w}$) coefficients determined for (1 Fe)–rubredoxin are 1.5×10^{-13} second and 6.6×10^{-7} cm^2 sec^{-1}, respectively; these values are not significantly different for the *S*-carboxymethylated apoprotein. These constants, taken with the molecular weight, suggest that the polypeptide chain of rubredoxin may have an unusual shape or be highly unfolded.

Spectrally, the forms containing one and two atoms of iron are similar, except that (1 Fe)–rubredoxin has an absorption maximum at 497 nm with $A_{280}:A_{497} = 6.3$, and (2 Fe)–rubredoxin has a maximum at 495 nm with $A_{280}:A_{495} = 3.7$, as shown in Fig. 2. Rubredoxins isolated from anaerobes have very similar absorption spectra with an absorption maximum at 490 nm and $A_{280}:A_{490} = 2.4$ (Lovenberg and Sobel, 1965). The absorption spectra of the reduced forms of (1 Fe)– and (2 Fe)–rubredoxin are also highly similar, with maxima occurring at about 338, 312, and 280 nm (Peterson and Coon, 1968).

The intense and detailed circular dichroism (CD) and optical rotatory dispersion (ORD) spectra of *P. oleovorans* rubredoxin also closely resemble those of rubredoxins from anaerobes (Peterson and Coon, 1968). The marked Cotton effects occurring at about 565 nm (positive) and 500 nm (negative) in the ORD spectrum and at about 632 nm (negative), 565 nm (positive), 502 nm (negative), 285 nm (negative), and 225 nm (negative) in the CD spectrum indicate that the iron–ligand chromophores are in a highly asymmetric environment. The ORD and CD spectra above 350 nm are lost when the protein is reduced with dithionite.

The oxidized forms of *P. oleovorans* (1 Fe)– and (2 Fe)–rubredoxin give EPR signals at $g = 9.4$ and a complex resonance at $g = 4.3$ which consists of both a wide and a narrow component (Peterson and Coon, 1968). These signals are characteristic of high-spin ferric ion in a rhombic field. At 82°K, the EPR spectra of the rubredoxins from *C. pasteurianum* and *Peptostreptoccus elsdenii* and of (1 Fe)– and (2 Fe)–rubredoxin are highly similar (Lode and Coon, 1971). A recent study by Peisach *et al.* (1971a,b) of the temperature dependence of the EPR spectrum of *P. oleovorans* rubredoxin has shown that as the temperature is lowered from 10° to 1.5°K the intensity of the $g = 9.4$ signal increases while that of the $g = 4.3$ signal decreases (Fig. 3). The spectrum of both (1 Fe)– and (2 Fe)–rubredoxin is primarily that of mononuclear iron in a rhombic field ($E/D = 0.328$). The population of spins undergoing an excited ($g = 4.3$) and ground ($g = 9.4$) state transition was fitted to a Boltzmann distribution from which D, the axial splitting term of the spin Hamiltonian, was calculated. The study also indicated that the iron atoms in (1 Fe)– and (2 Fe)–rubredoxin are in highly similar environments, since the values found for D were 1.51 and 1.50 cm^{-1}, respectively. A comparison of D calculated from model compounds having all-oxygen or all-sulfur coordination of iron with D calculated from (1 Fe)– or (2 Fe)–rubredoxin indicates all-sulfur chelation of the iron atoms in these proteins. An EPR resonance at $g = 4.3$ has also been reported for *P. denitrificans* rubredoxin (Kusunose *et al.*, 1967b, 1968).

A quantitative comparison of the loss in visible absorbance at 495 nm with the loss in EPR resonance at $g = 4.3$ upon the reduction of rubredoxin by NADPH indicated that one electron is accepted per atom of iron (Peterson and Coon, 1968). Other studies (Lode and Coon, 1971) have also shown that only one electron is accepted per iron atom by (1

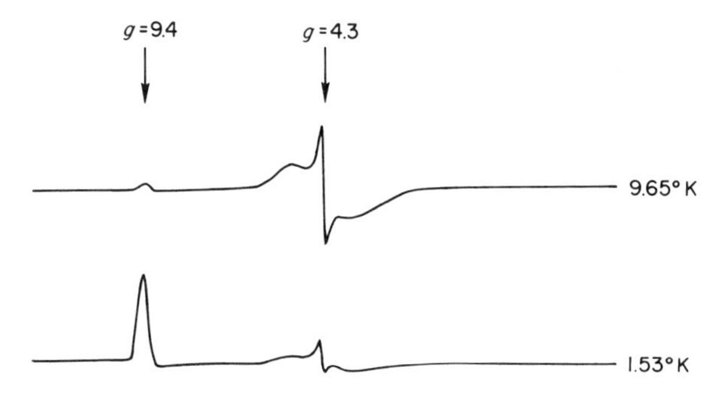

Fig. 3. Effect of temperature on EPR spectrum of (1 Fe)–rubredoxin.

Fe)– or (2 Fe)–rubredoxin as determined by titrations with dithionite or with NADH in the presence of a catalytic amount of the bacterial NADH-rubredoxin reductase. The oxidation–reduction potential of rubredoxin at pH 7.0 (determined in the presence of indigotrisulfonate) is —0.037 V.

B. Structure and Chemical Properties

Although rubredoxins have also been isolated from two other pseudomonads (Kususose *et al.*, 1967b, 1968; van Eyk and Bartels, 1970), the *P. oleovorans* protein is the only one for which extensive purification and detailed properties have been reported. In addition to exhibiting similarities in spectral properties, rubredoxins from *P. oleovorans* and from anaerobes each contain iron atoms chelated by four cysteine residues (Lode and Coon, 1971). On the other hand, the *P. oleovorans* protein differs from rubredoxins isolated from anaerobes in its ability to bind either one or two atoms of iron and in its greatly decreased stability at low pH.

The form now isolated routinely, (1 Fe)–rubredoxin, is obtained in much higher quantity than rubredoxins from anaerobes. This probably reflects the higher concentration of rubredoxin present in the pseudomonad grown on hydrocarbon. The (2 Fe)–rubredoxin can be prepared from the (1 Fe) form by conversion of the latter to the apoprotein by precipitation with trichloroacetic acid in the presence of mercaptothanol, followed by incubation with ferrous ammonium sulfate and mercaptoethanol.

The amino acid composition of *P. oleovorans* rubredoxin is compared with that of rubredoxins isolated from several anaerobes in Table I. The *P. oleovorans* protein contains 174 amino acid residues, including histidine and arginine, which have not been found in rubredoxins from anaerobes, as well as a single methionine residue and ten cysteine residues. Anaerobic rubredoxins have so far been found to contain only four cysteine residues, which, as sequence studies on rubredoxin from *Micrococcus aerogenes* (Bachmayer *et al.*, 1967, 1968a), *P. elsdenii* (Bachmayer *et al.*, 1968b), and *C. pasteurianum* (Eaton and Lovenberg, 1970) have shown, are located in pairs near the NH_2- and COOH-terminal ends of the polypeptide chain. Two other amino acid residues separate the cysteines in each pair. Methionine (Bachmayer *et al.*, 1968a,b) or *N*-formylmethionine (McCarthy and Lovenberg, 1970) has been found as the NH_2-terminal residue in anaerobic rubredoxins. In contrast, sequence studies on *P. oleovorans* rubredoxin have revealed that the ten cysteine residues are located in groups of five, each located near an end of

TABLE I

COMPARISON OF AMINO ACID COMPOSITION OF *P. oleovorans* RUBREDOXIN WITH THAT OF RUBREDOXINS ISOLATED FROM VARIOUS ANAEROBIC BACTERIA

Amino acid	Residues per mole[a]						
	P. oleovorans	*C. pasteurianum*	*P. elsdenii*	*M. aerogenes*	*M. lactilyticus*	*D. gigas*	*D. desulfuricans*
Asp	19	11	10	8	5	8	7
Thr	12	3	2	2	1	2	2
Ser	12	0	1	1	1	4	2
Glu	18	6	3	8	7	5	8
Pro	13	5	2	4	3	5	6–7
Gly	15	6	5	5	5	6	6
Ala	12	0	7	3	2	5	6
Cys	10	4	4	4	4	4	4
Val	10	5	3	4	8	4	5
Met	1	1	2	1	2	1	1
Ile	6	2	2	1	1	3	2
Leu	11	1	1	3	1	2	0
Tyr	8	3	3	3	3	3	3
Phe	4	2	2	3	2	1	3
Lys	10	4	4	2	4	5	4
His	4	0	0	0	0	0	0
Arg	3	0	0	0	0	0	0
Trp	6	2	1	1	2	1	1
Total	174	55	52	53	51	60	61

[a] The data are taken from the following sources: *C. pasteurianum*, Lovenberg and Williams (1969); *P. elsdenii*, Bachmayer *et al.* (1968b); *M. aerogenes*, Bachmayer *et al.* (1968a); *M. lactilyticus*, Lovenberg (1966); *D. gigas*, Laishley *et al.* (1969); *D. sulfuricans*, Newman and Postgate (1968); and *P. oleovorans*, Benson *et al.* (1971).

the polypeptide chain (Benson *et al.*, 1970, 1971). As shown in Fig. 4, each group consists of a "pair" and a "triplet." The amino acid sequences of residues 1–53 and 119–172 are each homologous with the sequences of anaerobic rubredoxins. Possibly, during evolution, the *P. oleovorans* protein arose from anaerobic rubredoxin via gene duplication. Since the first four residues at the NH_2 terminus of *P. oleovorans* rubredoxin are the same as those found by Keresztes-Nagy *et al.* (1969) in alfalfa ferredoxin, there may also be an evolutionary relationship between the ferredoxins and rubredoxins (Volume II, Chapter 2).

Two possible general ways in which the cysteine residues might provide two iron-binding sites are shown schematically in Fig. 5. Recent evidence indicates that model B is correct. The two peptides prepared from

1 10 20

NH_2 - Ala - Ser - Tyr - Lys - Cys - Pro - Asp - Cys - Asn - Tyr - Val - Tyr - Asp - Glu - Ser - Ala - Gly - Asn - Val - His - Glu - Gly - Phe -

30 40

Ser - Pro - Gly - Thr - Pro - Trp - His - Leu - Ile - Pro - Glu - Asp - Trp - Asp - Cys - Pro - Cys - Cys - Ala - Val - Arg -

50 60

Asp - Lys - Leu - Asp - Phe - Met - Leu - Ile - Glu - Ser - Gly - Val - Gly - Glu - Lys - Gly - Val - Thr - Ser -

70 80

Thr - His - Thr - Ala - Glx - Ala - Val - Val - Ala - Pro - Thr - Thr - Ser - Pro - Asn - Leu - Ser - Glu - Val - Ser - Gly - Thr - Ser - Leu - Ser - Leu -

90 100 110

Glu - Lys - Leu - Pro - Ser - Ala - Asp - Val - Lys - Gly - Gln - Asp - Leu - Tyr - Lys - Thr - Glu - Pro - Pro - Arg - Ser - Asp - Ala - Glu -

120 130

Gly - Gly - Lys - Ala - Tyr - Leu - Lys - Trp - Ile - Cys - Ile - Thr - Cys - Gly - His - Ile - Try - Asp - Trp - Glu - Ala - Leu - Gly -

140 150

Asp - Glu - Ala - Glu - Gly - Phe - Thr - Pro - Gly - Thr - Arg - Phe - Glu - Asp - Ile - Pro - Asp - Trp - Asp -

160 170 174

Cys - Cys - Trp - Cys (Asx, Pro) Gly - Ala - Thr - Lys - Glu - Asn - Tyr - Val - Leu - Tyr - Glu - Glu - Lys - CO_2H

Fig. 4. Amino acid sequence of *P. oleovorans* rubredoxin (Benson *et al.*, 1971).

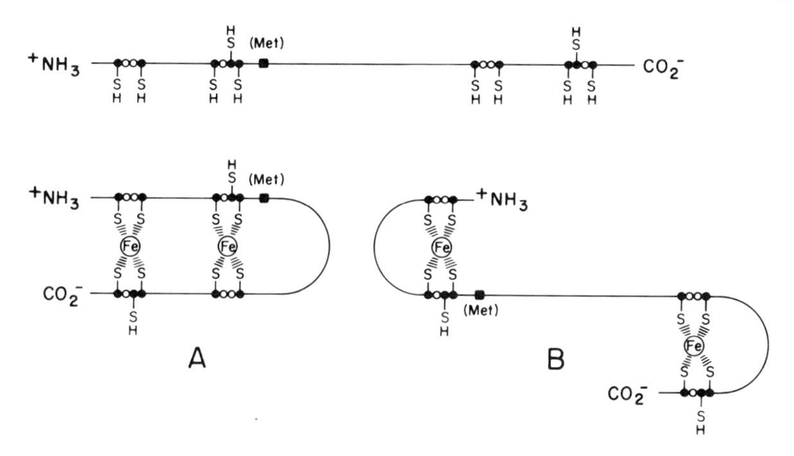

Fig. 5. Schematic diagram of *P. oleovorans* rubredoxin and possible models for iron-binding sites (Lode and Coon, 1971).

cyanogen bromide cleavage at the methionine residue of the aporubredoxin-mercaptoethanol mixed disulfide derivative can each be reconstituted in the presence of ferrous ammonium sulfate and mercaptoethanol to give an iron-containing form with a visible absorption spectrum characteristic of rubredoxin (Fig. 6). The iron atom in (1 Fe)–rubredoxin is probably chelated by the COOH-terminal group of cysteine residues, since, after alkylation with iodoacetamide and cleavage at the methionine residue, *S*-carboxymethylcysteine was found primarily in the NH_2-terminal peptide. Presumably those cysteine residues involved in iron binding were not available for alkylation. Had the iron atoms in (1 Fe)–rubredoxin been distributed between the two binding sites, or had each binding site consisted of portions of each group of cysteine residues, as in model

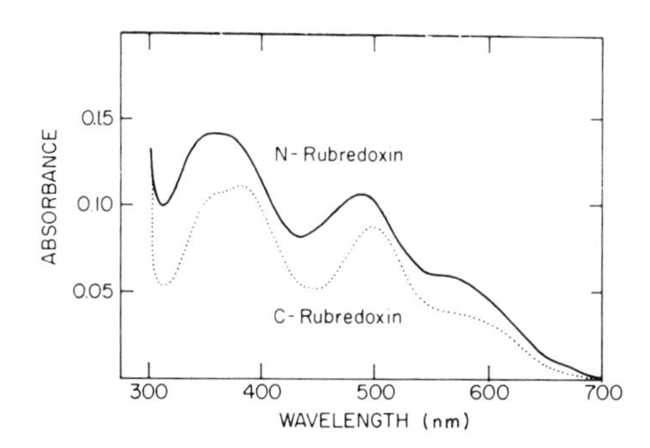

Fig. 6. Absorption spectra of the oxidized forms of N–rubredoxin (0.14 mg per ml) and C–rubredoxin (0.23 mg per ml) in 0.01 *M* tris-chloride buffer, pH 7.3 (from Lode and Coon, 1971).

A, then the ratio of carboxymethylcysteine in the NH_2-terminal peptide to that in the COOH-terminal peptide would have been much lower than the 6:1 ratio which was found (Lode and Coon, 1971).

IV. ACTIVITY OF IRON–SULFUR PROTEINS AS ELECTRON CARRIERS IN ω-HYDROXYLATION

A. *P. oleovorans* (1 Fe)–Rubredoxin and Its Derivatives: (2 Fe)–, C–, and N–Rubredoxin

As shown in Fig. 7, the (1 Fe) and (2 Fe) forms of *P. oleovorans* rubredoxin function equally well as electron carriers in the ω-hydroxylation system (Lode and Coon, 1971). Of the two peptides obtained from the cyanogen bromide cleavage of the protein and "reconstituted" with iron, C–rubredoxin, which was derived from the COOH-terminal portion of the original polypeptide, had about 30% of the activity of (1 Fe)–rubredoxin. On the other hand, N–rubredoxin, which was derived from the NH_2 portion, did not have significant activity. Since the iron atom in (1 Fe)–rubredoxin is chelated by the COOH-terminal group of cysteine residues, the partial activity of C–rubredoxin might have been predicted.

B. Rubredoxins from Anaerobic Bacteria and Other Iron–Sulfur Proteins

The rubredoxins isolated from *C. pasteurianum* and *P. elsdenii* were not active as electron carriers in ω-hydroxylation (Peterson *et al.*, 1966;

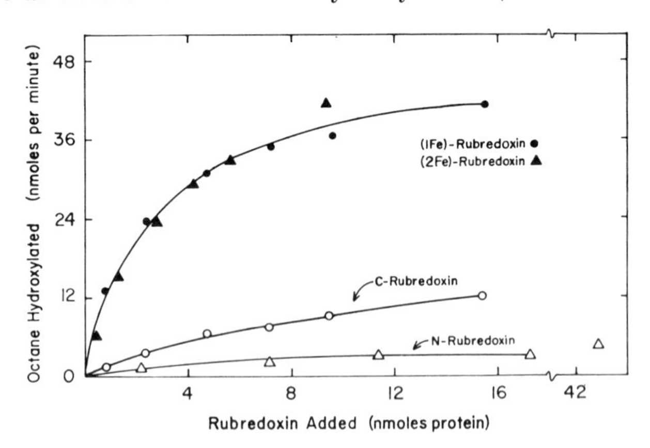

Fig. 7. Activity of (1 Fe)–, (2 Fe)–, C–, and N–rubredoxin in ω-hydroxylation system. NADH, radioactive octane, reductase, and ω-hydroxylase were present, and the assay was conducted as described elsewhere (Lode and Coon, 1971).

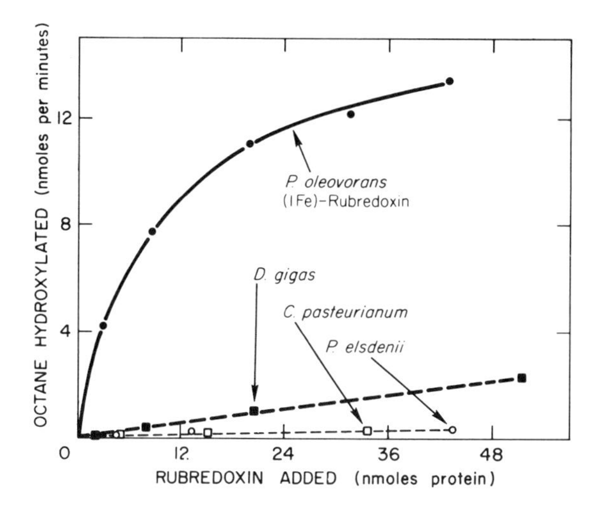

Fig. 8. Activity of various rubredoxins in ω-hydroxylation system. Rubredoxin from *D. gigas* was kindly furnished by Dr. J. LeGall, from *C. pasteurianum* by Dr. W. Lovenberg, and from *P. elsdenii* by Dr. S. G. Mayhew.

Lode and Coon, 1971), and in more recent experiments rubredoxin from *Desulfovibrio gigas* appeared to function only slightly (Fig. 8). Since each of these rubredoxins is rapidly reduced by NADH in the presence of the bacterial reductase (Ueda *et al.*, 1972), but not all function in ω-hydroxylation, certain unique structural features must be essential for electron transfer from rubredoxin to the hydroxylase. Ferredoxins from spinach and *C. pasteurianum* as well as the *Chromatium* high-potential iron protein and flavodoxin apparently do not function in the *P. oleovorans* system. On the other hand, spinach ferredoxin and adrenodoxin were reported to serve in the ω-hydroxylation system of *P. denitrificans* (Kusunose *et al.*, 1968).

V. ACTIVITY OF *P. oleovorans* RUBREDOXIN IN REDUCTION OF ALKYL HYDROPEROXIDES

As recently shown by Boyer *et al.* (1971), alkyl hydroperoxides are reduced to alcohols in the presence of NADH and homogenous rebredoxin and NADPH-rubredoxin reductase from *P. oleovorans*. The reaction does not occur when NADPH is substituted for NADH or when rubredoxins from anaerobic bacteria are substituted for the *P. oleovorans* nonheme iron protein. The reaction appears to have broad substrate specificity; 1-octyl, 2-octyl, cyclohexyl, and cumyl hydroperoxides are all reduced at significant rates. The reduction of alkyl hydroperoxides is inhibited by cyanide, as is the hydroxylation of alkanes and fatty acids in the same enzyme system supplemented with the ω-hydroxylase.

These results establish that *P. oleovorans* rubredoxin serves as an electron carrier for hydroperoxide reduction, but the significance of the reaction is not yet clear. Attempts to identify an alkyl hydroperoxide as a free intermediate in the hydroxylation of alkanes have so far proved negative. On the other hand, an enzyme-bound hydroperoxide could not have been detected. Some similarities in hydroperoxide reduction and in substrate hydroxylation in the *P. oleovorans* enzyme system should be emphasized. Both reactions are specific for rubredoxin from *P. oleovorans*, have the same pH optimum (7.4), and are inhibited to a similar extent by cyanide. Furthermore, the rates of the two reactions appear to be similar when rubredoxin is the limiting component. It should be noted that the reaction of an alkane with 0_2 to form the hydroperoxide, followed by the reduction reaction described in this chapter, would constitute a hydroxylation reaction having the expected overall stoichiometry. The alternative possibility may also be considered that the function of rubredoxin in aerobic cells is to decompose hydroperoxides by a reaction unrelated to substrate hydroxylation.

VI. SUMMARY

Rubredoxin functions as an essential electron carrier in fatty acid and hydrocarbon hydroxylation in an inducible enzyme system isolated from *P. oleovorans* grown under aerobic conditions. The electron transfer sequence is as follows: NADH → reductase (FAD) → rubredoxin → ω-hydroxylase. The hydroxylase catalyzes the following overall reaction: 2 rubredoxin (Fe^{2+}) + O_2 + RH → 2 rubredoxin (Fe^{3+}) + H_2O + ROH, where R is an alkyl or ω-carboxyalkyl residue.

The rubredoxin of *P. oleovorans* differs from rubredoxins isolated from anaerobic bacteria in its higher molecular weight (19,500) and ability to bind either one or two atoms of iron but is similar in its other properties, such as the EPR, CD, and absorption spectra. The amino acid sequence of the *P. oleovorans* rubredoxin differs in important respects from that of "anaerobic" rubredoxins, but the sequences around the cysteine residues which bind the iron atoms are homologous. Rubredoxins from anaerobic bacteria and other iron–sulfur proteins (ferredoxins and the high-potential iron protein) show slight activity, if any, when added to the hydroxylation system. Since all of these iron–sulfur proteins are readily reduced by NADH (or NADPH) in the presence of the appropriate reductase, the specificity must reside in the step in which electrons are transferred from reduced rubredoxin to the ω-hydroxylase.

More recent studies have shown that alkyl hydroperoxides are reduced to alcohols in the presence of NADH and homogeneous rubredoxin and NADH-rubredoxin reductase from *P. oleovorans*. Attempts to identify an alkyl hydroperoxide as a free intermediate in the hydroxylation of alkanes have so far proved negative. The possibilities may be considered that an enzyme-bound hydroperoxide is formed during hydroxylation or that one of the functions of rubredoxin in aerobic cells is to decompose hydroperoxides in a reaction unrelated to substrate hydroxylation.

REFERENCES

Abbott, B. J., and Casida, L. E., Jr. (1968). *J. Bacteriol.* **96,** 925.

Ali Khan, M. Y., Hall, A. N., and Robinson, D. S. (1963). *Nature (London)* **198,** 289.

Azoulay, E., and Heydeman, M. T. (1963). *Biochem. Biophys. Acta* **73,** 1.

Azoulay, E., and Senez, J. C. (1958). *C. R. Acad. Sci. Paris* **247,** 1251.

Azoulay, E., Chouteau, J., and Davidovics, G. (1963). *Biochim. Biophys. Acta* **77,** 554.

Bachmayer, H., Piette, L. H., Yasunobu, K. T., and Whiteley, H. R. (1967) *Proc. Nat. Acad. Sci. U.S.* **57,** 122.

Bachmayer, H., Benson, A. M., Yasunobu, K. T., Garrard, W. T., and Whiteley, H. R. (1968a). *Biochemistry* **7,** 986.

Bachmayer, H., Yasunobu, K. T., Peel, J. L., and Mayhew, S. (1968b). *J. Biol. Chem.* **243,** 1022.

Baptist, J. N., and Coon, M. J. (1959). *Abstr. Amer. Chem. Soc.* 135th Meeting, Boston, Massachusetts, 9C.

Baptist, J. N., Gholson, R. K., and Coon, M. J. (1963). *Biochim. Biophys. Acta* **69,** 40.

Benson, A. M., Haniu, M., Lode, E. T., Coon, M. J., and Yasunobu, K. T. (1970). *Fed. Proc.* **29,** 859.

Benson, A., Tomoda, K., Chang, J., Matsueda, G., Lode, E. T., Coon, M. J., and Yasunobu, K. T. (1971). *Biochem. Biophys. Res. Commun.* **42,** 640.

Boyer, R. F., and Coon, M. J. (1970). Unpublished data.

Boyer, R. F., Lode, E. T. and Coon, M. J. (1971). *Biochem. Biophys. Res. Commun.* **44,** 925.

Cardini, G., and Jurtshuk, P. (1968). *J. Biol. Chem.* **243,** 6071.

Cardini, G., and Jurtshuk, P. (1970). *J. Biol. Chem.* **245,** 2789.

Chouteau, J., Azoulay, E., and Senez, J. C. (1962). *Nature (London)* **194,** 576.

Coon, M. J., and Gunsalus, I. C. (1970). Unpublished data.

Coon, M. J., Peterson, J. A., Lu, A. Y. H., and Lode, E. T. (1967). *Abstr. 7th Int. Congr. Biochem. (Tokyo)* p. 793.

Cushman, D. W., Tsai, R. L., and Gunsalus, I. C. (1967). *Biochem. Biophys. Res. Commun.* **26,** 577.

Das, M. L., Orrenius, S., and Ernster, L. (1968). *Eur. J. Biochem.* **4,** 519.

Eaton, W. A., and Lovenberg, W. (1970). *J. Amer. Chem. Soc.* **92,** 7195.

Foster, J. W. (1962). *Antonie van Leeuwenhoek* **28,** 241.
Foust, G. P., Mayhew, S. G., and Massey, V. (1969). *J. Biol. Chem.* **244,** 964.
Fredricks, K. M. (1967). *Antonie van Leeuwenhoek* **33,** 41.
Fühs, G. W. (1961). *Arch. Mikrobiol.* **39,** 374.
Gholson, R. K., Baptist, J. N., and Coon, M. J. (1963). *Biochemistry* **2,** 1155.
Gillard, R. D., McKenzie, E. D., Mason, R., Mayhew, S. G., Peel, J. L., and Stangroom, J. E. (1965). *Nature (London)* **208,** 769.
Heringa, J. W., Huybregtse, R., and van der Linden, A. C. (1961). *Antonie van Leeuwenhoek* **27,** 51.
Herriot, J. R., Sieker, L. C., Jensen, L. H., and Lovenberg, W. (1970). *J. Mol. Biol.* **50,** 391.
Heydeman, M. T., and Azoulay, E. (1963). *Biochim. Biophys. Acta* **77,** 545.
Huybregtse, R., and van der Linden, A. C. (1964). *Antonie van Leeuwenhoek* **30,** 185.
Ichihara, K., Kusunose, E., and Kusunose, M. (1969). *Biochim. Biophys. Acta* **176,** 713.
Iizuka, H., Iida, M., and Fujita, S. (1969). *Z. Allgem. Mikrobiol.* **9,** 223.
Imilik, B. (1948a). *C. R. Acad. Sci. Paris* **226,** 922.
Imilik, B. (1948b). *C. R. Acad. Sci. Paris* **226,** 1227.
Imilik, B. (1948c). *C. R. Acad. Sci. Paris* **226,** 2082.
Katagiri, M., Ganguli, B. N., and Gunsalus, I. C. (1968). *J. Biol. Chem.* **243,** 3543.
Keresztes-Nagy, S., Perini, F., and Margoliash, E. (1969). *J. Biol. Chem.* **244,** 981.
Kester, A. S., and Foster, J. W. (1963). *J. Bacteriol.* **85,** 859.
Kimura, T., and Suzuki, K. (1965). *Biochem. Biophys. Res. Commun.* **19,** 340.
Kusunose, M., Kusunose, E., and Coon, M. J. (1964a). *J. Biol. Chem.* **239,** 1374.
Kusunose, M., Kusunose, E., and Coon, M. J. (1964b). *J. Biol. Chem.* **239,** 2135.
Kusunose, M., Ichihara, K., Kusunose, E., Nozaka, J., and Matsumoto, M. (1967a). *Agr. Biol. Chem.* **31,** 990.
Kununose, M., Matsumoto, J., Ichihara, K., Kusunose, E., and Nozaka, J. (1967b). *J. Biochem. (Tokyo)* **61,** 665.
Kusunose, M., Ichihara, K., Kusunose, E., and Nozaka, J. (1968). *Physiol. Ecol.* **15,** 45.
Laishley, E. J., Travis, J., and Peck, H. D., Jr. (1969). *J. Bacteriol.* **98,** 302.
Leadbetter, E. R., and Foster, J. W. (1960). *Arch. Mikrobiol.* **35,** 92.
Lebeault, J. M., Lode, E. T., and Coon, M. J. (1971). *Biochem. Biophys. Res. Commun.* **42,** 413.
LeGall, J., and Dragoni, N. (1966). *Biochem. Biophys. Res. Commun.* **23,** 145.
Lode, E. T., and Coon, M. J. (1970). *Fed. Proc.* **29,** 900.
Lode, E. T., and Coon, M. J. (1971). *J. Biol. Chem.* **246,** 791.
Lovenberg, W. (1966). *In* "Protides of the Biological Fluids, Proceedings of the 14th Colloquium, Bruges" (H. Peeters, ed.), Vol. 14, pp. 165–172, Elsevier, New York.
Lovenberg, W., and Sobel, B. (1965). *Proc. Nat. Acad. Sci. U.S.* **54,** 193.
Lovenberg, W., and Williams, W. M. (1969). *Biochemistry* **8,** 141.
Lu, A. Y. H., and Coon, M. J. (1968). *J. Biol. Chem.* **243,** 1331.
Lu, A. Y. H., Junk, K. W., and Coon, M. J. (1969a). *J. Biol. Chem.* **244,** 3714.
Lu, A. Y. H., Strobel, H. W., and Coon, M. J. (1969b). *Biochem. Biophys. Res. Commun.* **36,** 545.

Lu, A. Y. H., Strobel, H. W., and Coon, M. J. (1970). *Mol. Pharmacol.* **6,** 213.
McCarthy, R. D. (1964). *Biochim. Biophys. Acta* **84,** 74.
McCarthy, K. F., and Lovenberg, W. (1970). *Biochem. Biophys. Res. Commun.* **40,** 1053.
McKenna, E. J., and Kallio, R. E. (1965). *Ann. Rev. Microbiol.* **19,** 183.
McKenna, E. J., and Coon, M. J. (1970). *J. Biol. Chem.* **245,** 3882.
Mitchell, M. P., and Hübscher, G. (1968). *Eur. J. Biochem.* **7,** 90.
Nagai, J., and Bloch, K. (1966). *J. Biol. Chem.* **241,** 1925.
Newman, D. J., and Postgate, J. R. (1968). *Eur. J. Biochem.* **7,** 45.
Omura, T., Sato, R., Cooper, D. Y., Rosenthal, O., and Estabrook, R. W. (1965). *Fed. Proc.* **24,** 1181.
Payne, W. J. (1963). *Biotechnol. Bioeng.* **5,** 355.
Peisach, J., Blumberg, W. E., and Lode, E. T. (1971a). *Abstr. Biophys. Soc. Annu. Meeting, New Orleans,* February.
Peisach, J., Blumberg, W. E., Lode, E. T., and Coon, M. J. (1971b). *J. Biol. Chem.* **246,** 5877.
Peterson, J. A. (1969). *In* "Microsomes and Drug Oxidations" (J. R. Gillette *et al.,* eds.), pp. 167–171. Academic Press, New York.
Peterson, J. A. (1970). *J. Bacteriol.* **103,** 714.
Peterson, J. A., and Coon, M. J. (1968). *J. Biol. Chem.* **243,** 329.
Peterson, J. A., Basu, D., and Coon, M. J. (1966). *J. Biol. Chem.* **241,** 5162.
Peterson, J. A., Kusunose, M., Kusunose, E., and Coon, M. J. (1967). *J. Biol. Chem.* **242,** 4334.
Peterson, J. A., McKenna, E. J., Estabrook, R. W., and Coon, M. J. (1969). *Arch. Biochem. Biophys.* **131,** 245.
Senez, J. C., and Azoulay, E. (1961). *Biochim. Biophys. Acta* **47,** 307.
Senez, J. C., and Konovaltschikoff-Mazoyer, M. (1956). *C. R. Acad. Sci. Paris* **242,** 2873.
Stadtman, T. C. (1965). *In* "Nonheme Iron Proteins: Role in Energy Conversion" (A. San Pietro, ed.), pp. 439–445. Antioch Press, Yellow Springs, Ohio.
Stetten, D., Jr. (1943). *J. Biol. Chem.* **147,** 327.
Stewart, J. E., Kallio, R. E., Stevenson, D. P., Jones, A. C., and Schissler, D. O. (1959). *J. Bacteriol.* **78,** 441.
Stewart, J. E., Finnerty, W. R., Kallio, R. E., and Stevenson, D. P. (1960). *Science* **132,** 1254.
Thijsse, G. J. E., and van der Linden, A. C. (1961). *Antonie van Leeuwenhoek* **27,** 171.
Thijsse, G. J. E., and van der Linden, A. C. (1963). *Antonie van Leeuwenhoek* **29,** 89.
Treccani, V., and Canonica, L. (1953). *Annali Microbiol.* **5,** 162.
Ueda, T., Lode, E. T., and Coon, M. J. (1972). *J. Biol. Chem.* **247,** 2109.
Ullrich, V. (1969). *Hoppe-Seyler's Z. Physiol. Chem.* **350,** 357.
van der Linden, A. C. (1970). Personal communication.
van der Linden, A. C., and Huybregtse, R. (1969). *Antonie van Leeuwenhoek* **35,** 344.
van der Linden, A. C., and Thijsse, G. J. E. (1965). *Advan. Enzymol.* **27,** 469.
van Eyk, J., and Bartels, T. J. (1968). *J. Bacteriol.* **96,** 706.
van Eyk, J., and Bartels, T. J. (1970). *J. Bacteriol.* **104,** 1065.
Wagner, F., Zahn, W., and Bühring, U. (1967). *Angew. Chem.* **79,** 314.

CHAPTER 8

Adrenodoxin: An Iron-Sulfur Protein of Adrenal Cortex Mitochondria

RONALD W. ESTABROOK, KOJI SUZUKI, J. IAN MASON, JEFFREY BARON, WAYNE E. TAYLOR, EVAN R. SIMPSON, JOHN PURVIS, and JOHN McCARTHY*

I. INTRODUCTION

In addition to the classical energy-conserving electron transport sequence present in all mitochondria, the mitochondria of many stereoido-

* Virginia Lazenby O'Hara Professor of Biochemistry.

genic organs, such as the adrenal cortex, contain a second electron transport system which functions in the activation of oxygen required for hydroxylation of steroid molecules. In the adrenal cortex this second electron transport pathway consists of three proteins: a protoheme-containing oxygenase, cytochrome P-450; an iron–sulfur protein, adrenodoxin; and an FAD-containing flavoprotein which serves as a NADPH dehydrogenase and adrenodoxin reductase. Together, the flavoprotein dehydrogenase and the iron–sulfur protein function as a NADPH:cytochrome P-450 oxidoreductase, with adrenodoxin serving as the electron transfer intermediate between the reduced flavoprotein and oxidized cytochrome P-450 (Omura *et al.*, 1965a,b). The present review will summarize our knowledge of the iron–sulfur protein (adrenodoxin) with particular emphasis on its biological role associated with steroid hydroxylation reactions as catalyzed by mitochondria of the adrenal cortex. Since the chemistry of adrenodoxin has recently been thoroughly reviewed by Kimura (1968), no attempt will be made to reiterate this material. It should be emphasized that the iron–sulfur protein, adrenodoxin, is distinct from the iron–sulfur proteins associated with NADH dehydrogenase and succinate dehydrogenase of the classical mitochondrial respiratory pathway (Beinert and Lee, 1961), although many common properties are shared by this general class of electron carriers.

A. The Discovery of Adrenodoxin

The isolation and partial purification of an iron–sulfur protein from adrenocortical mitochondria occurred almost simultaneously in Japan and the United States with reports from Kimura and Suzuki (1965) and from Omura *et al.* (1965). Omura *et al.* (1966) isolated this iron–sulfur protein while seeking the reductase for cytochrome P-450 which was known to be present and presumed to be functional in steroid hydroxylation reactions associated with adrenocortical mitochondria. In their initial studies, Omura *et al.* (1967), using a minimal number of purification steps, were able to isolate from adrenocortical mitochondria a reddish-brown protein and a yellow protein which were required for electron transport from NADPH to either cytochrome c or P-450. At the time of these studies, San Pietro and Avron were studying, with Chance at the Johnson Research Foundation, University of Pennsylvania, the roles of PPNR (photosynthetic pyridine nucleotide reductase) and ferredoxin in photosynthetic reactions. A comparison of the optical absorption spectrum of the adrenal iron–sulfur protein with the spectrum of plant ferredoxin established the similarity of the two proteins (Omura *et al.*, 1965). Relatively simple tests subsequently characterized the adrenal protein as an iron–sulfur protein, i.e., the release of H_2S upon acidification, partial bleaching of the spectrum upon reduction of the protein with sodium dithionite, and nearly total bleaching upon the addition of a mercurial, such as mersalyl (Kimura, 1968).

B. Intracellular Distribution of Adrenodoxin

The report by Harding *et al.* (1964) that cytochrome P-450 is associated with adrenal cortex mitochondria initiated a series of studies by Omura *et al.* (1965) to determine the path of electron transport from reduced pyridine nucleotide to cytochrome P-450. These experiments—coupled with the isolation and purification of adrenodoxin—raised the question of the intracellular localization of adrenodoxin and the exclusiveness of its role in mitochondrial cytochrome P-450 reduction when compared to similar reactions catalyzed by cytochrome P-450 associated with the microsomal fraction. As illustrated in Table I, adrenodoxin is principally associated with adrenal cortex mitochondria with only traces present in the microsomal or cytosol fraction of the cell. The ability to detect small amounts of adrenodoxin in these fractions may be the conse-

TABLE I
DISTRIBUTION OF ADRENODOXIN IN BEEF ADRENAL CORTEX[a]

	Adrenodoxin concentration	
Cell fraction	(%)	(nmole/mg protein)
Homogenate	100	0.18
750 *g* supernate	27	0.16
Mitochondria	22	1.4
Microsomes	0.6	0.08
100,000 *g* supernate	3.0	0.03

[a] Bovine adrenal cortex was homogenized in 0.25 *M* sucrose to form a 10% homogenate (w/v). This homogenate was centrifuged for 10 minutes at 750 *g* to remove unbroken cells, nuclei, etc. The 750 *g* supernate was then centrifuged at 8500 *g* for 10 minutes to sediment mitochondria. The 8500 *g* supernate was recentrifuged at 18,000 *g* for 10 minutes to ensure removal of all mitochondrial fragments. The 18,000 *g* supernate was further centrifuged at 100,000 *g* for 1 hour to sediment the microsomal fraction of the cellular homogenate. Samples were sonicated for 2 minutes in 0.2 *M* tris-chloride buffer, pH 7.4, and the adrenodoxin was reduced by addition of sodium dithionite. The concentration of reduced adrenodoxin was then determined by EPR spectroscopy using samples cooled to the temperature of liquid nitrogen. (From J. Baron, W. Taylor, K. Suzuki and R. W. Estabrook, unpublished results.)

quence of some disruption of the mitochondrial structure during the initial preparation of the homogenate. Efforts to isolate and purify adrenodoxin from the 100,000 *g* supernatant fraction do not indicate any differences in spectral properties from adrenodoxin isolated from the mitochondrial fraction.

Of interest is the calculation that adrenodoxin represents approximately 0.3% of the protein of the beef adrenal gland and about 2% of the protein of beef adrenal cortex mitochondria. With isolated adrenal cortex mitochondria, adrenodoxin is present at a concentration equal to that of mitochondrial cytochrome P-450, i.e., a stoichiometric ratio of 1:1. The equivalence of adrenodoxin and cytochrome P-450 concentrations suggests an intramitochondrial association of these electron transport carriers to optimize the efficiency of steroid hydroxylation reactions (see below). The inability to solubilize adrenodoxin by simple osmotic swelling of adrenal cortex mitochondria supports the conclusion that adrenodoxin is not uniformly distributed in the mitochondrial matrix but is in a spatial proximity to the membrane-bound cytochrome P-450.

C. Purification and Crystallization of Adrenodoxin

A number of methods have been described (Kimura, 1968; Omura *et al.,* 1967) for the isolation and purification of adrenodoxin from adrenal glands. The methods generally differ in the manner of dissociating adrenodoxin from the mitochondria with techniques such as freezing and thawing of the tissue, preparation of acetone powders, and sonic treatment of homogenates employed. After release of adrenodoxin from the mitochondria, adsorption chromatography on DEAE cellulose followed by gel filtration on Sephadex results in a high degree of purification with a minimal amount of effort. A summary of one such preparation of adrenodoxin from whole beef adrenal glands (K. Suzuki *et al.,* unpublished experiment) is illustrated in Table II. A yield of approximately 25% of nearly a hundredfold purified adrenodoxin can be obtained by this method. Because of the apparent lability to oxygen of adrenodoxin when present in dilute solutions, the yield of purified material obtained is greatly enhanced if precautions are taken to decrease the oxygen content of solutions during chromatography and dialysis by gassing thoroughly with nitrogen.

Suzuki was able to extend this purification procedure by dialyzing the highly purified adrenodoxin against a 60% solution of ammonium sulfate at pH 8.1 under reduced pressure and obtained crystals of the protein (Fig. 1). Using this technique, small, brown-red crystals of adrenodoxin appeared as fine needles within a few hours. The subsequent storage of

TABLE II
SUMMARY OF PURIFICATION OF ADRENODOXIN FROM WHOLE ADRENAL GLANDS[a]

	Total volume (ml)	Total protein (mg)	Total iron (μmole)	Iron content (nmoles/mg protein)
Homogenate	2500	37,000	31.7	0.85
13,000 *g* supernatant	1950	23,600	23.7	1.0
First DEAE cellulose column chromatography	350	1,750	16.8	9.6
Third DEAE cellulose column chromatography	70	350	11.7	33.4
Sephadex G-75 gel filtration	150	80	6.7	84.0
Fourth DEAE cellulose column chromatography	28	79	6.4	81.0
First crystallization	4	18	1.9	105

[a] Beef adrenal glands were obtained from the slaughter house and dissected free of adhering fat and connective tissue. The trimmed glands were frozen at −20°C for at least 2 weeks prior to homogenization in a Waring blender (20% w/v) in 0.15 *M* KCl and 0.1 *M* tris-chloride buffer pH 7.4. The resulting homogenate was centrifuged at 13,000 *g* to remove particulate material and the supernatant adsorbed on DEAE cellulose. Chromatography on DEAE cellulose or Sephadex G-75 was as described by Omura *et al.* (1966). The purified adrenodoxin was crystallized as described in the text. The concentration of adrenodoxin was determined by EPR measurements of samples reduced with sodium dithionite and cooled to the temperature of liquid nitrogen. (From K. Suzuki, W. Taylor, J. Baron, and R. W. Estabrook, unpublished results.)

this suspension at 4°C for several weeks allowed the crystals to grow to 0.1–0.3 mm in length. The enzymatic and biochemical properties of recrystallized adrenodoxin examined thus far, such as absorption spectra, ability to reconstitute 11β-hydroxylase activity, and the iron and labile sulfide content, were found to be essentially identical to those observed for preparation of adrenodoxin from acetone powders or sonic extracts of adrenal mitochondria.

II. PHYSICAL PROPERTIES OF ADRENODOXIN

A. Optical Absorption Spectrum

Adrenodoxin, like many other iron–sulfur proteins, exhibits characteristic optical absorption bands (Fig. 2) which are associated with the oxidized and reduced forms of the pigment. The optical absorption spectrum of the oxidized form of adrenodoxin appears to be nearly identical with that of plant ferredoxin, except for a bathochromic shift of about 10 nm for the absorption band maxima at 415 and 453 nm. A second distinguishing feature of adrenodoxin is the hyperfine structure observed (Fig. 3) in the ultraviolet portion (250–280 nm) of the spectrum suggestive of a perturbation of electron densities associated with an aromatic amino acid near the iron–sulfur region of the molecule. In addition, Wilson (1967) has shown that iron–sulfur proteins, including adrenodoxin, have optical absorption characteristics in the near infrared portion of the spectrum (Fig. 4).

Reduction of adrenodoxin, either enzymatically by NADPH via the flavoprotein dehydrogenase (adrenodoxin reductase) or chemically with sodium dithionite, results in a loss of absorbance of approximately 50%

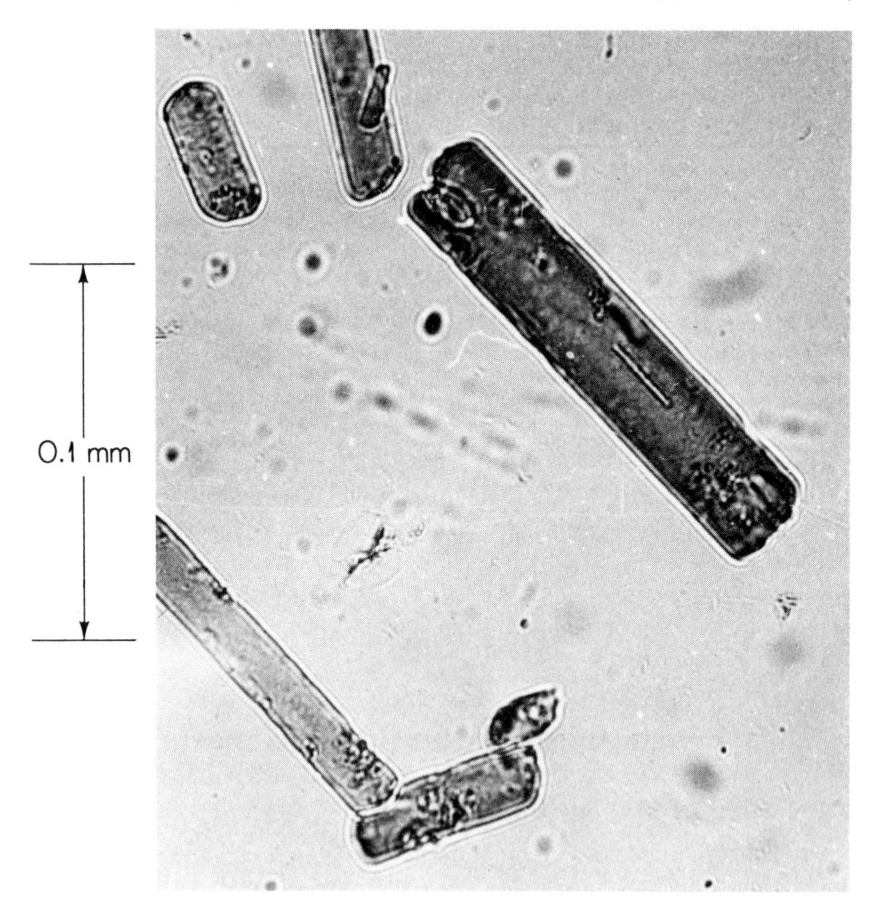

Fig. 1. Crystalline adrenodoxin. The purified iron–sulfur protein was isolated from beef adrenals and crystallized in the presence of 60% ammonium sulfate at pH 8.1 as described in the text. (From K. Suzuki, unpublished results.)

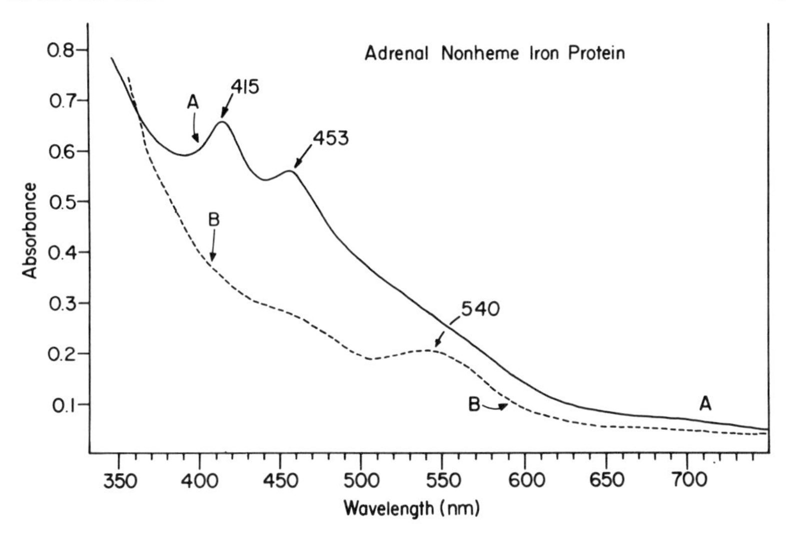

Fig. 2. The optical absorption spectrum of oxidized and reduced adrenodoxin. Purified beef adrenal cortex adrenodoxin was diluted in 0.035 *M* tris-chloride buffer, pH 7.4 to a final concentration of 67 μM, and the absorption spectrum of the oxidized pigment (A, solid line curve) recorded. The addition of NADPH in the presence of a catalytic concentration of the purified flavoprotein (NADPH:adrenodoxin oxidoreductase) was then added and the spectrum of reduced adrenodoxin was recorded (B, dashed line curve). (From K. Suzuki and Estabrook, unpublished results.)

between 400 and 500 nm concomitant with the appearance of a broad absorption band at about 540 nm. The absorption band associated with the reduced form of adrenodoxin is most clearly seen when the spectral characteristics of adrenodoxin are studied at the temperature of liquid nitrogen (Palmer *et al.* 1967).

Although many different approaches, including optical rotary dispersion (Kimura and Suzuki, 1967), circular dichroism (Palmer *et al.*, 1967), and Mössbauer spectra (Dunham *et al.*, 1971) have been taken in the study of purified adrenodoxin, the interpretation of the unique spectral properties of this iron–sulfur protein remains complex and unresolved (Orme-Johnson *et al.*, 1968; Tsibris and Woody, 1970; Beinert and Orme-Johnson, 1969).

B. Magnetic Properties

In the reduced state adrenodoxin exhibits a characteristic electron paramagnetic resonance (EPR) spectrum when examined at liquid-nitrogen temperatures as seen in Fig. 5 (Estabrook *et al.*, 1966). A striking

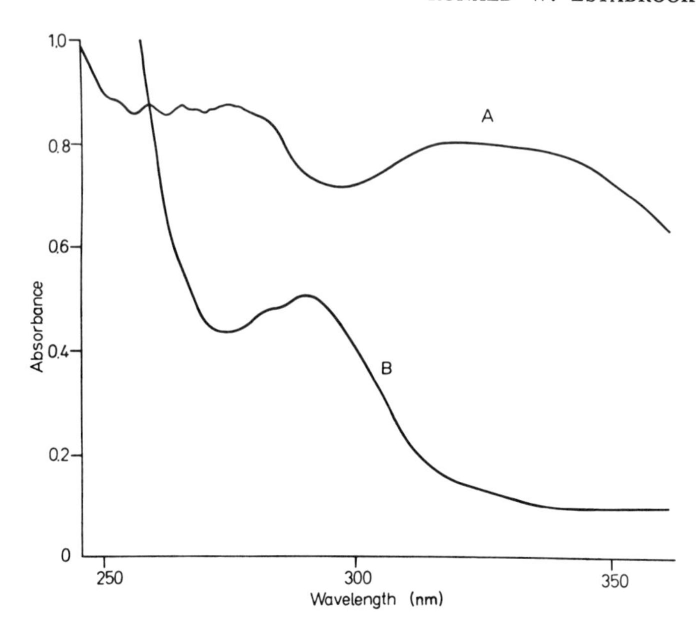

Fig. 3. The ultraviolet absorption spectrum of adrenodoxin. A solution of adrenodoxin was diluted in 0.003 *M* phosphate buffer, pH 7.4, to a concentration of 0.88 mg per ml and the absorption spectrum (curve A) was recorded. A comparable sample was diluted in 0.1 *N* NaOH and allowed to remain at room temperature for 24 hours before recording the absorption spectrum (curve B). (From Kimura, 1968.)

feature of this EPR spectrum is the so-called $g = 1.94$ signal which is characteristic of many electron transport iron–sulfur proteins. This signal is dependent upon both the state and environment of the iron in the protein molecule (DerVartanian *et al.*, 1967; Shethna *et al.*, 1964; Palmer *et al.*, 1967) and is suggested to result from the association of sulfur with rhombic iron (Palmer *et al.*, 1967). Kimura *et al.* (1970) have shown by magnetic susceptibility measurements that only one of the iron atoms of reduced adrenodoxin contributes to the $g = 1.94$ signal. At present the exact nature of the species contributing to this characteristic EPR signal of reduced adrenodoxin is not known, although this signal probably arises from either high-spin ferric iron or from high-spin ferrous iron (Kimura *et al.*, 1970; Dunham *et al.*, 1971) in association with sulfur ligands of undefined valency.

Since adrenodoxin is the major iron–sulfur protein in adrenocortical mitochondria, EPR determinations have become invaluable in studying the redox changes of adrenodoxin during steroid hydroxylation reactions (see below). With the present available instrumentation it is possible to

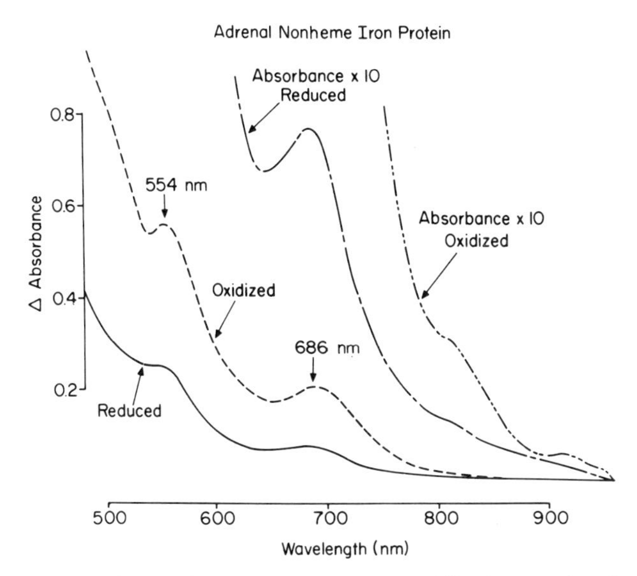

Fig. 4. The near infrared absorption spectrum of oxidized and reduced adrenodoxin at 86°K. Adrenodoxin was diluted to a final concentration of 95 μM Fe in a medium containing 0.9 M sucrose and 25 mM potassium phosphate buffer, pH 8.0. (From Wilson, 1967.)

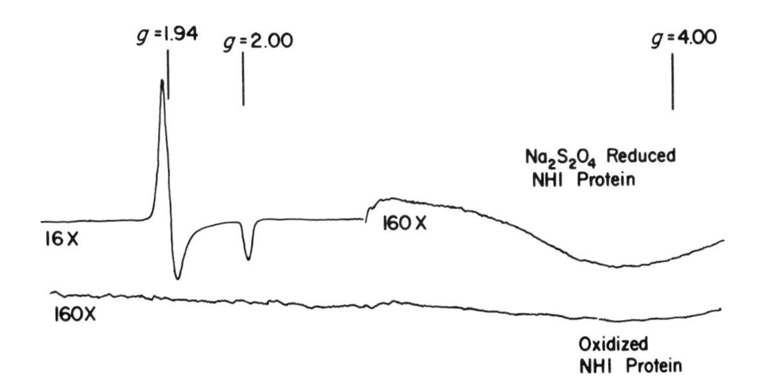

Fig. 5. The EPR spectra of oxidized and reduced adrenodoxin. Purified adrenodoxin (about 10 mg/ml) was suspended in 20 mM potassium phosphate buffer, pH 7.2. Samples containing oxidized adrenodoxin and adrenodoxin reduced by the addition of a few crystals of solid sodium dithionite were placed in matched quartz capillary tubes, frozen by immersion in liquid nitrogen, and examined in a Varian E4 EPR spectrometer equipped with a variable temperature attachment. The appropriate instrument parameters were modulation amplitude, 20 G; microwave power, 10 mW; and temperature, 82°K. (Reprinted from Estabrook *et al.*, 1966.)

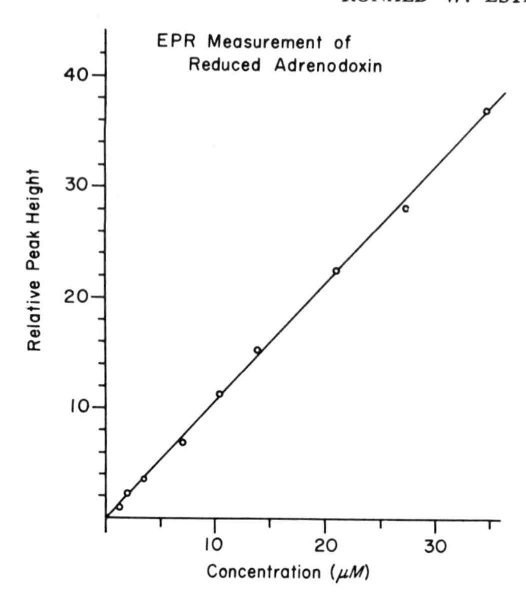

Fig. 6. Relationship between adrenodoxin concentration and magnitude of the $g = 1.94$ signal at 100°K. Purified adrenodoxin was diluted to various concentrations in 0.15 *M* tris-chloride buffer, pH 7.4. The concentration of adrenodoxin was determined from the absorbance of the oxidized protein at 415 nm using a millimolar extinction coefficient of 10 mM^{-1} cm^{-1}. Adrenodoxin in the various samples was reduced by the addition of a few crystals of solid sodium dithionite, and the samples were frozen in matched quartz capillary tubes by immersion in liquid nitrogen. The magnitude of the $g = 1.94$ signal of reduced adrenodoxin was determined using a Varian E4 EPR spectrometer equipped with a variable temperature attachment. The appropriate instrument parameters were: modulation amplitude, 12.5 G; modulation frequency, 100 kHz; microwave power, 50 mW; and temperature, 100°K. (From W. Taylor and R. W. Estabrook, unpublished results.)

measure quantitatively by EPR techniques concentrations of reduced adrenodoxin as low as 1 nmole/ml of solution (Fig. 6). This sensitivity permits the determination of changes in concentration of reduced adrenodoxin in small samples of tissue culture cells or in a *single* rat adrenal gland (see below).

C. Redox Potential

The oxidation–reduction potential (E_0') of adrenodoxin was first determined by Kimura and Suzuki (1967) using a potentiometric titration method with sodium dithionite and was reported to be +164 mV at pH 7.4° and 26°C with an n value of 2. This determination of a positive oxidation–reduction potential for adrenodoxin as well as its supposed ability to serve as a two electron acceptor indicated it was markedly different from other known iron–sulfur proteins (Tagawa and Arnon, 1962). This discrepancy prompted Suzuki and Estabrook (unpublished results) to redetermine the oxidation–reduction potential of adrenodoxin, and strikingly different results were obtained. In these later studies, a value for E_0' of approximately -360 mV at pH 7.4 with an n value of 1 was obtained. The reason for the discrepancy of values obtained remains unexplained but may result from side reactions occurring with sodium dithionite in the studies of Kimura and Suzuki (1967).

The value of n (the number of electron equivalents transferred per mole of adrenodoxin) was redetermined by Suzuki and Estabrook (unpublished results) by first establishing equilibrium conditions of adrenodoxin and NADPH in the presence of a catalytic concentration of the purified flavoprotein, NADPH dehydrogenase-adrenodoxin reductase. For these studies strict anaerobic conditions were established using glucose and glucose oxidase in the presence of catalase with an evacuated anaerobic sample. The extent of adrenodoxin reduced and NADPH oxidized was then determined spectrophotometrically using varying concentrations of NADPH and adrenodoxin. As shown in Table III, an n value of 1 is obtained indicating that only 1 mole of iron per mole of adrenodoxin is reduced enzymatically. Using a different approach of reductively titrating adrenodoxin with sodium dithionite and then evaluating the magnitude of the $g = 1.94$ signal observed by EPR spectroscopy, Orme-Johnson and Beinert (1969) also found (Fig. 7) that adrenodoxin accepts only one electron equivalent on reduction.

The oxidation–reduction potential of adrenodoxin was reexamined by

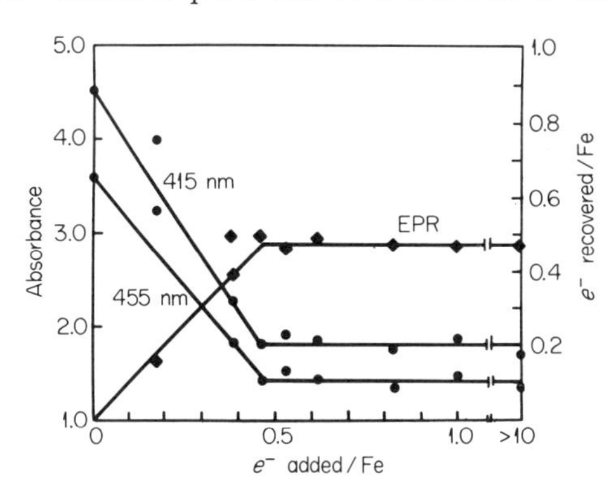

Fig. 7. Titration of pig adrenal iron–sulfur proteins with sodium dithionite. Samples of adrenodoxin (315 ng atoms of iron) were diluted in potassium phosphate buffer, pH 7.4, together with the dye methyl viologen. Optical absorption changes at 415 and 455 nm were determined and compared with the double integral of the EPR spectrum using the technique described by Orme-Johnson and Beinert (1969a). The optical scale is on the left and that for the EPR data is on the right. (Reprinted from Orme-Johnson and Beinert, 1969.)

TABLE III

DETERMINATION OF THE EQUIVALENTS (n) OF ELECTRONS TRANSFERRED PER MOLE OF ADRENODOXIN[a]

No.	Initial optical density		Optical density (after equilibrium)		Decreased absorbance at				NADPH oxidation (nmole)	Adrenodoxin reduced (nmole)	n
					340 nm			455 nm			
	340 nm	455 nm	340 nm	455 nm	Total	Due to NADPH oxidation	Due to adrenodoxin reduced				
1	0.770	0.350	0.679	0.268	0.091	0.048	0.043	0.082	15.4	32.7	0.94
2	0.870	0.500	0.750	0.400	0.120	0.067	0.053	0.100	21.5	39.2	1.1
3	0.897	0.345	0.792	0.250	0.105	0.055	0.050	0.095	17.7	37.0	0.96

[a] Various concentrations of NADPH and adrenodoxin were incubated in an anaerobic cuvette in an incubation mixture containing 0.035 M tris-chloride buffer, pH 7.4, 0.06 M glucose, 0.3 mg of glucose oxidase, and 0.1 mg of catalase in a final volume of 2 ml. To the chamber of the plunger, 0.05 ml of an 8 mg/ml solution of adrenodoxin reductase was added and the contents of the cuvette were gassed with prepurified argon for 15 minutes. The oxidation of NADPH and the reduction of adrenodoxin were determined from the changes in absorbance at 340 and 455 nm, respectively. The concentrations of NADPH and adrenodoxin used were: experiment No. 1, 88.4 nmoles of NADPH and 81.3 nmoles of adrenodoxin; experiment No. 2, 64.5 nmoles of NADPH and 114 nmoles of adrenodoxin; and experiment No. 3, 136 nmoles of NADPH and 78 nmoles of adrenodoxin. (From K. Suzuki and R. W. Estabrook, unpublished results.)

TABLE IV
DETERMINATION OF OXIDATION–REDUCTION POTENTIAL OF ADRENODOXIN[a]

No.	Initial optical density at		Optical density (after equilibrium) at		Adrenodoxin reduced (nmole)	Benzyl viologen reduced (nmole)	E_0' (V)
	455 nm	600 nm	455 nm	600 nm			
1 (pH 8.0)	0.307	0.081	0.263	0.302	34.9	54.8	−0.376
2 (pH 8.0)	0.615	0.165	0.490	0.364	62.8	54.0	−0.381
3 (pH 7.4)	0.386	0.102	0.305	0.275	47.6	44.8	−0.363

[a] Various concentrations of NADPH, adrenodoxin, and benzyl viologen were incubated in an anaerobic cuvette in an incubation mixture containing 0.035 *M* tris-chloride buffer at the indicated pH, 0.06 *M* glucose, 0.3 mg of glucose oxidase, and 0.1 mg of catalase in a final volume of 2 ml. To the chamber of the plunger, 0.05 ml of an 8 mg/ml solution of adrenodoxin reductase was added and the contents of the cuvette were gassed with prepurified argon for 15 minutes. The reduction of benzyl viologen and of adrenodoxin were calculated from the absorbance changes at 600 and 455 nm. The concentrations of NADPH, benzyl viologen, and adrenodoxin used were: experiment No. 1, 70.7 nmoles of adrenodoxin, 82.5 nmoles of benzyl viologen, and 196 nmoles of NADPH; experiment No. 2, 140 nmoles of adrenodoxin, 82.5 nmoles of benzyl viologen, and 196 nmoles of NADPH; and experiment No. 3, 90 nmoles of adrenodoxin, 82.5 nmoles of benzyl viologen, and 192 nmoles of NADPH. (From K. Suzuki and R. W. Estabrook, unpublished results.)

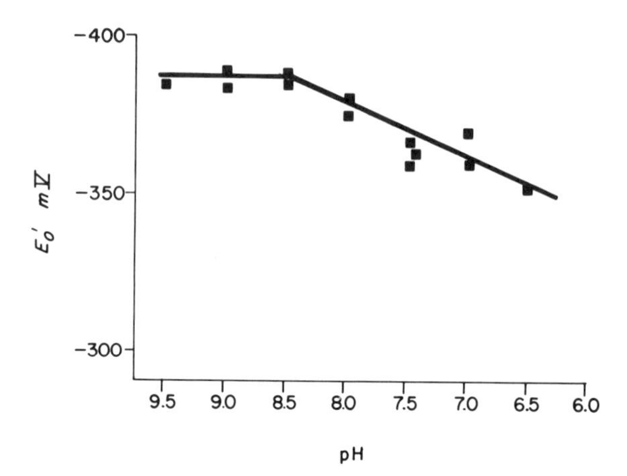

Fig. 8. The influence of pH on the oxidation–reduction potential of adrenodoxin. A series of experiments were carried out using benzyl viologen, NADPH:adrenodoxin oxidoreductase, and the reaction mixture to ensure anaerobiosis as described in Table IV. The extent of adrenodoxin or benzyl viologen reduction was determined spectrophotometrically in anaerobic cuvettes. (From K. Suzuki and R. W. Estabrook, unpublished results.)

Suzuki and Estabrook (unpublished results), since it was apparent from the anaerobic titration experiments with NADPH described above that the oxidation–reduction potential was more negative than the E_0' for NADP/NADPH and significantly different from the value determined earlier by Kimura and Suzuki (1967). Using the dye benzyl viologen and anaerobic conditions, the oxidation–reduction potential of beef adrenodoxin was shown to be —0.36 to 0.38 V (Table IV). Figure 8 shows the influence of pH on the oxidation–reduction potential as determined spectrophotometrically using the electromotively active dye benzyl viologen and the reductant NADPH in the presence of the flavoprotein, adrenodoxin reductase. These results indicate that adrenodoxin is similar in many respects to plant ferredoxin as reported by Tagawa and Arnon (1962).

III. THE BIOLOGICAL FUNCTION OF ADRENODOXIN

A. Role in Steroid Hydroxylation Reactions

The iron–sulfur protein, adrenodoxin, functions (Omura *et al.*, 1965) in the mitochondrial fraction of adrenal cortex by mediating electron transfer from a NADPH flavoprotein dehydrogenase to the hemoprotein,

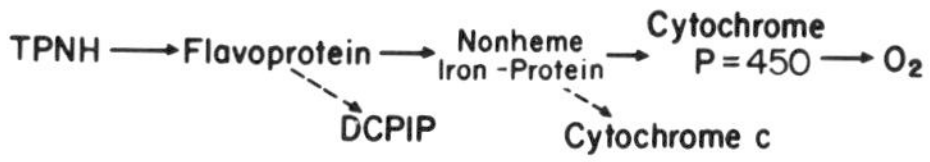

TPNH → Flavoprotein → Nonheme Iron-Protein → Cytochrome P=450 → O_2
DCPIP   Cytochrome c

Fig. 9. Electron transport sequence of mixed-function oxidations involving adrenodoxin of adrenal cortex mitochondria. In these reactions, NADPH donates an electron to a flavoprotein which, in turn, transfers an electron to adrenodoxin. Reduced adrenodoxin can then donate an electron to cytochrome P-450. The pathways for transfer of electrons *in vitro* from NADPH to DCPIP and cytochrome c are also included.

cytochrome P-450 (Fig. 9). Similar iron-sulfur proteins function in a comparable manner in other steroidogenic organs such as testis and ovary (Kimura and Ohno, 1968). In addition to the reaction of reduced adrenodoxin with cytochrome P-450, reduced adrenodoxin can also donate reducing equivalents to cytochrome c *in vitro*.

Our present understanding of the cyclic process of cytochrome P-450 reduction and oxidation indicates that reduced adrenodoxin participates at *two* distinct sites in the cycle as illustrated in Fig. 10. The series of reactions occurring during cytochrome P-450 interaction with the steroid substrate, oxygen, and reduced adrenodoxin are now recognized (Estabrook *et al.*, 1968) to consist of at least six distinct steps, although all have not been fully characterized.

REACTION 1

The low-spin form of ferric cytochrome P-450 reacts with a substrate, such as deoxycorticosterone, to form a complex (Cammer *et al.*, 1967)

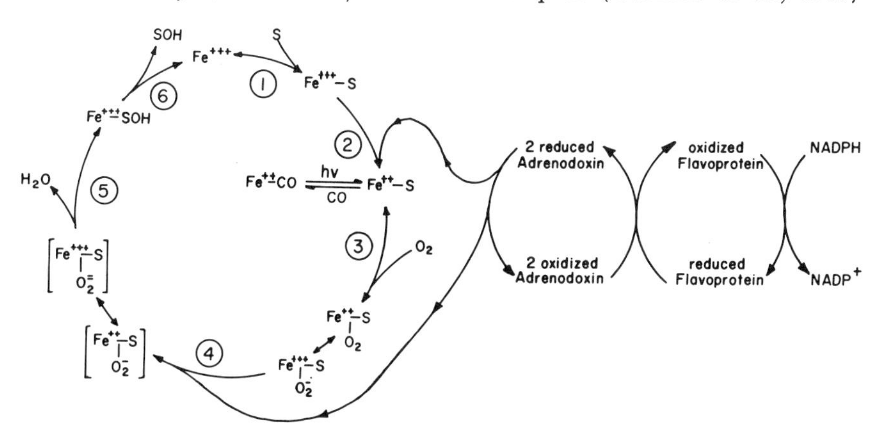

Fig. 10. Proposed roles for adrenodoxin in cytochrome P-450-mediated reactions. The donation of electrons to cytochrome P-450 by reduced adrenodoxin are indicated at steps **2** and **4** in the cyclic reduction and oxidation transitions of cytochrome P-450 as described in the text.

of substrate with high-spin ferric cytochrome P-450. The interaction can be readily measured by determining the perturbation of the optical absorption spectrum of oxidized cytochrome P-450, i.e., associated with substrate interaction there is a loss of absorbance at about 420 nm concomitant with the appearance of an absorbance band at about 385 nm (Narashimhulu *et al.,* 1965).

REACTION 2

The high-spin ferric cytochrome P-450·substrate complex undergoes a *one electron* reduction by interaction with reduced adrenodoxin forming a ferrous cytochrome P-450·substrate complex. This form of reduced cytochrome P-450 can react with carbon monoxide to form the light dissociable CO adduct characterized by an absorbance band maximum at about 450 nm, hence the name cytochrome P-450 (Omura and Sato, 1964).

REACTION 3

In the presence of oxygen, the ferrous cytochrome P-450·substrate complex can react to form an "oxygenated cytochrome P-450" (Ishimura *et al.,* 1971; Estabrook *et al.,* 1971) which may be present as an oxygen·substrate·ferrous cytochrome P-450 ternary complex in equilibrium with a superoxide anion·substrate·ferric cytochrome P-450 complex (Peterson *et al.,* 1972a). At this time it is not known which form of oxygenated cytochrome P-450 is functional during steroid hydroxylation reactions catalyzed by adrenal cortex mitochondrial cytochrome P-450.

REACTION 4

Analogous to studies carried out (Gunsalus *et al.,* 1972; Peterson *et al.,* 1972a,b) with the purified bacterial cytochrome P-450, it is presumed that the "oxygenated cytochrome P-450" interacts with reduced adrenodoxin to form an intermediary stage of oxygen reduction such as a superoxide anion·substrate complex with ferrous cytochrome P-450 in equilibrium with a hydroperoxide substrate complex of ferric cytochrome P-450.

REACTION 5

Through a rearrangement of the "activated oxygen" present in the substrate complex of cytochrome P-450, one atom of molecular oxygen is inserted into the steroid substrate molecule while the other atom of molecular oxygen is reduced to water. This proposed concerted reaction results in the formation of a complex of the product with ferric cytochrome P-450.

REACTION 6

The cycle of cytochrome P-450 function is completed by dissociation of the product molecule from ferric cytochrome P-450 regenerating the low-spin form of the ferric hemoprotein now poised for reaction with another molecule of the steroid substrate.

Reduced adrenodoxin therefore participates as an electron donor at two specific sites in this cycle, i.e., reactions 2 and 4. Although the direct demonstration of these reactions has not as yet been accomplished with the cytochrome P-450 of adrenal cortex mitochondria, sufficient similarities exist with the purified bacterial cytochrome P-450 system, which has been examined in great detail (Gunsalus *et al.*, 1971; Peterson and Ishimura, 1971) to assume that a similar reaction sequence is operative. The validity of this extrapolation of the mechanism developed for the bacterial cytochrome P-450 to the steroid hydroxylating mitochondrial cytochrome P-450 system is supported by recent studies by Baron *et al.* (1972) using an antibody to adrenodoxin, and the studies of Schleyer *et al.* (1972) who measured product formation on addition of reduced adrenodoxin to the complex of deoxycorticosterone and reduced adrenal mitochondrial cytochrome P-450 in the presence of oxygen (see below).

B. Studies with the Resolved Steroid Hydroxylating System of Adrenal Cortex Mitochondria

In the late 1950's and early 1960's, studies by Tompkins (1959), Sweat (1962), and Sharma *et al.* (1962) showed that the electron transport complex functional in adrenal cortex for the 11β-hydroxylation of deoxycorticosterone could be resolved into a particulate fraction and a soluble fraction. It was not until 1965 when Kimura and Suzuki (1965) as well as Omura *et al.* (1965) isolated and purified the iron–sulfur protein, adrenodoxin, from adrenal cortex mitochondria that it was realized that this electron transport carrier working together with an NADPH-specific flavoprotein were the required components present in the soluble fractions isolated earlier. With the purified adrenodoxin as well as the flavoprotein available, studies were initiated (Omura *et al.*, 1966) to reexamine the reconstitution of steroid hydroxylation reactions using the membrane-bound cytochrome P-450 associated with fragments derived from adrenal cortex mitochondria. As shown in Table V, little or no steroid hydroxylation was detectable unless *both* the purified flavoprotein and the purified adrenodoxin were present together with cytochrome P-450 in the reaction mixture. In the presence of excess flavoprotein as well as the particle-bound cytochrome P-450, it was possible to demonstrate directly the in-

TABLE V
RECONSTITUTION OF THE 11β-HYDROXYLASE SYSTEM[a,b]

No.	Nonheme iron (nmole)		Flavoprotein (nmole)		11β-Hydroxylated product (nmole min^{-1})
	46	92	1.8	3.6	
1	−	−	−	−	0.1
2	+	−	−	−	1.6
3	−	−	+	−	0.2
4	−	+	−	−	1.6
5	−	−	−	+	0.2
6	+	−	+	−	11.2
7	+	−	−	+	10.8
8	−	+	+	−	13.6
9	−	+	−	+	13.4

[a] 1.69 nmole cytochrome P-450 per vessel.
[b] Conditions were similar to those described for Fig. 11 (from Simpson *et al.*, 1969).

fluence of varying concentrations of adrenodoxin (Fig. 11) on the overall rate of deoxycorticosterone hydroxylation. Similar results demonstrating the role of adrenodoxin in the 11β- and 18-steroid hydroxylation reactions of rat adrenal mitochondria have been reported by Nakamura *et al.* (1966), while Ichii *et al.* (1967) and Simpson and Boyd (1967) demonstrated the need for adrenodoxin for the oxidative reactions associated with the side chain cleavage of cholesterol.

Of interest is the observation that rather large amounts of adrenodoxin were required in order to obtain maximal rates of activity in the reconstituted system. As discussed by Cooper *et al.* (1968), a ratio of adrenodoxin:flavoprotein:cytochrome P-450 of 55:1.1:1.0 was necessary in order to obtain maximal activity for the hydroxylation of deoxycorticosterone. The need for this large excess of adrenodoxin relative to the other components functional in this electron transport system, under the conditions for the reconstitution of hydroxylating activity, remains unknown. Consideration of the stoichiometric ratio of about one for adrenodoxin to cytochrome P-450 in intact adrenal cortex mitochondria indicates that a spatial organization of these compounds might exist within the mitochondria or that the artificial conditions for reconstitution of activity introduces changes in the association of the needed electron transport carriers. Studies by Foust *et al.* (1969) have shown that the iron–sulfur protein, ferredoxin, can form a complex with flavoproteins which is readily dissociated in the presence of high salt concentrations. During the isolation and purification of adrenodoxin, it has been observed (Omura and Estabrook, unpublished observations) that a salt sensitive complex of adrenodoxin with the NADPH flavoprotein dehydrogenase can occur, suggesting that the salts present in the reaction mixture used in the above described reconstitution experiments may be partially responsible for the large excess of adrenodoxin required to obtain maximal activity. Recent results reported by Mitani and Horie (1969) indicate that even higher ratios of adrenodoxin to cytochrome P-450 are required for reconstitution of steroid 11 β-hydroxylase activity when a detergent-treated, partially purified preparation of adrenal cortex cytochrome P-450 is employed. Comparable studies have been carried out by Bryson and Sweat (1968) as well as Young and Hall (1969) demonstrating the

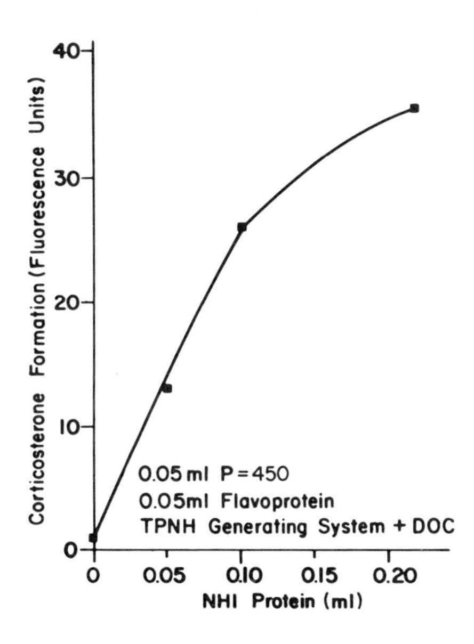

Fig. 11. The influence of varying concentrations of adrenodoxin on the rate of formation of corticosterone. The main compartment of a Warburg reaction vessel contained 20 mg of DOC suspended in 2.1 ml of a buffer mixture consisting of 0.047 *M* KCl, 0.42 m*M* $MgCl_2$, 0.0115 *M* glycylglycine buffer pH 7.4, 0.058 *M* NaCl, 17.8 mg of crystalline bovine serum albumin, 0.05 ml of P-450, 0.05 ml of adrenodoxin reductase, and various concentrations of adrenodoxin. After equilibration for 5 minutes at 25°C with a gas mixture containing 4% oxygen and 96% nitrogen, the reaction was initiated by the addition from the sidearm of the Warburg vessel 0.3 ml of a NADPH-generating system containing 0.5 m*M* NADP, 0.13 *M* glucose 6-phosphate and 0.5 Kornberg units of glucose-6-phosphate dehydrogenase. Corticosterone formation was measured fluorometrically. (From Omura *et al.*, 1966.)

need for a large excess of adrenodoxin for the reconstitution of hydroxylase activity associated with the side-chain cleavage of cholesterol as catalyzed by cytochrome P-450 of adrenal cortex mitochondria.

C. Immunochemical Studies of Adrenodoxin

The availability of pure adrenodoxin permits the formation of inhibitory antibodies useful for demonstrating the role of this pigment in a number of electron transport reactions. Using homogeneous adrenodoxin purified from bovine adrenal cortex mitochondria, Baron *et al.* (1972) and Masters *et al.* (1971) prepared γ-globulin fractions from immunized rabbits which interacted with adrenodoxin. This interaction could be demonstrated by formation of a precipitin reaction with either purified adrenodoxin or adrenodoxin present in sonic extracts of adrenal mitochondria. Of interest is the observation (Fig. 12) that the anti-adrenodoxin γ-globulin fraction prevented the reduction of purified adrenodoxin

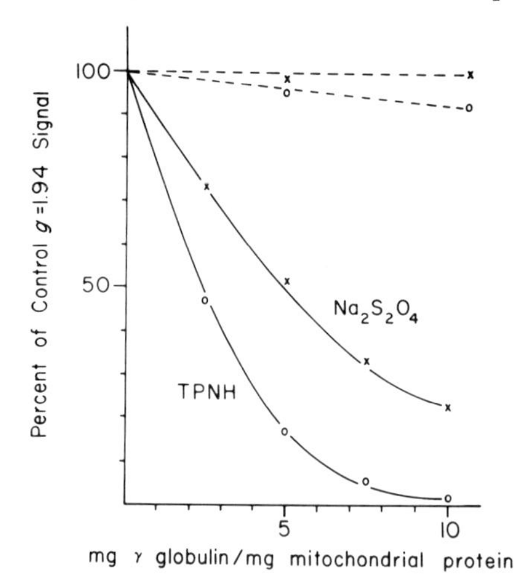

Fig. 12. Effect of antibody to adrenodoxin on the extent of reduction of adrenodoxin by NADPH and dithionite. Bovine adrenocortical mitochondria were sonicated for 5 minutes and were suspended to a protein concentration of 2 mg/ml in 0.2 *M* tris-chloride buffer, pH 7.4, without γ-globulin or containing the specified concentrations of either immune or preimmune γ-globulin. The samples were then gassed with argon for 10 minutes, after which either 200 μM. NADPH (○——○) or a few crystals of sodium dithionite (×——×) were added. The EPR spectrum were recorded as described in Fig. 6. The solid lines represent the effect of immune γ-globulin, while the dashed lines represent the effect of preimmune γ-globulin. In the absence of γ-globulin, NADPH reduced 0.73 nmoles adrenodoxin/mg mitochondrial protein, while sodium dithionite reduced 0.91 nmoles/mg protein. (Reprinted from Baron *et al.*, 1972.)

TABLE VI
COMPARISON OF THE INHIBITORY EFFECTS OF ANTIBODY TO ADRENODOXIN ON NADPH-DEPENDENT ACTIVITIES IN BOVINE ADRENOCORTICAL MITOCHONDRIA[a]

γ-Globulin per mg mitochondria protein (mg)	NADPH:cytochrome c reductase		NADPH-cytochrome P-450 reductase		11β-Hydroxylase	
	(nmoles/mg/minute)	(% control)	(nmoles/mg in the initial 15 seconds)	(% control)	(nmoles/mg/minute)	(% control)
0	20.4	100	0.43	100	1.70	100
20 (immune)	7.1	35	0.17	37	0.77	45
20 (preimmune)	20.1	99	0.42	98	1.77	104

[a] Bovine adrenocortical mitochrondia were frozen and thawed five times prior to the determination of enzymic activities. (From Baron *et al.*, 1972.)

TABLE VII
INHIBITION OF NADPH:CYTOCHROME c REDUCTASE ACTIVITY BY ANTIBODY TO ADRENODOXIN AS A FUNCTION OF MITOCHONDRIAL INTEGRITY[a]

Condition of mitochondria[b]	Percent of maximal inhibition[c]
Intact	3
Frozen and thawed	
1×	38
2×	51
3×	86
4×	90
5×	95
6×	96
Sonicated (5 minutes)	100

[a] From Baron *et al.* (1972).

[b] Mitochondria were suspended in distilled water to a concentration of 10 mg protein/ml. The activity of NADPH:cytochrome c reductase was determined in the presence of both NADPH and malate. The mitochondria were frozen and thawed using a dry ice–acetone mixture.

[c] The inhibition produced by a ratio of 10 mg γ-globulin/mg mitochondria protein using sonicated mitochondria (resulting in 70% inhibition as compared to a control with preimmune γ-globulin) was taken to be equal to 100%. The activity of NADPH: cytochrome c reductase in the absence of γ-globulin was 28.7 nmoles cytochrome c reduced/minute/mg protein.

either by sodium dithionite or by NADPH in the presence of the required NADPH dehydrogenase flavoprotein. This result suggests that the antibody forms a complex with adrenodoxin affecting either directly or indirectly the active site of the iron–sulfur protein.

Studies of electron transport reactions of adrenal cortex mitochondria showed that the antibody was an effective inhibitor of NADPH:cytochrome c oxidoreductase activity, NADPH:cytochrome P-450 oxidoreductase activity, as well as the 11β-hydroxylation of deoxycorticosterone (Table VI). In contrast, the antibody to adrenodoxin had *no effect* on any of the electron transport activities associated with drug oxidative metabolism as catalyzed by liver microsomes or the 21-hydroxylation of 17-hydroxyprogesterone as catalyzed by adrenal cortex microsomes. Thus, these experiments are conclusive evidence for the role of this unique iron–sulfur protein in a specific electron transport sequence functional in adrenal cortex mitochondria for steroid hydroxylation reactions.

Of interest are the observations that the antibody to adrenodoxin did not inhibit the NADPH:cytochrome c oxidoreductase activity of adrenal cortex mitochondria (Table VII) unless the structural integrity of these mitochondria was modified by a sequence of freeze–thaw treatments or by sonication of the mitochondria. Such a result suggests that adrenodoxin is *not* associated with the outer membrane structure of adrenal cortex mitochondria. This result is consistent with earlier observations (McCarthy and Peron, 1967; Omura *et al.*, 1966) that demonstrated the need to modify adrenal cortex mitochondrial structure by calcium ion treatment or sonic disruption in order to observe the NADPH-supported reduction of cytochrome P-450 for the overall reaction of 11β-hydroxylation of steroids.

D. Steady-State Reduction of Adrenodoxin during Steroid Hydroxylation Reactions

The ability to readily detect reduced adrenodoxin by its characteristic electron paramagnetic resonance signal permits a series of experiments to determine changes in steady state reduction associated with altered electron transport flux during adrenal steroid hydroxylation reactions. For these experiments adrenal cortex mitochondria were isolated and examined by the oxygen electrode technique to ensure optimal conditions for the 11β-hydroxylation of deoxycorticosterone (Fig. 13). A comparable experimental protocol was used at more concentrated protein concentrations and samples withdrawn for examination by electron paramagnetic resonance spectroscopy. As shown in Fig. 14 the presence of endogenous substrates serving as reducing equivalent donors in adrenal cortex mitochondria results in only a partial (about 25%) reduction of adrenodoxin. The addition of malate to activate the NADP$^+$-dependent malic enzyme of mitochondria or the presence of succinate which may generate NADPH via energy-linked, reversed electron transport reactions (Cam-

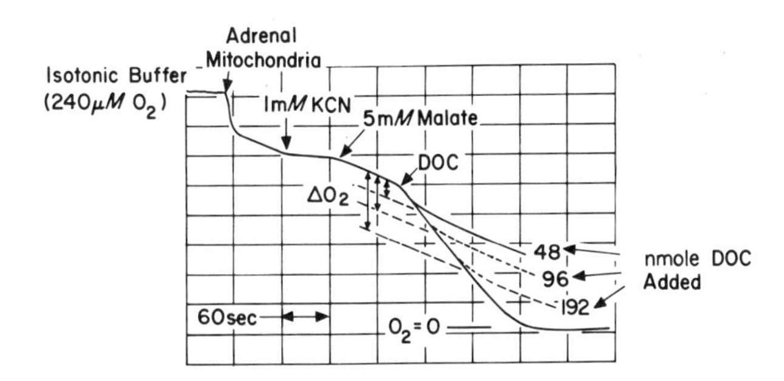

Fig. 13. The respiratory activity of adrenal cortex mitochondria oxidizing malate in the presence of limiting concentrations of deoxycorticosterone. (Reprinted from Cammer *et al.*, 1967.)

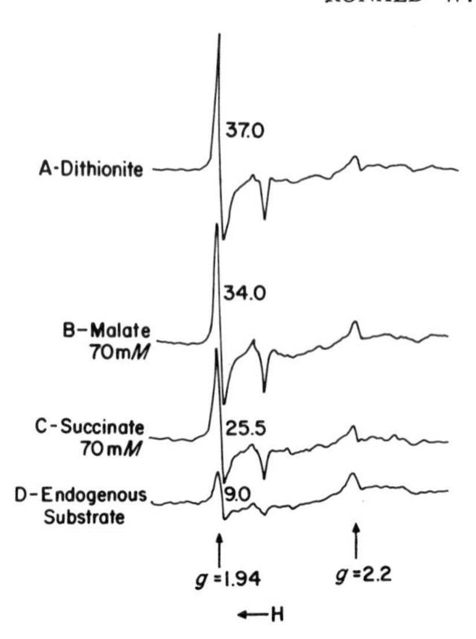

Fig. 14. EPR measurements of the aerobic steady-state reduction of adrenodoxin of adrenal cortex mitochondria in the presence of various substrates. Mitochondria were suspended in an isotonic buffer at a concentration of about 20 mg of protein per ml in the presence of the substrates indicated and samples withdrawn for examination by EPR spectroscopy. (Reprinted from Cammer *et al.*, 1967.)

mer *et al.*, 1967) markedly enhance the extent of adrenodoxin reduced during the aerobic steady state. Indeed, adrenodoxin is about 90% in the reduced form (when compared to the magnitude of the EPR signal observed in the presence of the chemical reductant sodium dithionite) in the presence of malate. The subsequent addition of deoxycorticosterone, to initiate the cytochrome P-450 catalyzed hydroxylation reaction, results in a dramatic change in the extent of adrenodoxin reduced (Fig. 15) in the presence of malate. Comparable shifts in the steady-state reduction of adrenodoxin were observed when deoxycorticosterone was added to adrenal cortex mitochondria pretreated with sodium succinate. This alteration in the extent of steady-state reduction of adrenodoxin suggests that the reduction of adrenodoxin by the NADPH-flavoprotein dehydrogenase may play a key role in dictating the overall rate of steroid hydroxylation as catalyzed by bovine adrenal cortex mitochondria.

E. Reaction of Reduced Adrenodoxin with Oxygenated Cytochrome P-450

The scheme describing the sequence of reactions associated with cytochrome P-450 function during steroid hydroxylation reactions presented above (Fig. 10) predicts that reduced adrenodoxin serves at two sites as a one electron donor to cytochrome P-450. Recently, experiments have been reported by Huang and Kimura (1971) and Schleyer *et al.* (1972) demonstrating the ability of reduced adrenodoxin to interact with the substrate complex of reduced cytochrome P-450 in the presence of oxygen resulting in a stoichiometric formation of the hydroxylated steroid product. An example of the results of such an experiment is illustrated in Table VIII taken from the work of Schleyer *et al.* (1972). For this study, cytochrome P-450 resolved from beef adrenal cortex mitochondria was interacted with deoxycorticosterone and then reduced with a limiting concentration of sodium dithionite. After oxygenation of this reaction mixture, reduced adrenodoxin was added and the amount of product formed (corticosterone) determined. As shown in Table VIII, only reduced adrenodoxin served as a competent electron donor to the proposed oxygenated form of the reduced cytochrome P-450 substrate complex. As stated by Huang and Kimura (1971): "From these observations it is clear that reduced adrenodoxin is *absolutely* required for the hydroxylation reaction of steroids." The failure to observe product formation even when a dye (benzyl viologen) of sufficient negative E_0' is employed indicates the specificity of this reaction for reduced adrenodoxin.

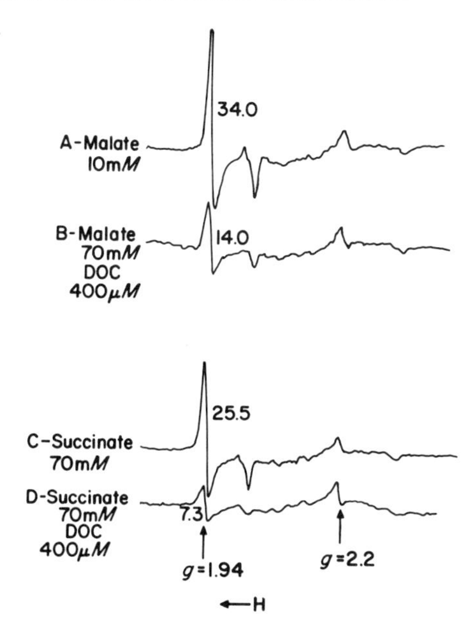

Fig. 15. The change in steady-state reduction of adrenodoxin upon addition of deoxycorticosterone to adrenal cortex mitochondria. Conditions are similar to those described for Fig. 15. (Reprinted from Cammer *et al.*, 1967.)

TABLE VIII
FUNCTION OF REDUCED ADRENODOXIN IN THE 11β-HYDROXYLATION OF DEOXYCORTICOSTERONE BY OXYGENATED CYTOCHROME P-450[a]

Experiment	Reduced cytochrome P-450 plus oxygen (nmoles)	ISP_{red} (nmoles)	Other additions (nmoles)		Corticosterone formed (nmoles)	Preparative yield (percent of theoretical)
A	20	200	—	—	16.8	83.9
B	20	—	ISP_{ox}	200	0.58	2.88
C	20	—	Benzyl viologen	200	0.45	2.26
D	—	200	P-450(Fe^{3+})	20	0.79	3.97
E	—	200	Hb · O_2	20	0.01	0

[a] The reactions were carried out in 2.5 ml incubation mixtures containing 0.05 *M* potassium phosphate buffer, pH 7.4, 600 nmoles of deoxycorticosterone and other components as indicated. Corticosterone formation was determined fluorometrically. (Reprinted from Schleyer *et al.*, 1972.)

F. *In Vivo* Biological Variation in Adrenodoxin Concentration as Influenced by ACTH

The steroidogenic effect of ACTH on adrenal tissue *in vivo* has been extensively studied in a number of laboratories. In particular Griffith and Glick (1966) have reported the increased activity of the 11β-hydroxylase of adrenal in response to long-term stimulation of intact animals with ACTH. Since the 11β-hydroxylase activity is intimately associated with the mitochondrial fraction of adrenal cortex and involves a sequence of electron carriers involving adrenodoxin as well as cytochrome P-450, a series of studies were initiated to determine any changes in adrenodoxin concentration during ACTH stimulation. The initial experiments to determine if adrenodoxin did respond to ACTH treatment of adrenal cells were carried out by Kowal *et al.* (1970) using monolayer cultures of cells derived from mouse adrenal tumors (Kowal and Fiedler, 1968). As shown in Table IX, prolonged exposure of these tissue culture cells resulted in a 50% increase in the concentration of adrenodoxin and cytochrome P-450 with no detectable change in the concentration of other cytochromes associated with mitochondria. For these studies the concentration of adrenodoxin was determined in cell suspensions by electron paramagnetic resonance spectroscopy (Fig. 16). The ability to demonstrate a specific increase by ACTH of the electron carriers required for

TABLE IX
EFFECT OF ACTH ON INTRACELLULAR CONCENTRATION OF CYTOCHROME P-450, ADRENODOXIN, AND MITOCHONDRIAL RESPIRATORY CHAIN CYTOCHROMES IN TISSUE CULTURE CELLS[a]

Source	Cytochromes (pmoles/mg protein)				Adrenodoxin (pmoles/mg protein)
	$a + a_3$	b	$c + c_1$	P-450	
Whole cell extract					
T_{422} control	69	90	200	22	136
T_{422} ACTH	60	78	180	46	180

[a] From Kowal *et al.* (1970).

steroid 11β-hydroxylation in the absence of a general stimulation of cell growth undoubtedly represents specific protein synthesis associated with the requirement to maintain a continual high level of steroid synthesis during prolonged stimulation with ACTH.

Success with experiments showing an influence of ACTH on the concentration of adrenodoxin in adrenal cells in tissue culture suggested the need to examine comparable effects on the pigments associated with the functional adrenal in a living animal. The sensitivity of electron paramagnetic resonance spectroscopy is sufficient to permit the quantitative measurement of reduced adrenodoxin concentration in a single rat adrenal gland. A series of studies were initiated by Purvis *et al.* (1972a) using hypophysectomized rats and followed the change in cytochrome P-450

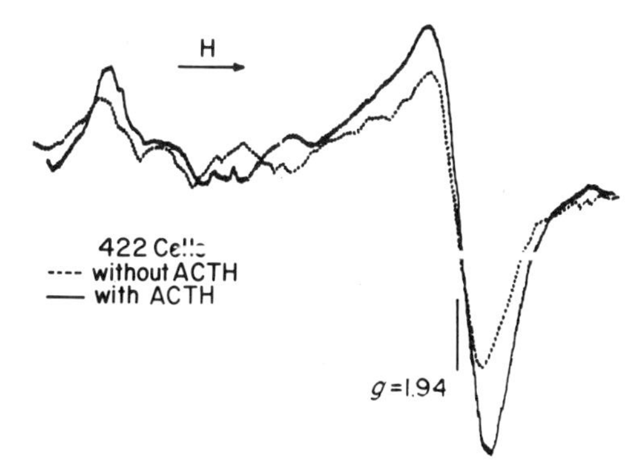

Fig. 16. The EPR spectra of control and ACTH stimulated tissue culture cells derived from rat adrenal cortex tumors. Concentrated suspensions of cells (approximately 20 mg of protein per ml in each case) were treated with a small sample of sodium dithionite and placed in calibrated quartz EPR tubes. (Reprinted from Kowal *et al.*, 1970.)

TABLE X
RAT ADRENAL HOMOGENATE[a]

Experiment	Adrenal weight (mg)	P-450 (nmoles/mg wet wt.)	Adrenodoxin (nmoles/mg wet wt.)	P-450/adrenodoxin	11β-Hydroxylase (nmoles/mg wet wt./30 minutes)
Control	25 (5)	0.123	0.116	1.07	9.6
$SHAM_{11}$	24 (5)	0.078	0.095	0.82	11.1
H_4	17.1 (4)	0.060	0.080	0.75	5.6
H_7	14.0 (4)	0.050	0.073	0.69	2.9
H_{11}	13.1 (4)	0.028	0.051	0.55	3.9
H_{14}	12.0 (6)	0.020	0.040	0.50	2.0
$H_{14}A_2$	13.3 (3)	0.033	0.045	0.73	7.2
$H_{14}A_4$	11.8 (4)	0.045	0.053	0.77	9.0
$H_{14}A_7$	16.9 (4)	0.060	0.071	0.84	10.8

[a] Adrenal glands were excised from rats at various times after hypophysectomy (H) and after subsequent ACTH (A) treatment (10 international units daily). The subscripts refer to the number of days elapsed in the hypox and ACTH therapy phases. The number of animals examined at each sampling is included in parentheses. The adrenodoxin concentration of the adrenal homogenates was determined by EPR spectroscopy as described in Fig. 6. The homogenate cytochrome P-450 content was determined according to the method of Estabrook *et al.* (1972).

and adrenodoxin concentrations concomitant with the decrease in steroid 11β-hydroxylase activity in homogenates of the adrenal gland. As summarized in Table X, hypophysectomy results in a decrease to nearly one-third of the original levels in the tissue concentration of adrenodoxin at the end of 14 days. Subsequent treatment of the hypophysectomized animals with ACTH restores the 11β-hydroxylase activity and initiates a synthesis of adrenodoxin as well as cytochrome P-450 comparable to that observed with the adrenal tissue culture cells. From studies of this type, it is possible to estimate the half-life of adrenodoxin in the mitochondria of rat adrenal cortex as 5.3 days (Purvis *et al.*, 1972b). Of interest is the observation that the half-life of mitochondrial cytochrome P-450 is significantly different (3.5 days) from that for adrenodoxin, suggesting that their synthesis is not coordinately controlled in response to ACTH treatment.

IV. CONCLUSION

Of all the mixed-function oxidations catalyzed by cytochrome P-450 in mammalian tissues, only those reactions which occur in the mitochondrial fractions of steroidogenic tissues require an iron–sulfur protein. More specifically, in the hydroxylations of steroids in adrenal cortex mitochondria, the iron–sulfur protein, adrenodoxin, serves as an electron transport intermediate between a flavoprotein dehydrogenase and cytochrome P-450, the terminal oxidase. In the present chapter, the properties of adrenodoxin have been briefly reviewed in light of our present knowledge of the biology and chemistry of iron–sulfur proteins. At this time, it is proposed that adrenodoxin functions by donating one electron to cytochrome P-450 at two discrete steps in hydroxylation reactions catalyzed by adrenal cortex mitochondria. Further experiments will, however, undoubtedly yield greater insight into the molecular role of iron–sulfur proteins in these reactions.

Of greatest interest is the unanswered question concerning the regulation of synthesis of adrenodoxin in the adrenal cortex in response to hormones such as ACTH. The mechanism of transmission of genetic information for the synthesis of this unique intramitochondrial protein undoubtedly will provide new and exciting fundamental principles at the frontier of biology of the future.

ACKNOWLEDGMENT

The work reported in this chapter was supported in part by U.S. Public Health Service Grant No. 1 P11 GM16488.

REFERENCES

Baron, J., Taylor, W. E., and Masters, B. S. S. (1972). *Arch. Biochem. Biophys.* **150,** 105.

Beinert, H., and Lee, W. (1961). *Biochem. Biophys. Res. Commun.* **5,** 40.

Beinert, H., and Orme-Johnson, W. H. (1969). *Ann. N.Y. Acad. Sci.* **158,** 336.

Bryson, M. J., and Sweat, M. L. (1968). *J. Biol. Chem.* **243,** 2799.

Cammer, W., Cooper, D. Y., and Estabrook, R. W. (1967). *In* "Functions of The Adrenal Cortex" (K. McKerns, ed.), Vol. 2, p. 943. Appleton, New York.

Cooper, D. Y., Schleyer, H., Estabrook, R. W., and Rosenthal, O. (1968). *In* "Progress in Endocrinology," Proc. 3rd. Int. Congr. Endocrinol. (C. Gual, ed.), p. 784. Excerpta Medica Foundation, Mexico, D.F.

Der Vartanian, D. V., Orme-Johnson, W. H., Hansen, R. E., and Beinert, H. (1967). *Biochem. Biophys. Res. Commun.* **26,** 569.

Dunham, W. R., Palmer, G., Sands, R. H., and Bearden, A. J. (1971). *Biochim. Biophys. Acta* **253,** 373.

Estabrook, R. W. Gonze, J., and Omura, T. (1966). *In* "Flavins and Flavoproteins" (E. C. Slater, ed.), Biochim. Biophys. Acta Library, Vol. 8, p. 78. Elsevier, Amsterdam.

Estabrook, R. W., Hildebrandt, A. G., Remmer, H., Schenkman, J. B., Rosenthal, O., and Cooper, D. Y. (1968). *In* "Colloquium Gesellschaft fuer der Biologie Chemie" (B. Hess and Hj. Staudinger, eds.), p. 142. Springer-Verlag, Berlin.

Estabrook, R. W., Hildebrandt, A. G., Baron, J., Netter, K. J., and Leibman, K. (1971). *Biochem. Biophys. Res. Commun.* **42,** 132.

Estabrook, R. W., Peterson, J. A., Baron, J., and Hildebrandt, A. G. (1972). *In* "Methods in Pharmacology" (C. F. Chignell, ed.), Vol. 2, p. 303. Appleton, New York.

Foust, G. P., Mayhew, S. G., and Massey, V. (1969). *J. Biol. Chem.* **244,** 964.

Griffith, K., and Glick, D. (1966). *J. Endocrinol.* **35,** 1.

Gunsalus, I. C., Tyson, C. A., Tsai, R., and Lipscomb, J. D. (1971). *Chem. Biol. Interactions* **4,** 75.

Gunsalus, I. C., Tyson, C. A., and Lipscomb, J. D. (1972). *In* "Second International Symposia on Oxidases and Related Redox Systems" (T. E. King, H. S. Mason, and M. Morrison, eds.), p. 583. Univ. Park Press, Baltimore, Maryland.

Harding, B. W., Wong, S. H., and Nelson, D. H. (1964). *Biochim. Biophys. Acta* **92,** 415.

Huang, J. J., and Kimura, T. (1971). *Biochem. Biophys. Res. Commun.* **44,** 1065.

Ichii, S., Omata, S., and Koboyashi, S. (1967). *Biochim. Biophys. Acta* **139,** 308.

Ishimura, Y., Ullrich, V., and Peterson, J. A. (1971). *Biochem. Biophys. Res. Commun.* **42,** 140.

Kimura, T. (1968). *In* "Structure and Bonding" (C. K. Jorgensen, J. B. Neilands, R. S. Nyholm, D. Reiner and R. J. P. Williams, eds.), Vol. 5, p. 1. Springer-Verlag, Berlin.

Kimura, T., and Ohno, H. (1968). *J. Biochem. (Japan)* **63,** 716.

Kimura, T., and Suzuki, K. (1965). *Biochem. Biophys. Res. Commun.* **20,** 373.

Kimura, T., and Suzuki, K. (1967). *J. Biol. Chem.* **242,** 485.

Kimura, T., Tasaki, A., and Watari, H. (1970). *J. Biol. Chem.* **245,** 4450.

Kowal, J., and Fiedler, R. P. (1968). *Arch. Biochem. Biophys.* **128,** 406.

Kowal, J., Simpson, E. R., and Estabrook, R. W. (1970). *J. Biol. Chem.* **245,** 2438.

Masters, B. S. S., Baron, J., Taylor, W. E., Isaacson, I. L., and LoSpalluto, J. (1971). *J. Biol. Chem.* **246,** 4143.

McCarthy, J. L., and Peron, F. G. (1967). *Biochemistry* **6,** 25.

Mitani, F., and Horie, S. (1969). *J. Biochem. (Japan)* **66,** 139.

Nakamura, Y., Otsuka, H., and Tamaoki, B. (1966). *Biochim. Biophys. Acta* **122,** 34.

Narasimhulu, S., Cooper, D. Y., and Rosenthal, O. (1965). *Life Sci.* **4,** 2101.

Omura, T., and Sato, R. (1964). *J. Biol. Chem.* **239,** 2370.

Omura, T., Sanders, E., Cooper, D. Y., Rosenthal, O., and Estabrook, R. W. (1965a). *In* "Non-heme Iron Proteins: Role In Energy Conservation" (A. San Pietro, ed.), p. 401. Antioch Press, Yellow Springs, Ohio.

Omura, T., Sato, R., Cooper, D. Y., Rosenthal, O., and Estabrook, R. W. (1965b). *Fed. Proc.* **24,** 1181.

Omura, T., Sanders, E., Estabrook, R. W., Cooper, D. Y., and Rosenthal, O. (1966). *Arch. Biochem. Biophys.* **117,** 660.

Omura, T., Sanders, E., Cooper, D. Y., and Estabrook, R. W. (1967). *In* "Methods in Enzymology" (R. W. Estabrook and M. Pullman, eds.), Vol. 10, Oxidation and Phosphorylation, p. 362. Academic Press, New York.

Orme-Johnson, W. H., and Beinert, H. (1969). *J. Biol. Chem.* **244,** 6143.

Orme-Johnson, W. H., and Beinert, H. (1969a). *Anal Biochem.* **32,** 425.
Orme-Johnson, W. H., Hansen, R. E., Beinert, H., Tsibris, J. C. M., Bartholomaus, R. C., and Gunsalus, I. C. (1968). *Proc. Nat. Acad. Sci. U.S.* **60,** 368.
Palmer, G. (1967). *Biochem. Biophys. Res. Commun.* **27,** 315.
Palmer, G., Brintzinger, H., and Estabrook, R. W. (1967). *Biochemistry* **6,** 1658.
Peterson, J. A., and Ishimura, Y. (1971). *Chem. Biol. Interactions* **3,** 300.
Peterson, J. A., Ishimura, Y., and Griffin, B. W. (1972a). *Arch. Biochem. Biophys.* **149,** 197.
Peterson, J. A., Ishimura, Y., Baron, J., and Estabrook, R. W. (1972b). *In* "Second International Symposia on Oxidases and Related Redox Systems" (T. E. King, H. S. Mason, and M. Morrison, eds.), p. 565. Univ. Park Press, Baltimore, Maryland.
Purvis, J. L., Mason, J. I., Canick, J. A., McCarthy, J. L., and Estabrook, R. W. (1972a). *Fed. Proc.* **31,** 294.
Purvis, J. L., Canick, J. A., Mason, J. I., McCarthy, J. L., and Estabrook, R. W. (1972b). *Int. Congr. Endocrinol., 4th, Washington.*
Schleyer, H., Cooper, D. Y., and Rosenthal, O. (1972). *In* "Second International Symposia on Oxidases and Related Redox Systems" (T. E. King, H. S. Mason, and M. Morrison, eds.), p. 469. Univ. Park Press, Baltimore, Maryland.
Sharma, D. C., Forchielli, E., and Dorfman, R. I. (1962). *J. Biol. Chem.* **237,** 1495.
Shethna, Y. I., Wilson, P. W., Hansen, R. E., and Beinert, H. (1964). *Proc. Nat. Acad. Sci. U.S.* **52,** 1263.
Simpson, E. R., and Boyd, G. S. (1967). *Biochem. Biophys. Res. Commun.* **28,** 945.
Simpson, E. R., Cooper, D. Y., and Estabrook, R. W. (1969). *Recent Progr. Hormone Res.* **25,** 523.
Sweat, M. L. (1962). *Fed. Proc.* **21,** 189.
Tagawa, K., and Arnon, D. I. (1962). *Nature (London)* **195,** 537.
Tompkins, G. (1959). *Proc. Int. Congr. Biochem., 4th, Vienna* **13,** 153.
Tsibris, J. C. M., and Woody, R. W. (1970). *Coordination Chem. Rev.* **5,** 417.
Wilson, D. F. (1967). *Arch. Biochem. Biophys.* **122,** 254.
Young, D. G., and Hall, P. F. (1969). *Biochemistry* **7,** 2987.

CHAPTER 9

Iron–Sulfur Flavoprotein Dehydrogenases

THOMAS P. SINGER, M. GUTMAN, and VINCENT MASSEY

I. INTRODUCTION

Some 14 years have passed since two of the present authors reviewed the chemistry and catalytic involvement of nonheme iron in metal–flavoproteins (Singer and Massey, 1957). In the intervening years knowledge in this field has been accruing at an ever-increasing pace, so that a comparative review of the subject appears to be both necessary and timely, although it is recognized that by the time these articles reach print, some of the information presented may be outdated.

Major impetus for the surge of experimental work on complex iron–sulfur proteins was provided by two important discoveries: that of the EPR signal given by the nonheme iron upon reduction (Beinert and Sands, 1960; Sands and Beinert, 1960) and the fact that "labile" sulfide is structurally and functionally interrelated with the nonheme iron in this group of enzymes (Massey, 1957; Fry and San Pietro, 1962; Gewitz and Volker, 1962). The former, in conjunction with Bray's rapid freezing technique (Bray, 1961), provided a powerful tool for the study of the electron transport functions of nonheme iron, while the latter laid the theoretical basis for the eventual understanding of the environment of nonheme iron in proteins of this class.

This and the following chapter provide a survey of the so-called "complex iron–sulfur proteins." The definition of this class of enzymes is somewhat unsatisfactory. The term was originally devised to distinguish enzymes, such as xanthine oxidase and succinate dehydrogenase, from low molecular weight proteins containing functional iron–sulfur moieties, e.g., the ferredoxins, which were thought to be characterized by low molecular weight, low potential, similar catalytic properties, and amino acid sequence (Buchanan and Arnon, 1970). However, it has become clear that ferredoxins cover a wide range of potential and molecular weight (Buchanan and Arnon, 1970) and a number of low molecular weight iron–sulfur proteins which are not ferredoxins have been discovered. There remain the distinctions that all known "complex iron–sulfur proteins" have molecular weight in excess of 100,000 in the native state and that they contain additional prosthetic groups (flavin or Mo or both) besides iron and sulfur.

This and the next chapter deal with five enzymes in this class: NADH, dihydroorotate, and succinate dehydrogenases, and aldehyde and xanthine oxidases, since these are the only ones for which extensive information is available concerning the content, reactivity, and functions of the nonheme iron constitutents. Although primary emphasis of this volume is placed on the chemistry and role of the nonheme iron centers, in the case of the complex iron–sulfur proteins the interaction of these centers with other redox components in a given enzyme has been of major interest to investigators. Accordingly, these reviews include detailed consideration of such interactions as well as background material on the properties of the individual enzymes necessary for the understanding of their functioning.

For ease of presentation, this topic has been divided into two chapters. The first one deals with the iron–sulfur-containing dehydrogenases (NADH, succinate, and dihydroorotate dehydrogenases), the second with the hydroxylases (xanthine and aldehyde oxidases).

II. SUCCINATE DEHYDROGENASE

A. Biological Functions

Succinate dehydrogenase plays very important roles in the metabolism of all living cells, but its functions are different in aerobic and anaerobic cells (Singer, 1971). In typical aerobic cells, succinate dehydrogenase is a membrane-bound enzyme which is part of the respiratory chain, and its function is the oxidation of succinate to fumarate during the operation of the Krebs cycle, linking this event to the energy-conservation system. In aerobic cells, the kinetic properties of the enzyme are admirably suited to this purpose. In obligate anaerobic cells, in contrast, fumarate usually acts as a terminal oxidant, replacing O_2, and thus the function of succinate dehydrogenase becomes the reduction of fumarate to succinate, thereby providing a means of reoxidizing the reduced pyridine nucleotides generated in anaerobic fermentations. Succinate, either as such, or after conversion to short-chain fatty acids, is excreted. The properties of succinate dehydrogenase (or fumarate reductase, as the "anaerobic" enzyme is often called) in anaerobic cells are excellently adapted to catalysis of the reductive reaction.

Certain facultative anaerobes, such as *E. coli* (Hirsch *et al.*, 1963) and *S. cerevisiae* (Hauber and Singer, 1967; Tisdale *et al.*, 1968) are capable of synthesizing both the membrane-bound aerobic enzyme and the cytoplasmic anaerobic one; the two forms of succinate dehydrogenase are under separate genetic control. In such instances, fermentative growth conditions repress formation of the aerobic enzyme, and high O_2 tension represses synthesis of the anaerobic one. The biological function of the anaerobic enzyme in certain organisms, such as *E. coli* and yeast, is not known, since in these cells fumarate is not thought to function as an oxidant of NADH and thus the electron donor of the enzyme is not known (Singer, 1971).

B. Aerobic Succinate Dehydrogenases

1. Molecular Properties

Although succinate dehydrogenase has been isolated in highly purified form from animal tissues as well as from aerobic yeast and the properties of the enzymes in these two sources have been shown to be substantially identical (Singer, 1971), the mammalian enzyme has been more extensively investigated; for this reason the present section discusses only the properties of the mammalian enzyme.

It is uncommonly difficult to prepare succinate dehydrogenase in homogeneous form and to maintain purified preparations in the active state. One reason is the very tight linkage to the respiratory chain, which requires special treatments, such as exposure to organic solvents and alkaline pH or concentrated perchlorate, for extraction of the enzyme. The second reason is that, notwithstanding occasional reports to the contrary (Baginsky and Hatefi, 1968), the experience of most workers is that soluble preparations are very unstable, so that preservation even for a short period requires very low temperatures, anaerobiosis, and the presence of protective agents. The instability of the soluble enzyme may well be the consequence of modification during the extraction procedure. The fact that, despite the extraordinary amount of experimental work devoted to this enzyme during the past two decades, there is little agreement even on such fundamental properties as its molecular weight or turnover number is probably the result of these difficulties and of the fact that there does not appear to be any completely satisfactory method available as yet for isolation of the pure enzyme.

Available isolation procedures fall into one of four categories according to the method of extraction. The first preparation reported to be homogeneous was isolated from alkaline extracts of acetone powders of beef heart mitochondria (Singer and Kearney, 1954; Singer *et al.*, 1956a). The method was later revised (Bernath and Singer, 1962), and more recently an acetone powder of ETP has been used as the source material (Coles *et al.*, 1972). The second method utilizes extraction of cyanide-treated membrane preparations with aqueous *n*-butanol at alkaline pH (Wang *et al.*, 1956) but differs little from the first reported preparations in subsequent steps. When it was later discovered (Keilin and King, 1960) that neither preparation is active in reconstitution tests (cf. below) because of the absence of succinate during extraction and because of the cyanide treatment in the second procedure, a variant of the latter, omitting exposure to cyanide and including succinate during extraction, was devised (King, 1963), followed by a number of refinements of the butanol method (Veeger *et al.*, 1963; Hucko and King, 1964; Bruni and Racker, 1968). It should be noted that both the acetone and the butanol methods yield "reconstitutionally active" preparations if the precautions mentioned are observed (King, 1962). The reconstitution test referred to here involves combining a soluble preparation of the dehydrogenase with membrane fragments containing the cytochrome system in which succinate dehydrogenase was inactivated by alkali, and thereby reestablishing electron transport from succinate, via the respiratory chain, to O_2 (Keilin and King, 1960).

The third method utilizes extraction with cyanide (Wu and King,

1967), the rationale being that the enzyme is thought to be joined to the respiratory chain by way of nonheme irons which react with cyanide (Giuditta and Singer, 1958). The main limitation of this procedure is the growing evidence (Giuditta and Singer, 1958; Rossi *et al.*, 1970; Cerletti *et al.*, 1971) that cyanide treatment modifies the dehydrogenase. The fourth and most recent method involves extraction of a purified, particulate form of the enzyme (complex II) with concentrated perchlorate (Davis and Hatefi, 1971). The method is rapid and simple, although the isolation of complex II is not, and yields highly active preparations, which appear to be, however, modified as far as regulatory properties are concerned (cf. below).

a. Molecular Weight and Composition. On the basis of sedimentation–diffusion analysis, the molecular weight appears to be 150,000 (Singer *et al.*, 1956a), while bound flavin and iron content, after correction for impurities from electrophoretic and ultracentrifugal analysis, gave a provisional molecular weight of 200,000 (Singer *et al.*, 1956a; Kearney, 1960). A value of 140,000 to 160,000 was calculated for the butanol preparation from flavin and iron analyses (Wang *et al.*, 1956), but the purity of the preparations used does not appear to have been determined by physical methods. More recently, Davis and Hatefi (1971) calculated a molecular weight of 110,000 from chromatography on Agarose and bound flavin content and a somewhat lower value (97,000) from subunit analysis on acrylamide gels in SDS for the perchlorate-extracted preparations.

Values in agreement with this lower molecular weight have also been reported for the butanol-extracted enzyme by Cerletti and co-workers (Righetti and Cerletti, 1971; Cerletti *et al.*, 1972) (110,000 MW by density gradient ultracentrifugation and 99,000 MW by subunit analysis on polyacrylamide SDS). Despite this agreement, the results are not unambiguous. Thus the preparation used by these workers showed at least two catalytically active components on DEAE–Sephadex and in density gradients, confirming the earlier report of Kimura and Hauber (1963) that "reconstitutionally active" butanol preparations of succinate dehydrogenase are resolved into several molecular forms on Sephadex. Second, the position of the enzyme in density gradients does not agree with the bound flavin content, which indicates molecular weights of about 180,000 and 500,000 for the two enzymic components detected. Molecular heterogeneity on Sephadex G-200 and in density gradients has also been found in this laboratory (Kimura and Hauber, 1963).

Recent studies (Coles *et al.*, 1972) resolved these discrepancies and provided a firm value for the molecular weight of the enzyme. The enzyme extracted from acetone powders and the perchlorate-extracted preparation were shown to have identical molecular weights. Chemical determination of the minimum molecular weight (covalently bound flavin content, summation of the weights of subunits) gave a value of 100,000 for both preparations, while physical measurements (gradient centrifugation, gel chromatography) at the protein concentrations usually used in such studies (1 to 10 mg/ml) indicated a weight average molecular weight of 175,000 ± 10,000. The enzyme was shown to exist in solution as a self-associating system governed by temperature and protein concentration, with a $K_d = 1.2 \times 10^{-6}$ *M* at 5°C, pH 7.5. Thus at concentrations above 1 mg/ml it exists predominantly as a dimer.

The composition of the enzyme is usually expressed as molar ratios of bound flavin:iron:sulfur. The first apparently homogeneous preparations, isolated by the acetone or butanol methods, were reported to contain four irons per flavin (Singer *et al.*, 1956a; Wang *et al.*, 1956). Preparations obtained by butanol extraction in the presence of succinate were found to contain eight irons per flavin (King, 1964). Complex II (Lusty *et al.*, 1965; Davis and Hatefi, 1971) as well as "reconstitutionally active" butanol preparations (King, 1964) has been reported to contain labile S in approximately equimolar proportions to iron, but one report claims a flavin:iron:sulfur ratio of 1:8:4 for the latter (Zeijlemaker *et al.*, 1965).

Treatment of succinate dehydrogenase (perchlorate or butanol preparations) with SDS, perchlorate, urea, guanidine, or thiocyanate results in dissociation into subunits with complete loss of catalytic activity (Davis and Hatefi, 1971; Righetti and Cereletti, 1971). Two subunits have been detected, one with a molecular weight of 70,000, containing flavin:iron:sulfur in the approximate ratio 1:4:4, and a so-called "iron–protein" (molecular weight, 30,000 by gel electrophoresis) containing traces of bound flavin but equimolar amounts of iron and labile sulfide. This subunit ratio is readily ascertained in particulate preparations, but all soluble preparations yield a subunit ratio significantly in excess of 1:1 in favor of the 30,000 MW subunit. The reason for this may be that on extraction some dissociation into subunits and differential loss of the more insoluble 70,000 MW subunit occurs. The excess 30,000 MW subunit may then recombine with the native enzyme (Coles *et al.*, 1972).

b. Covalently Bound Flavin and Peptide Sequence at the Active Center. The fact that the FAD prosthetic group of the enzyme is covalently linked to the peptide chain has been known since the mid-1950's (Kearney and Singer, 1955; Singer *et al.*, 1956a). Kearney isolated a pure flavin peptide and showed that the attachment of the flavin to the protein

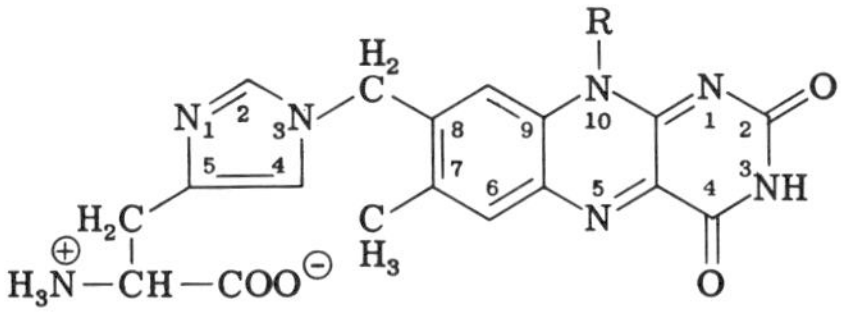

Fig. 1. Structure of histidyl-8α-FAD.

is through the isoalloxazine ring system, but not at the 10 position (Kearney, 1960). On the basis of ESR, ENDOR, and absorption spectra of the peptide, the point of attachment was identified as the 8α-CH_3 group of riboflavin (Hemmerich *et al.*, 1969; Walker *et al.*, 1969; Singer *et al.*, 1971a; Salach *et al.*, 1972) and shortly thereafter the complete structure was determined to be 8α-histidyl FAD (Walker and Singer, 1970) (Fig. 1) and the peptide sequence at the active center shown to be

Ser-His-Thr-Val-Ala
|
Flavin

More recent work has resulted in the determination of a sequence of some 30 amino acids around the flavin (Kenney *et al.*, 1972).

Attachment of the imidazole N to the 8α-CH_2 group results in an unusual pH–fluorescence profile: maximum fluorescence is at pH 3.2 and above this pH the fluorescence is quenched almost completely with a pK = 4.6 (Kearney, 1960). This property has been the basis of a highly sensitive and specific method for chemically determining the concentration of the enzyme regardless of physical form, purity, or activity (Singer *et al.*, 1971b).

c. Iron–Sulfur Components. Starting with Beinert and Sands (1960; Sands and Beinert, 1960), many laboratories (King *et al.*, 1961; Beinert, 1965, 1966; Griffin *et al.*, 1967; Van Voorst *et al.*, 1967; Dervartanian *et al.*, 1969; Baginsky and Hatefi, 1969) have studied the EPR signals ascribed to the iron–sulfur components of succinate dehydrogenase. This component has been detected after reduction with substrate, dithionite, or other artificial reductants in membrane-bound preparations as well as highly purified soluble ones, isolated by various procedures.

Besides the free radical signal at $g = 2.003$, the reduced enzyme shows an asymmetric signal at $g = 1.94$. Some preparations but not others also show an additional signal at $g = 4.3$ in the oxidized state, which is slowly reduced by substrate and is typical of Fe^{3+} in the high-spin state, but oxidation–reduction of the iron represented by this signal is not considered to be of significance in the catalytic mechanism (Dervartanian *et al.*, 1969).

Available information on the kinetics of the oxidation–reduction of the iron–sulfur species represented by the $g = 1.94$ signal is summarized in the section dealing with mechanism of action. At this point it should be emphasized, however, that reductive titrations indicate that at most one unpaired electron per mole of bound flavin is accounted for by the $g = 1.94$ signal detected at 77°K at maximal development, and thus at the most one pair of iron–sulfur moieties of the total of eight appears to be responsible for this signal (Devartanian *et al.*, 1969).

The opinion has been frequently voiced (e.g., King, 1967a) that fresh preparations containing 8 gm-atoms each or iron and sulfur per mole of flavin are active in reconstitution tests, whereas those containing 4 are not. The situation is not quite as simple as this formula would suggest, since a "reconstitutionally active" preparation was reported to contain 4, not 8, moles of labile sulfide (Zeijlemaker *et al.*, 1965), and the enzyme extracted from complex II by freeze thawing at alkaline pH (Baginsky and Hatefi, 1969) is inactive in this test, although it contains eight iron and sulfur, but, interestingly, the reconstitution activity is regained on incubation with ferrous ammonium sulfate, sodium sulfide, and mercapto-ethanol in the manner previously used to reconstitute ferredoxin from apoferredoxin (Malkin and Rabinowitz, 1966). The composition of such reactivated preparations is not known. Thus the interesting question remains open whether such reconstituted preparations are double-headed, containing both the catalytically inactive iron–sulfur centers previously

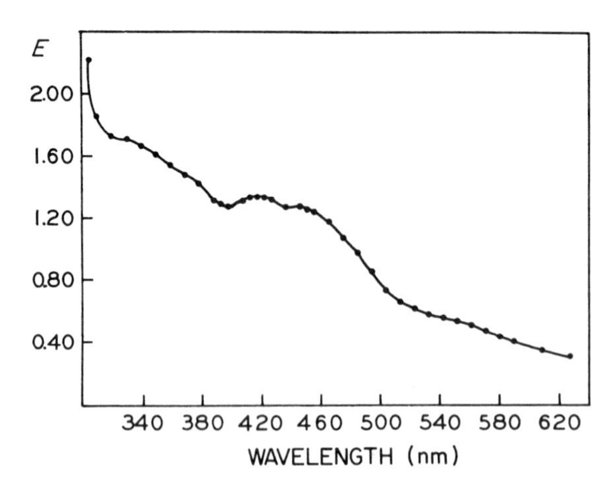

Fig. 2. Absorption spectrum of succinate dehydrogenase from beef heart. From Singer *et al.* (1956a).

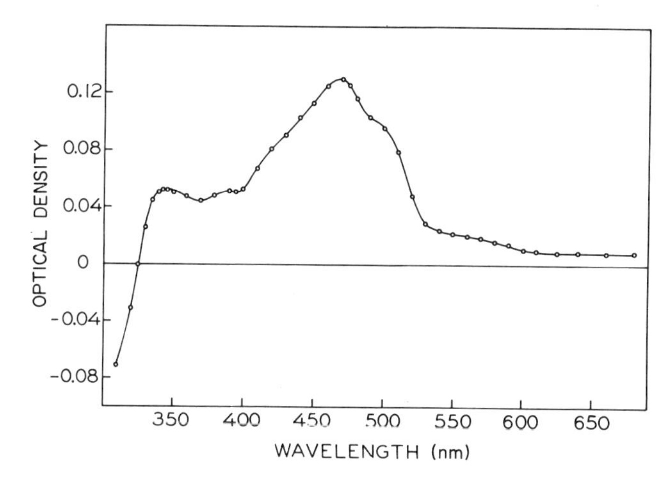

Fig. 3. Difference spectrum resulting from reduction of fully activated enzyme by succinate. Positive values denote bleaching. From Kearney (1957).

present, as well as the newly acquired active ones, in the manner seen in succinoxidase preparations reconstituted from alkali-treated particles and purified succinate dehydrogenase, which contain twice the amount of dehydrogenase as untreated ones (Kimura *et al.*, 1963).

The absorption spectrum of the oxidized enzyme (Fig. 2), showing a broad absorption in the entire visible range, with peaks at 420 and 460 nm and a shoulder at 550 nm, as well as the difference spectrum obtained on reduction with substrate (Fig. 3), indicated many years ago (Singer *et al.*, 1956a) that much of the color is due to iron–protein linkages, rather than to flavin. These early studies also provided evidence for the presence of labile sulfide in the enzyme, for when purified preparations were denatured, the odor of H_2S was detected. This observation of Massey (1957), in fact, led to the discovery of iron–sulfur linkages in proteins. Further evidence that iron–sulfur centers are responsible for the 420 nm peak and most of the absorbance above 500 nm came from treatment of the dehydrogenase with *p*-mercuriphenylsulfonate (Fig. 4), which caused considerable bleaching and the appearance of a recognizable flavoprotein spectrum (Singer and Massey, 1957; Massey, 1958). As in the case of ferredoxin, the mercurial appears to detach the iron–sulfur moieties. The same type of experiment has been performed with preparations isolated from complex II, leading to the conclusion that the absorbance bleached by the mercurial corresponds, per gm-atom of iron, to the same extinction coefficient as in ferredoxins (Baginsky and Hatefi, 1969; Davis and Hatefi, 1971). It has not been possible, however, to pre-

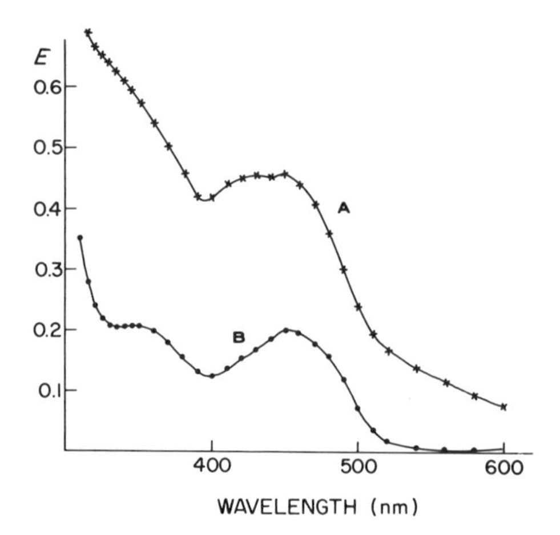

Fig. 4. Effect of *p*-mercuriphenylsulfonate on the spectrum of succinate dehydrogenase. A, enzyme alone; B, same after addition of 2.06 μmoles of PCMS/mg of enzyme. From Massey (1958).

pare an iron–sulfur-free flavoprotein in this manner from which the catalytically active holoenzyme may be reconstituted.

d. Action of —SH Reagents. The action of —SH reagents on the enzyme is varied and complex. Since most of them produce conformational alterations in the enzyme, often involving the iron–sulfur centers, it seems appropriate to discuss them under molecular properties.

The presence of —SH groups in succinate dehydrogenase which are essential for catalytic activity was discovered in 1938 (Hopkins and Morgan, 1938; Hopkins *et al.*, 1938), and it was shown at that time that substrates and competitive inhibitors protect the dehydrogenase from —SH reagents. This observation led to the hypothesis that the substrate binding site either involves or is in the immediate vicinity of certain thiol groups. Since that time the dehydrogenase has been a favorite test material for studies on —SH reagents.

Although inhibition by oxidizing agents, arsenoxides, and mercurials is reversible if the treatment is brief and the inhibitor concentration low, it is difficult to separate reversible effects from irreversible structural alterations, particularly with purified, soluble preparations. Thus, when *N*-ethylmalemide (NEM) or bromopyruvate (Sanborn *et al.*, 1971) were tested as potential active-site-directed irreversible inhibitors in order to label the substrate binding site with ^{14}C-marked inhibitors and subsequently determine its sequence, it was found (Kenney and Singer, 1971), in accord with studies in Hollocher's laboratory (Sanborn *et al.,* 1971), that succinate and malonate only delayed inactivation of the enzyme moderately but did not prevent it, provided that the enzyme is used as isolated (Bernath and Singer, 1962). This is because this form of the enzyme contains tightly bound oxalacetate (cf. below), which protects the —SH groups at the substrate site. If this oxalacetate is removed with Br^- (cf. below), the enzyme is readily inactivated by NEM, and malonate protects against this inactivation and prevents combination at the substrate site. Thus NEM and bromopyruvate do act as substrate site-directed inhibitors under appropriate conditions and —SH groups do play a role in substrate binding (Kenney, 1973). According to Sanborn *et al.,* (1971), NEM and bromopyruvate act differently from mercurials, since they do not attack the iron–sulfur centers, as judged from the minor spectral alterations they produce and other evidence (Felberg and Hollocher, 1972).

One of the unexpected changes that alkylating —SH reagents produce in the highly purified enzyme is illustrated in Fig. 5. In this experiment the perchlorate-extracted preparation of Davis and Hatefi (1971) was used which appears homogeneous in chromatography on Sephadex G-200 (Fig. 5, left panel), and moves just ahead of lactate dehydrogenase (MW = 142,000). After treatment with 0.3 mM [^{14}C]NEM in the presence of succinate, the elution profile changes remarkably: The band corresponding to the untreated enzyme broadens and develops an extra peak on the leading edge, and a substantial part of the protein is now excluded from G-200, as judged by either ^{14}C counts or absorbance at 280 nm. The same behavior is observed with [^{14}C]iodoacetamide. In general, the higher the concentration of the inhibitor and the longer the treatment, the more prominent these changes become (Kenney and Singer, 1971). This suggests that alkylation of —SH groups initiates polymerization or aggregation of the protein.

Indications that —SH inhibitors cause structural changes in succinate dehydrogenase under appropriate conditions were obtained many years ago in Massey's studies (Singer and Massey, 1957; Massey, 1958). He reported that the extremely slow reaction of the iron components of four-iron preparations with iron chelating agents (*o*-phenanthroline, α,α'-bipyridyl, Tiron) is greatly accelerated by prior treatment with organic mercurials. The effect of the mercurial was interpreted as a conformation change, since (1) it mimics the action of urea in this regard, (2) prolonged contact with mercurial resulted in irreversible inactivation, (3) changes in the absorption spectrum and fluorescence accompanied the inactivation, and (4) ultimately nonheme iron was split off in dialyzable form.

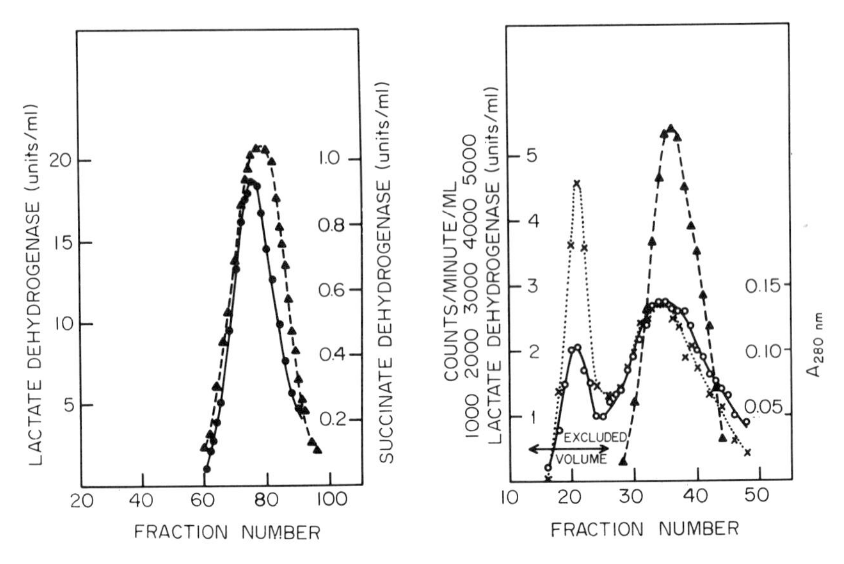

Fig. 5. Effect of NEM treatment on chromatographic behavior of succinate dehydrogenase on Sephadex G-200. Left side, SDB preparation of Davis and Hatefi, as isolated (1.44 mg) plus 1 mg of pig heart lactate dehydrogenase in 1 ml volume were chromatographed on a 1.1 × 55 cm Sephadex G-200 column at 0°C under N_2, equilibrated and eluted with 0.1 *M* NaCl—0.05 *M* tris—0.02% (w/v) Na azide, pH 7.4. Fraction size, 0.46 ml. Right side, chromatographic conditions as above except that a 1.1 × 23 cm column was used. The solution chromatographed contained 0.125 mg lactate dehydrogenase plus succinate dehydrogenase [SDB (Davis and Hatefi, 1971)] pretreated as follows. Perchlorate and ammonium sulfate were removed on Sephadex G-25 (equilibrated with 50 m*M* tris-20 m*M* succinate, pH 7.4) at 0°C under N_2. The sample (2.4 mg of protein/ml) was then incubated with 0.3 m*M* [^{14}C]NEM (1.7 mCi/mmole) for 45 minutes at 0°C, then for 43 minutes at 23°C under N_2, resulting in 50% loss of activity. Mercaptoethanol (0.5 μl) was added to react with excess NEM; the solution was passed through Sephadex G-25 (equilibrated with 0.05 *M* tris, pH 7.4) and applied to the Sephadex G-200 column. Symbols: ●, succinate dehydrogenase activity as 15°C in arbitrary units; ▲, lactate dehydrogenase activity at 25°C; ○, absorbance at 280 nm; ×, radioactivity in counts/minute/ml.

e. Action of Iron Complexing Agents. As discussed above, the iron components of the enzyme react only sluggishly with most iron chelators, unless the enzyme is previously treated with mercurials, urea, or hydrosulfite, all of which appear to produce conformation changes which render the iron components reactive toward these chelators. Thenoyltrifluoroacetone (TTF) appears to be an exception: Although it has been reported to be without effect on soluble (four- or eight-iron type) preparations (Ziegler, 1961; Takemori and King, 1964a,b; King, 1967a; Cerletti *et al.*, 1967; Baginsky and Hatefi, 1969), in the particle bound form electron transport to and from the respiratory chain is completely inhibited at relatively low concentrations. Baginsky and Hatefi (1969) state that the reduction of phenazine methosulfate is unaffected by TTF even in particulate preparations and from this conclude that this dye reacts with the flavin, rather than the iron–sulfur components, a view not shared by other investigators (Massey and Singer, 1957; Giuditta and Singer, 1958). As seen in Fig. 6, TTF acts as a noncompetitive inhibitor in the phenazine assay in untreated membrane preparations; the inhibition becomes competitive with respect to phenazine methosulfate in CoQ-depleted samples, so that it is not detectable at the high fixed dye concentrations used in the studies of Baginsky and Hatefi (1969), and the original inhibition characteristics are stored when the CoQ content is restored (Rossi *et al.*, 1970). Recently Nelson *et al.*, (1971) reported another unusual effect of TTF on the iron–sulfur components of succinate dehydrogenase. Addition of TTF to CoQ-depleted submitochondrial particles in which cytochrome b had been reduced by treatment with succinate plus KCN (and/or antimycin A), caused rapid and complete reoxidation of cytochrome b. These authors suggest that TTF activates an electron sink (probably nonheme iron) by conformational modification of a nonheme iron component of the succinate dehydrogenase–cytochrome b system.

Another iron complexing agent which behaves similarly is cyanide. It was noted by Tsou many years ago (Tsou, 1951) that succinate dehydrogenase activity (as measured by methylene blue reduction) is abolished on prolonged incubation of particulate preparations with cyanide. An ex-

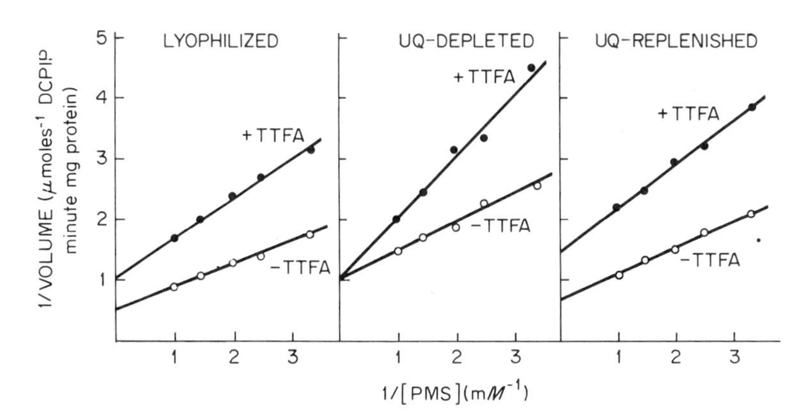

Fig. 6. Effect of thenoyltrifluoroacetone on the succinate dehydrogenase activity of lyophilized, UQ-depleted and UQ-replenished particles. Where indicated, 100 μM thenoyltrifluoroacetone (TTFA) was added to the cuvette, before the addition of particles. The final concentration of lyophilized, UQ-depleted, and UQ-replenished particles was 15.7, 16.7, and 16.7 μg protein/ml, respectively. From Rossi *et al.* (1970).

tensive study of the phenomenon (Guiditta and Singer, 1958) brought to light many interesting characteristics and led to the suggestion that inactivation by cyanide is the result of conformation changes initiated by combination of cyanide with the iron components.

Cyanide does not inhibit the activities of the soluble enzyme (Giuditta and Singer, 1958), although the enzyme is reported to bind [^{14}C]cyanide (Lee and King, 1962). In particles the inactivation progresses as a first order reaction, the rate constant being the same regardless of which activity is being followed, and depends only on temperature and cyanide concentration, but the final extent of inactivation depends on the assay system used (Guiditta and Singer, 1958) (Fig. 7). While reactivity with the respiratory chain or methylene blue is completely lost, only some 50–70% of succinate–phenazine methosulfate or succinate–ferricyanide activities disappear, while the fumarate–$FMNH_2$ reaction (at V_{max} with respect to $FMNH_2$) is unaffected. Accompanying these events there is a major increase in the K_m for $FMNH_2$ and phenazine methosulfate. The rate of inactivation increases considerably with pH in the range 7–9. Reducing agents (succinate, NADH, dithionite) prevent the inactivation and reverse the effects on K_m for electron carriers but not the inactivation at infinite dye concentration (Giuditta and Singer, 1958). It is interesting that in CoQ-depleted preparations little inactivation is noted at V_{max}^{PMS} but the change in K_m is still apparent, and on restoration of the CoQ content the preparation becomes again inactivable by cyanide (Rossi *et al.*, 1970), similarly to the behavior with respect to TTF (Fig. 8).

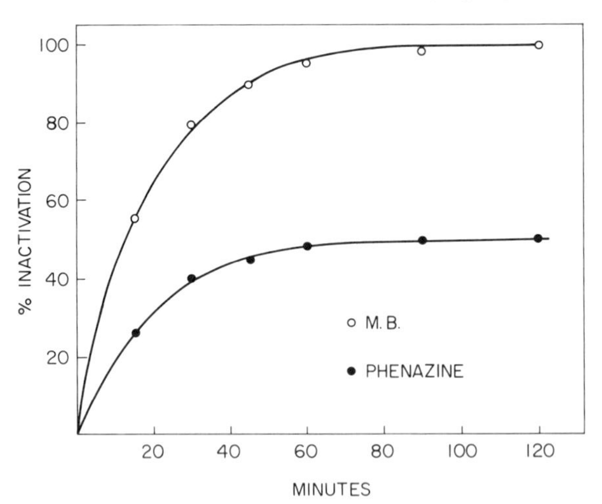

Fig. 7. Time course of the inactivation by 0.02 *M* cyanide at pH 7.85, 30°C. Activity determinations were based on V_{max} values with both dyes at 38°C. M.B., methylene blue. From Giuditta and Singer (1958).

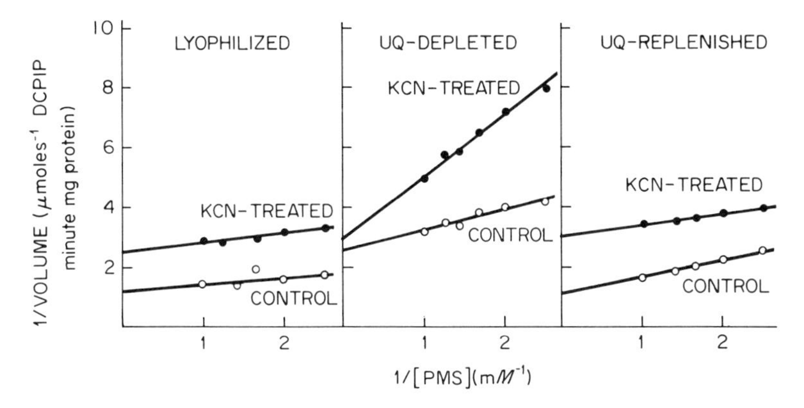

Fig. 8. Effect of pretreatment with cyanide on the succinate dehydrogenase activity of lyophilized, UQ-depleted and UQ-replenished particles. The particles were incubated at 30°C in the presence of 167 m*M* sucrose, 50 m*M* tris-acetate, and 20 m*M* KCN, final pH = 8.5. After 30 minutes, 20 m*M* succinate was added and the incubation carried on for 30 minutes more. In the control sample succinate was added before KCN, thus preventing the inactivating effect of the latter. The final concentrations of lyophilized, UQ-depleted and UQ-containing particles were 16.8, 40.5, and 17.8 μg protein/ml, respectively. From Rossi *et al.* (1970).

Giuditta and Singer (1958) interpreted these effects as representing complex conformation changes consequent to combination of the cyanide with some of the nonheme iron. On the basis of the kinetics of the reversible and the irreversible effects of cyanide treatment and the high energy of activation of the process, they postulated that combination of cyanide with a reactive site in the protein is followed by a series of intramolecular transformations, as represented in Eqs. (1) and (2). In Eq. (1) the reversible reaction of succinate dehydrogenase (S.D.) with cyanide (which may be rapid) results in the formation of the complex succinate dehydrogenase—CN

$$\text{S.D.} + \text{CN}^- \xrightleftharpoons{\text{fast (?)}} \text{S.D.—CN} \xrightleftharpoons{\text{slow}} \text{S.D.—CN}^* \rightarrow \text{S.D.—CN}^{**} \quad (1)$$

$$\text{S.D.} + \text{CN}^- \xrightleftharpoons{\text{slow}} \text{S.D.—CN}^* \rightarrow \text{S.D.—CN}^{**} \quad (2)$$

retaining the original activity and affinity for dyes. Secondary changes in protein structure lead to the slow formation of S.D.—CN*, which is still active but has a lowered affinity for dyes, and, eventually, by an irreversible step to succinate dehydrogenase—CN**, an inactive form. The second scheme [Eq. (2)] differs from the first one in that a slow reversible combination of the enzyme with cyanide is visualized as leading directly to S.D.—CN*, the form which is still active but has lost its high affinity for dyes. In either scheme the enzyme–cyanide complex may have a high dissociation constant, and high concentrations of cyanide may be needed to drive the slow reaction.

A corollary of these studies is the speculation that extraction of the dehydrogenase in soluble form or treatment of particles with cyanide may render half of the nonheme iron required for the electron transport from succinate to phenazine methosulfate or ferricyanide (or one of the two reaction sites of phenazine methosulfate) nonfunctional (Giuditta and Singer, 1958). This idea was advanced to explain the odd coincidence that (a) cyanide inactivation proceeds to a maximum of a little over 50% loss of succinate–phenazine methosulfate activity, (b) that even the highest activities reported for soluble preparations (Bernath and Singer, 1962; Hanstein *et al.*, 1971) represent only 50–60% of the turnover number of the membrane-bound enzyme in the phenazine assay [18,000 moles succinate/minute/mole of bound flavin at 38°C (Cerletti *et al.*, 1963; Singer, 1966)], although (c) neither extraction nor cyanide treatment changes significantly the turnover number in the fumarate–$FMNH_2$ assay, which is not thought to involve the iron–sulfur centers (Massey and Singer, 1957a). The idea, which lay dormant for many years, was recently revived by Rossi *et al.* (1970) to explain the disappearance of sensitivity to TTF and cyanide on removal of CoQ from the particles (Figs. 6 and 8). They have proposed that extraction of the CoQ modifies the reactivity of the nonheme irons in a manner analogous to extraction of the enzyme. This, in turn, has been the basis of the conclusion that CoQ is not only an electron acceptor but also a modulator of succinate dehydrogenase, as discussed later in this chapter.

f. Reconstitution. Keilin and King (1960) discovered that if membrane fragments from heart muscle are exposed to pH about 9.3 until all or nearly all of the succinate dehydrogenase activity is inactivated, the resulting particles will recombine with soluble, purified succinate dehydrogenase to form a catalytically competent succinoxidase system. As noted above, only very fresh preparations of the enzyme, extracted in the presence of succinate and not exposed to cyanide are functional in this reconstitution test. It was later shown (King, 1961) that on aging of soluble preparations reconstitution activity decays faster than catalytic activity toward phenazine methosulfate, and this finding was used as the basis of the dictum that only the reconstitution test is a reliable criterion of the native or unmodified state of a given preparation (King, 1961). Actually, the observation of differential loss of activities of succinate dehydrogenase on aging was not novel, since many years earlier it had been shown that activity in the succinate–phenazine methosulfate assay decays much faster than in the fumarate–$FMNH_2$ test (Massey and Singer, 1957a). The reconstitution test, by itself, may also lead to misleading conclusions regarding the physiological state of the enzyme in a given preparation, for several reasons. First, it has been shown that "reconstitutively active" preparations may have a lowered turnover number in the phenazine assay or an artificially increased fumarate–$FMNH_2$ activity, which are not detected in the reconstitution test (Kimura *et al.*, 1963). Second, such preparations contain several molecular forms of the dehydrogenase, which undoubtedly represent modifications during isolation, one of which is incapable of combination with alkali-treated particles, another which combines but does not confer catalytic activity, a third which combines and reconstitutes dehydrogenase but not oxidase activity, and a fourth which combines and reconstitutes all activities; the latter appears to be preferentially taken up by the alkali-treated particles (Kimura and Hauber, 1963). Third, it was recently discovered (Singer *et al.*, 1972a; Coles *et al.*, 1973) that perchlorate-extracted preparations of the enzyme (Davis and Hatefi, 1971), although endowed with high catalytic activity and competent in the reconstitution test, have nevertheless altered regulatory properties, so that after activation by succinate and removal of the activator, they do not become spontaneously deactivated. Since the regulatory properties of an enzyme are certainly physiologically relevant, it would seem reasonable to conclude that in considering the extent of modification of a succinate dehydrogenase preparation one should examine several criteria, including turnover number, reconstitution activity, regulatory properties, and molecular heterogeneity.

The mechanism of the reconstitution is viewed by King (1967a) as a dissociation of the dehydrogenase from the respiratory chain during the alkali treatment, resulting in a particle deficient in the flavoprotein, so that when the added soluble enzyme recombines, the original stoichiometry of flavoprotein to cytochrome system is restored. While at highly alkaline pH (at pH 10), the dehydrogenase is indeed dissociated, under these conditions the resulting particle is not capable of taking up soluble succinate dehydrogenase, and at pH 9.3, where the preparation is routinely prepared, little or no succinate dehydrogenase is dissociated (Kimura *et al.*, 1963; Singer, 1966). Hence, on titration with the flavoprotein, when full succinoxidase activity is restored, twice the amount of bound flavin is found in the particles than was originally present (Table I). This finding has been debated by King (1967a) but confirmed in another laboratory (Hanstein *et al.*, 1971).

2. Catalytic Properties

a. Turnover Number. The turnover numbers reported in the literature for the enzyme even in relatively intact preparations (mitochondria,

TABLE I
CATALYTIC ACTIVITIES AND TURNOVER NUMBERS IN ALKALI-TREATED AND REACTIVATED PREPARATIONS[a]

Soluble S.D. added (μmoles/mg Keilin–Hartree preparation)	Bound flavin in resulting complex (nmoles/mg)	Specific activity	
		Succinic dehydrogenase	Succinoxidase
None (untreated)	0.129	1.47	0.897
None (AT)	0.115	0.198	0.041
0.132	0.145	0.556	0.232
0.263	0.175	0.926	0.406
0.395	0.197	1.22	0.582
0.652	0.205	1.39	0.655
0.980	0.211	1.39	0.607
1.31	0.207	1.31	0.625
2.61	0.209	1.45	0.614

[a] Each sample was assayed after isolation of the reaction product of the alkali-treated preparation plus soluble dehydrogenase by repeated washing in the ultracentrifuge. Succinic dehydrogenase activity refers to the spectrophotometric phenazine assay. Specific activity is μmoles/minute/mg protein. AT, alkali-treated Keilin–Hartree preparation; S.D., succinic dehydrogenase. Data from Kimura *et al.* (1963).

inner membrane preparations) cover an appreciable range. Part of the reason for this is that prior to 1963 the limitations of the manometric phenazine assay (Arrigoni and Singer, 1962) were not recognized; other reasons are that some values refer to an arbitrary dye concentration, rather than V_{max}, or the preparation was not fully activated before assay. For the past five to six years, however, two laboratories (Cerletti *et al.*, 1963; Singer, 1966) have reported consistent values of 17,000 to 18,000 moles succinate/minute/mole of covalently bound flavin in the spectrophotometric phenazine assay at V_{max}, 38°C, for beef, rat, and pig heart mitochondria, ETP, ETP_H, and rat brain and aerobic yeast mitochondria. The same laboratories reported lower turnover numbers for liver preparations: this is because the covalently bound flavins of other enzymes (monoamine oxidase, sarcosine, and dimethylglycine dehydrogenases) contribute to the bound flavin content.

Soluble preparations show a wide range of turnover numbers, from about 4000 to 10,000–11,000 under the same assay conditions (Singer *et al.*, 1962; Veeger *et al.*, 1963; Singer, 1966; Baginsky and Hatefi, 1968; Hanstein *et al.*, 1971). All these are considerably lower than the activity measured in heart mitochondria or ETP, which may be related to the possible loss of one of the two reaction sites for phenazine methosulfate on extraction (Giuditta and Singer, 1958). It is important to note in this connection that unless the turnover number is compared with that of mitochondria, one may arrive at the conclusion that no loss of activity was incurred on extraction. Thus in the recent studies of Hanstein *et al.* (1971) both the material extracted (complex II) and the soluble enzyme showed turnover numbers of 10,000, but it has been known for some years that the enzyme suffers some modification during the preparation of complex II, as judged by its lowered turnover number (Singer *et al.*, 1962; Singer, 1966).

b. SPECIFICITY. Besides succinate itself, the enzyme has been reported to oxidize slowly methylsuccinate, ethylsuccinate, and very slowly D- and L-malate (Thunberg, 1933; Franke and Siewerdt, 1944; Gawron *et al.*, 1961; Dervartanian and Veeger, 1965). More recently, Tober *et al.* (1970) reported that DL-monofluorosuccinate is oxidized by the soluble as well as particulate enzyme to monofluorofumarate with a V_{max} of about 45% that of succinate.

The K_m for succinate has been determined in many laboratories but, not surprisingly, considerable variation has been reported, due in large part to the fact that different acceptors were used and lack of recognition that the K_m value for succinate contains kinetic terms dependent on the acceptor. Some of the very low values reported in the early literature may have been due to the electron acceptor being rate limiting, as in assays using succinoxidase or of succinate-cytochrome c reductase activity or a fixed concentration of dye as electron acceptor, or that the enzyme was not fully activated. Data in the literature for K_m (succinate) vary from 0.3 to 1.3 mM at 38°C and from 0.06 to 1.3 mM at 22° to 25°C (Singer, 1965, 1966; King, 1967b; Zeijlemaker *et al.*, 1969, Rossi *et al.*, 1970; Tober *et al.*, 1970). The value consistently obtained by the authors is 1.3 and 0.5 mM at the two respective temperatures at V_{max}^{PMS} (Singer, 1965). Interestingly the K_m at 30°C is reported to decline from 250 to 60 μM on depletion of the CoQ content of membrane preparations and to rise to 278 μM on reincorporating CoQ (Rossi *et al.*, 1970). This apparent dependence of the K_m for succinate on endogenous quinone has been taken as an indication that CoQ may be a modulator of the dehydrogenase.

The rate constants for the partial reactions and K_D, the dissociation constant, were studied by Dervartanian *et al.* (1966) and Zeijlemaker *et al.* (1969) and more recently by Tober *et al.* (1970). The values reported by these authors again cover a considerable range (35 to 250 μM), although the enzyme preparations used were comparable.

The specificity of the enzyme for electron acceptors is rather restricted, N-alkylphenazonium salts and ferricyanide being the only rapidly reacting electron acceptors. In early studies, where reactivity with ferricyanide was measured manometrically in bicarbonate buffer, the enzyme was thought to react more slowly with this oxidant (at V_{max}) than with phenazine methosulfate (Massey and Singer, 1957b). After the demonstration that bicarbonate is a competitive inhibitor of the enzyme (Zeijlemaker *et al.*, 1970), Zeijlemaker *et al.* found that at V_{max} the same velocity is measured with either oxidant in particulate or soluble preparations. Careful reexamination of the question, however, revealed (Singer, 1973) that the reaction of ferricyanide with the enzyme gives biphasic curves in Lineweaver–Burke plots and at V_{max}, under most conditions, the reaction of the enzyme with phenazine methosulfate is several times faster than with ferricyanide, in agreement with earlier data (Massey and Singer, 1957b).

The action of the enzyme is readily reversible, and a number of reductants may be used in measuring the rate of reduction of fumarate, among which $FMNH_2$ is preferred.

c. Inhibitors. Since the discovery of the inhibition of succinate oxidation by malonate (Quastel and Wooldridge, 1928), this has been the classical enzyme for the study of the phenomenon of competitive inhibition. In order of effectiveness the most commonly used competitive inhibitors are oxalacetate > malonate > pyrophosphate > fumarate. Their K_i values have been determined in a number of laboratories with varying results. More recently, di-, tri-, and perfluorosuccinate and mono- and difluorofumaric acid have been added to the list (Tober *et al.*, 1970). Other competitive inhibitors are listed in a review (Singer and Kearney, 1963).

The enzyme forms colored complexes with competitive inhibitors, which have been useful as a tool in determining their dissociation constants (Dervartanian and Veeger, 1964). Kearney, who first noted this effect with malonate, thought that the color change resulted from activation of the enzyme (Kearney, 1957). Dervartanian and Veeger (1964) reported that the formation of the colored complexes was unrelated to activation, although it was later shown that activation is a prerequisite for formation of the complexes (Kimura *et al.*, 1967). The relevance of these complexes to the reaction mechanism of the enzyme remains obscure.

3. Regulation

It has been known for over 15 years (Kearney *et al.*, 1955; Kearney, 1957) that incubation of succinate dehydrogenase with substrates or competitive inhibitors results in conversion from a low (or no) activity form to the fully active enzyme. The transformation is spontaneously reversible on removal of the activating agent (Kimura *et al.,* 1967). The regulatory significance of these observations was not appreciated until much later, when it was discovered that besides substrate, a number of physiologically occurring substances, including reduced CoQ_{10} and ATP, activate the enzyme, for while the concentration of succinate does not fluctuate dramatically in various metabolic states, that of $CoQ_{10}H_2$ and ATP does, so that it is now clear that the activation of the enzyme is an important control mechanism (Gutman *et al.,* 1971a,d,e; Singer *et al.,* 1972b,c).

Current knowledge of the regulation of the enzyme is too extensive to review here in detail, so that only a brief summary is presented. More extensive treatments will be found in recent reviews on the subject (Singer *et al.,* 1972b,c).

The most important features of the activation by substrates and substrate analogues (malonate, P_i) are as follows (Kearney, 1957): the activation has been observed in intact mitochondria, membranes, and all types of soluble preparations and is characteristic of aerobic-type succinate dehydrogenase (Singer, 1971). It is characterized by a very high activation energy (about 35 kcal/mole) and ΔS value (52 eu), suggesting that a conformation change is involved (Kearney, 1957; Gutman *et al.,* 1971a). The process is accompanied by some changes in kinetic constants (Kearney, 1957) and, according to a recent report (Sanborn *et al.,* 1971), results in the exposure of —SH groups, as judged by a greatly increased sensitivity to inhibition by NEM and bromopyruvate. Of particular importance in the context of this chapter is that the conformation change involved has been suggested to affect the environment of the iron–sulfur moieties, rather than the substrate-binding site (Kearney, 1957), since the K_m for succinate does not change appreciably nor the fumarate-$FMNH_2$ activity which is thought to involve only the flavin prosthetic group, but activities thought to require intact iron–sulfur residues (reactions with ferricyanide, phenazine methosulfate, and the respiratory chain) are all affected (Kearney, 1957).

The process is visualized as a free equilibrium between the deactivated (or unactivated) enzyme (E_U) and its activated form (E_A), the latter being stabilized by the activator (C) (Kimura *et al.,* 1967; Gutman *et al.,* 1971a; Thorn, 1962):

$$E_U \rightleftharpoons E_A \underset{-C}{\overset{+C}{\rightleftharpoons}} E_AC$$

Gutman *et al.* (1971a) recently demonstrated that the reduced form of CoQ_{10} activates the enzyme. Figure 9 demonstrates that while the oxi-

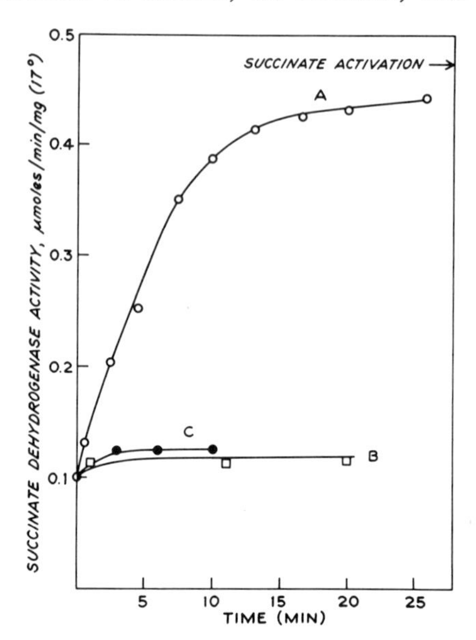

Fig. 9. Activation of succinate dehydrogenase by reduced CoQ_{10}. ETP_H, washed with 0.25 *M* sucrose—0.025 *M* tris—5 m*M* Mg (STM) buffer, pH 7.4, was resuspended in 0.18 *M* STM buffer, pH 7.4, at 4 mg of protein/ml. Antimycin A (1 nmole/mg) and cyanide (1 m*M*) were added and the sample placed under an atmosphere of N_2 to prevent autooxidation of $CoQ_{10}H_2$. CoQ_{10} was reduced with borohydride, neutralized with dilute acetic acid, and shaken till the first appearance of the yellow color of oxidized CoQ_{10} to ensure removal of unreacted borohydride, all at 0°C. Activation of succinate dehydrogenase was started by adding 50 μl of either $CoQ_{10}H_2$ (curve A) or of CoQ_{10} (curve B) in absolute ethanol to 3 ml of enzyme, giving 175 m*M* final concentration of the quinone. Curve C, no addition. Samples were withdrawn at intervals and assayed immediately at 17°C. The horizontal arrow indicates the maximal activation reached with succinate as activator. From Gutman *et al.* (1971a).

dized form of CoQ_{10} does not activate, the reduced form does, to the same maximal extent as succinate. In line with this, substances which reduce CoQ_{10} in mitochondria (Gutman *et al.*, 1971d,e) or inner membrane preparations (Gutman *et al.*, 1971a), such as NADH or α-glycerophosphate, activate the enzyme and when $CoQ_{10}H_2$ is reoxidized, the enzyme is spontaneously deactivated (Fig. 10). Since the ratio of CoQ_{10}/$CoQ_{10}H_2$ fluctuates extensively between metabolic states in mitochondria (Kröger and Klingenberg, 1966) and reflects the phosphate potential as well as the NAD^+/NADH ratio, it is clear that these factors, taken together, provide a rapid control mechanism for succinate dehydrogenase activity in mitochondria. It was also shown that in metabolic states where succinate and succinyl CoA are known to accumulate (e.g., state 3)

(VonKorff, 1967, Schäfer *et al.*, 1967; La Noue *et al.*, 1970) succinate dehydrogenase is in the deactivated state, primarily because the redox ratio $CoQ_{10}/CoQ_{10}H_2$ is high, while the opposite is true in state 4, where this ratio is low, the dehydrogenase is in the activated state (Gutman *et al.*, 1971e), and succinate is rapidly oxidized.

A third type of regulator of succinate dehydrogenase is ATP. Activation of the enzyme in yeast mitochondria by ATP was suggested several years ago (Gregolin and Scalella, 1965), but this and other demonstrations of increased succinoxidase activity induced by ATP in animal tissues were usually attributed (Greville, 1966; Pappa *et al.*, 1968) to the removal of oxalacetate. Recently, however, activation of the enzyme in mammalian mitochondria was demonstrated under conditions which preclude oxalacetate removal or changes in $CoQ_{10}H_2$ concentrations (Gutman *et al.*, 1971d,e). Figure 11 illustrates activation of the enzyme in rat heart mitochondria. It is seen that the same level of activity is reached as in activation by succinate and that, since oligomycin does not interfere with the process, the phosphorylation system is not involved.

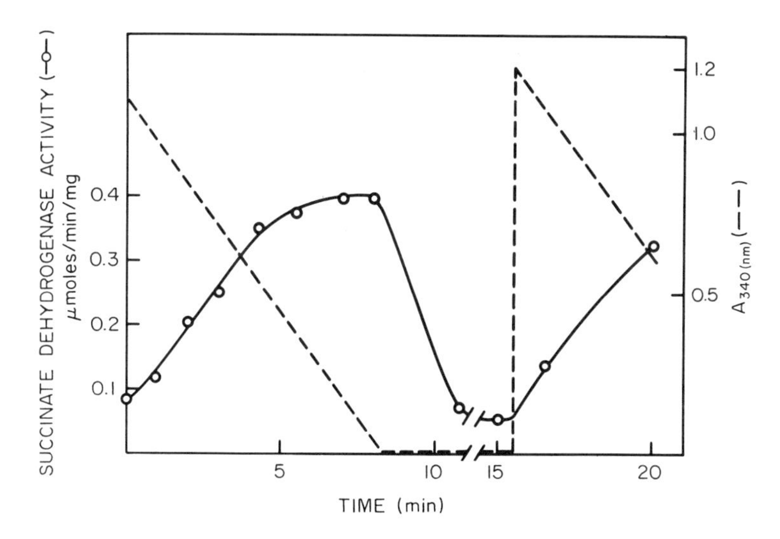

Fig. 10. Activation of succinate dehydrogenase by NADH. An ETP_H preparation (succinoxidase activity = 1.18 μmoles succinate/minute/mg at 30°C) was washed by centrifugation in STM buffer and resuspended in the same buffer to 1 mg of protein/ml. Antimycin A (1 nmole/mg of protein) was added to slow the rate of aerobic NADH oxidation, followed by 0.25 m*M* NADH. Oxidation of the latter at 23°C was monitored spectrophotometrically (dashed line). Samples were removed periodically and assayed immediately for succinate dehydrogenase activity in the presence of 0.33 mg phenazine methosulfate/ml (solid line). At 16 minutes a second aliquot of 0.25 m*M* NADH was added. From Gutman *et al.* (1971a).

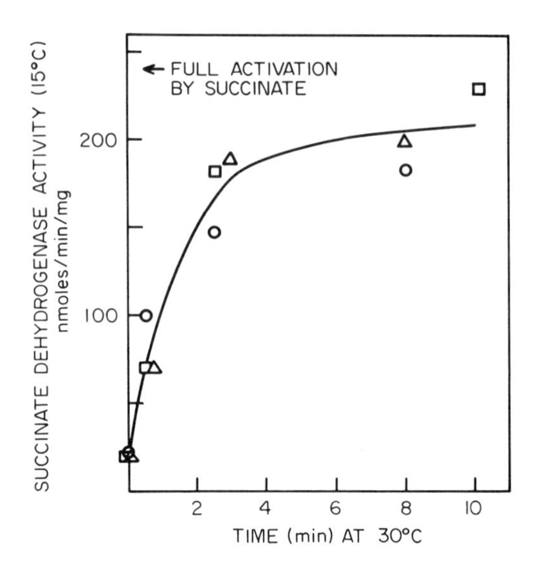

Fig. 11. Activation of succinate dehydrogenase in rat heart mitochondria initiated by ATP. The mitochondria were suspended in the buffer of Pande and Blanchaer at 1 mg protein/ml at 30°C and 1 mM ATP was added at 0 time. Aliquots were assayed for succinate dehydrogenase activity as in Fig. 10. Without ATP no change in succinate dehydrogenase activity occurred. Symbols: △, ATP alone; □, ATP + oligomycin (2 μg/mg of protein); ○, ATP + piericidin A (1.25 nmoles/mg of protein). From Gutman *et al.* (1971e).

In this system concentrations of ATP as low as 6 μM produce major activation.

It should be noted that the action of ATP may not involve a direct combination with the enzyme, since in membrane and soluble preparations ATP does not activate. Thus a compound produced from ATP might be the actual activator.

In addition to the many activators discussed above, in soluble or membrane preparations ITP and IDP activate but not IMP, cyclic IMP, 5′-AMP, 2′,3′- or 3′,5′-cyclic AMP (Gutman *et al.*, 1971e; Singer *et al.*, 1972b). Moreover, in studies with soluble and submitochondrial preparations (Kearney *et al.*, 1972a, 1973) it was discovered that the enzyme spontaneously (i.e., without activators) is converted to the active form on lowering the pH and returns to the deactivated state on neutralization. This activation is potentiated by certain anions. Anions, such as Br^-, Cl^-, I^-, NO_3^-, SO_4^{2-}, formate, etc., activate the enzyme even at neutral pH but considerably higher concentration of the activating anion is required for full activation than at or below pH 6. Although the enzyme activated by pH and anions seems to have the same turnover number and kinetic constants as that activated by other agents, the E_A value for pH anion activation is much lower ($\sim$25 kcal/mole) than for activation by substrates or $CoQ_{10}H_2$, showing that a different conformational change may be involved. High concentrations of ClO_4^-, as used in the extraction procedure of Davis and Hatefi (1971) also activate the enzyme even at 0°C (Coles *et al.*, 1973). This is the reason why these authors found that their preparation does not require activation (Davis and Hatefi, 1971). While activation by perchlorate is reversible on removal of this ion, activation of the perchlorate-extracted enzyme by succinate is exceedingly difficult to reverse (Coles *et al.*, 1973).

The question of how many of these recently discovered activators play a regulatory role in physiological conditions has not yet been sorted out, but it is clear that activation by ATP, by substances which reduce CoQ_{10}, and probably by succinate are of major importance and fit the general pattern of the activation-deactivation cycle observed in mitochondria (Gutman *et al.*, 1971e; Singer *et al.*, 1972b,c). It is also evident that succinate dehydrogenase is an important control point in the Krebs cycle (Gutman *et al.*, in press; VonKorff, 1967; Schäfer *et al.*, 1967) and that it is a highly regulated enzyme. A hypothesis to account for the existence of multiple regulatory mechanisms in terms of energy conservation has been presented (Singer *et al.*, 1971c, 1972b), but it is beyond the scope of this chapter to discuss it here.

Very recently important findings came to light concerning the chemical changes underlying the activation (Kearney *et al.*, 1972b; Ackrell and Kearney, 1973). It was found that succinate dehydrogenase in both membranal and soluble preparations, as isolated, contains extremely tightly bound oxalacetate in the deactivated form of the enzyme in equimolar ratio to the flavin in particulate preparations but in 1:2 ratio to the flavin in soluble ones. Fully activated preparations contain no oxalacetate. Evidence was obtained that the activation–deactivation cycle, under many conditions, involves movement of this oxalacetate in and out of this tight-binding site. The presence of oxalacetate in the deactivated form is not an instance of the long known inhibition of enzyme by oxalacetate, for conditions expected to dissociate it when acting as an inhibitor fail to do so when bound to the deactivated form. It takes incubation at elevated temperatures in the presence of activators (succinate, malonate, IDP, anions, etc.) to release the tightly bound oxalacetate; the latter may then be removed by gel exclusion. Although activation by Br^- and dissociation of this oxalacetate have the same activation energy, the rates of dissociation and of activation are markedly different, suggesting that activation is not merely a consequence of the dissociation of oxalacetate but that both are consequences of a conformational change. Moreover, the oxalacetate-free enzyme can be deactivated at alkaline pH and reactivated at neutral pH. Thus not all activation mechanisms of this enzyme involve oxalacetate dissociation.

The mechanism of the process requires considerable further study. At least some of the activators discovered fit the general pattern of allosteric activation and almost certainly cause conformation changes, although there is no evidence that they alter the K_m for the substrate significantly, nor is there evidence for cooperativity. The facts listed in previous sections (appearance of reactive —SH groups on activation, colored complex formation with competitive inhibitors in the activated but not deactivated state, and high energy or activation) are compatible with a conformation change. From measurements of the activation energies of the process in both directions, it has been calculated that the free energy change is minimal but a major entropy change is involved (Gutman *et al.*, 1971a).

4. Reaction Mechanism

Many attempts have been made to determine the reaction mechanism and intramolecular electron transfer sequence of the enzyme, including kinetic, stopped-flow, EPR, and isotope studies, but the problem remains largely unresolved. The results of isotope studies are particularly conflicting (Tober *et al.*, 1970) and will not be reviewed here.

From kinetic studies with activated, soluble preparations Zeijlemaker *et al.* (1969) proposed the reaction sequence shown in Eqs. (3)–(6).

$$E + S \underset{k_{-1}}{\overset{k_1}{\rightleftharpoons}} ES_I \tag{3}$$

$$ES_I \underset{k_{-2}}{\overset{k_2}{\rightleftharpoons}} ES_{II} \tag{4}$$

$$ES_{II} + A_{ox} \xrightarrow{k_3} EP + A_{red} \tag{5}$$

$$EP \underset{k_{-4}}{\overset{k_4}{\rightleftharpoons}} E + P \tag{6}$$

where ES_I and ES_{II} are two different enzyme–substrate complexes, perhaps enzyme–succinate and reduced enzyme–fumarate complexes, respectively, S is substrate, and A the electron acceptor. ES_{II} is proposed to be the species detected by EPR and stopped-flow studies on adding succinate to the enzyme.

A number of papers have appeared on the EPR behavior of the enzyme, but only in a few of the studies were the signals measured at a sufficiently low temperature to permit monitoring of the iron–sulfur components, and in only one case (Dervartanian *et al.*, 1969) was an attempt made to correlate the kinetics of the EPR signals with the catalytic constants obtained by conventional means. Even in that instance a formidable obstacle was posed by the regulatory properties of the enzyme: with deactivated preparations on adding substrate to the oxidized enzyme the activation, rather than electron transport, is rate limiting, and with preparations activated by substrates or competitive inhibitors the enzyme is in the reduced or inhibited form at the outset. For these reasons the results of Dervartanian *et al.* (1969), as recognized by them, must be considered very preliminary in nature as far as kinetics of the reduction of the EPR-active components of the enzyme is concerned because of the uncertain state of activation of the preparations used. However, their results on the reoxidation of the succinate-reduced enzyme are more significant, since activation probably occurred during the reductive phase. With the discovery that a number of substances activate the enzyme which are neither substrates nor competitive inhibitors (cf. above), the problem seems to have been overcome.

With the reservations mentioned, the results of Dervartanian *et al.* (1969) may be summed up as follows. As already pointed out, on anaerobic titration with NADH + phenazine methosulfate, at the most one unpaired electron per mole of enzyme flavin could be accounted for by the EPR signal at $g = 1.94$. Of this only about 16% was rapidly reduced by succinate in kinetic experiments. In anaerobic titrations the first reducing equivalent caused appearance of the $g = 1.94$ signal, and the free radical signal of the flavin became maximal only after the iron–sulfur center was fully reduced. This may be taken to suggest that the potential of the iron–sulfur is higher than that of the flavin and militates against an earlier proposal that flavin and nonheme iron react as a single electron acceptor unit.

While the kinetics and extent of reduction of the $g = 1.94$ iron–sulfur by succinate were not materially different at 0.5 and 50 m*M* substrate concentration, the extent of free radical formation was much greater at the higher substrate concentration. If the only effects of substrate concentration in these experiments were on the state of activation of the enzyme, the opposite result would be expected. Since the activation does not seem to influence the reactivity of the flavin (as judged by fumarate reductase activity) but affects the iron–sulfur components (as judged by catalytic activities thought to involve iron–sulfur), a greater extent of activation should be reflected in the behavior of the $g = 1.94$ components, not of the flavin. It should be pointed out, however, that 0.5 m*M* succinate is near the K_m at low enzyme concentrations and may be well below the apparent K_m at the high concentrations required for EPR studies, so that the predominant effect of succinate concentration in these experiments may have been one of substrate saturation. [In analogous studies on

NADH dehydrogenase (Beinert *et al.*, 1965), a major increase in apparent K_m has been noted at the high enzyme concentrations required for EPR experiments.]

Reoxidation of the $g = 2.00$ and $g = 1.94$ components of the enzyme, reduced by succinate, was rapid enough to be compatible with the overall kinetics, and also a larger fraction of the EPR-active iron reacted rapidly than in the reductive phase. In contrast, reoxidation was slow when NADH + phenazine methosulfate was the reductant; the difference may have been due to lesser activation in the latter case.

Although it is clearly desirable to extend these studies to fully activated preparations and to correlate them with the catalytic behavior under identical conditions, certain preliminary conclusions appear permissible to the present authors. (1) The free radical of the bound flavin seems to be a catalytically significant intermediate, although it is not certain whether the enzyme cycles between oxidized and free radical or oxidized and fully reduced flavin normally. (2) Iron–sulfur becomes reduced during catalysis but probably not simultaneously with flavin. (3) The question of how many of the eight iron–sulfur components participate in the catalytic cycle and are recognizable by their $g = 1.94$ signal remains entirely open, since in all studies hitherto reported the state of activation was uncertain and since the enzyme preparations used were not fully competent catalytically, as judged by their lower turnover number than that seen in intact membrane preparations. As emphasized earlier in this chapter, the lower catalytic activity of soluble preparations and their anomalous behavior toward TTF and cyanide may reflect alterations in the environment or reactivity of some of the iron–sulfur centers.

The availability of methods of activating succinate dehydrogenase without the use of substrates or competitive inhibitors and the recent discovery that the enzyme may be prepared with and without tightly bound oxalacetate pave the way to systematic studies of the role of the Fe–S centers in the mechanism of action of the enzyme by EPR methods. Added impetus to renewed attack on the problem has been provided by the recent discovery of Orme-Johnson and Beinert (1972) that at liquid helium temperature two EPR signals, differing in size and shape and corresponding probably to two different iron–sulfur centers of succinate dehydrogenase, may be detected in complex II preparations.

5. Anaerobic Succinate Dehydrogenases

Warringa *et al.* (1958) and Warringa and Giuditta (1958) isolated a succinate dehydrogenase from the obligate anaerobe *Micrococcus lactilyticus*, which is in just about every important respect different from the aerobic enzyme. At 30°C the turnover number is about 10^5 in the direction of fumarate reduction but only 2.6×10^3 in the direction of succinate oxidation. As in other anaerobic-type succinate dehydrogenases, malonate and fumarate inhibit succinate oxidation only poorly; the K_m for fumarate is very low and the K_m for succinate high. Since succinate increases the rate of fumarate reduction, it may be a regulator of the enzyme.

The enzyme is not membrane bound and thus is readily extractable. Preparations apparently homogeneous in electrophoresis and ultracentrifugation are deep brown in color, suggestive of high iron content. Indeed, 40 gm-atoms of nonheme iron were found present per mole of flavin. The molecular weight is not known for certain: the FAD content is 1 mole/4.6×10^5 gm protein, but the very high sedimentation constant ($s_{w,20^\circ} = 54$ S) suggests an aggregate or complex, rather than a molecular unit.

The flavin is noncovalently bound, being liberated completely on denaturation, but, curiously, mineralization of most of the nonheme iron requires proteolytic digestion. This behavior is not expected for the iron–sulfur prosthetic groups dealt with elsewhere in this chapter and, in fact, labile sulfide is not known to be present in the enzyme, since the work quoted antedated recognition of the importance of iron–sulfur units in oxidizing enzymes. For the same reason no EPR data are available for this interesting variant of succinate dehydrogenase. The probable involvement of nonheme iron in the catalytic action of the enzyme is nonetheless suggested by the rapid bleaching of the brown color by succinate and its reappearance on addition of fumarate.

A more recently studied and equally unusual enzyme in this class is the fumarate reductase of yeast cytoplasm (Hauber and Singer, 1967; Tisdale *et al.*, 1968). Although present in both anaerobic and aerobic yeast, its concentration is much higher in the former, particularly if grown at high glucose concentration. The enzyme is aptly named fumarate reductase, for no catalytic test has been devised to demonstrate the net oxidation of succinate by the enzyme (although succinate bleaches the spectrum and thus reacts with the enzyme), while the rate of reduction of fumarate to succinate is relatively high.

Several isoenzymes of fumarate reductase occur in yeast, which differ in charge, molecular weight, kinetic constants, and specificity for electron donors. The predominant type (type I) has a molecular weight of about 62,000 and five isoenzymes of this type differing only in charge have been detected. Type II, with different properties, shows molecular weights of 32,000 and 112,000. Neither type appears to be regulable nor contains covalently bound flavin. Type I enzyme has been extensively purified and shown to contain FAD and nonheme iron in approximately 1:1 ratio but no labile sulfide. Although there is also an equimolar amount of copper present, this may be an impurity.

The absorption spectrum shows a typical flavin peak in the 450 nm region and appreciable absorbance above 500 nm. Dithionite causes extensive bleaching in the flavin region and the appearance of a long wavelength band between 610 and 730 nm. Succinate causes less extensive but rapid bleaching below 500 nm and slower bleaching above 500 nm which is not fully reversed by fumarate. The interpretation of these changes in terms of oxidation–reduction of the flavin or of a metal component is difficult at this time. EPR examination revealed the presence of high-spin ferric iron and Cu (II) signals, neither of which responded to the addition of substrate (Tisdale *et al.*, 1968).

III. MAMMALIAN NADH DEHYDROGENASE

A. Different Forms of the Enzyme and Their Interrelations

The NADH dehydrogenase of the respiratory chain (NADHD) occurs in the inner membrane of mitochondria in a tightly bound form. Its primary physiological role is the initiation of the oxidation of all intramitochondrially generated NADH, passing electrons via CoQ_{10} to the cytochrome system. Thus oxidations initiated by NADHD are responsible for the majority of the ATP synthesized by oxidative phosphorylation. In addition, by influencing the redox state of CoQ_{10}, NADHD indirectly regulates the activity of succinate dehydrogenase (Gutman *et al.*, 1971a).

The enzyme has been isolated in two molecular forms (the so-called "high molecular weight" and "low molecular weight" forms) with different iron–sulfur content and varying properties. In order to understand differences in the behavior of the iron in different preparations, it is necessary to summarize their properties and interrelations. More extensive discussions on this subject will be found in recent reviews (Singer, 1968; Singer and Gutman, 1971).

The high molecular weight form has been purified both as a particle [the complex I of Hatefi *et al.* (1962)] and as a soluble enzyme, extracted with phospholipase A (Ringler *et al.*, 1963; Cremona and Kearney, 1964). The molecular and catalytic properties of the soluble and particulate preparations appear to be identical and are also indistinguishable from those seen in membrane preparations or mitochondria, with one exception: the soluble enzyme cannot reduce coenzyme Q homologues (Machinist and Singer, 1965a). There is a growing body of evidence, however, that the reaction of NADHD with external CoQ in a fully rotenone- and Amytal-sensitive manner requires the presence of lipids, probably including endogenous CoQ_{10}, which are removed during extraction of the soluble enzyme (Machinist and Singer, 1965a,b). Moreover, the specific and stoichiometric binding of rotenone and piericidin involves both the NADH dehydrogenase protein and lipids, so that on removal of the lipids specific binding and the resulting inhibition disappear (Gutman *et al.*, 1970c). Hence, the fact that the soluble enzyme is not sensitive to these inhibitors and does not reduce CoQ_{10} is most readily interpreted in terms of separation of the protein from its lipid environment, rather than as evidence of modification.

It should be noted that modification of the enzyme by heat (Machinist and Singer, 1965a) or by heat–acid–ethanol (Salach *et al.*, 1967) may lead to the appearance of artifactual CoQ reductase activity which is either insensitive to these inhibitors or is only partly inhibited by amytal and rotenone in a narrow range of concentrations, and the inhibition, curiously, disappears completely when more rotenone is added (Salach *et al.*, 1967). Therefore, inhibitor-sensitive NADH-CoQ reductase activity cannot be taken as an indication that the enzyme in a particular preparation is in the native state nor its absence as evidence of enzyme modification.

The low molecular weight form of the enzyme (MW = 78,000) has been prepared by the treatment of membrane preparations, particles, or of the highly purified high molecular weight form of the soluble enzyme with heat-acid-ethanol (Mahler *et al.*, 1952; Mackler, 1961; Watari *et al.*, 1963; Pharo *et al.*, 1966), heat plus phospholipase A (King and Howard, 1962; Cremona *et al.*, 1963), urea (Cremona *et al.*, 1963; Hatefi and Stempel, 1969), thiourea (Chapman and Jaggannathan, 1963), or proteolytic enzymes (Cremona *et al.*, 1963; Kaniuga, 1963). The soluble enzyme is very sensitive to all of these treatments, so that extensive or complete conversion to the low molecular weight form occurs under any of these conditions. The properties of the low molecular weight form of the enzyme extracted under these varying conditions are very similar, in several cases identical (Cremona *et al.*, 1963; Watari *et al.*, 1963).

Conversion of NADH dehydrogenase to the low molecular weight form is accompanied by extensive changes in composition and molecular and catalytic properties. The nonheme iron content is lowered from 17 ± 1 (Ringler *et al.*, 1963; Watari *et al.*, 1963) to 4, and the labile sulfide content from 27 to 2–4 per mole of flavin (Lusty *et al.*, 1965). Actually, no fixed stoichiometry can be assigned to flavin:Fe:S in the 78,000 molecular weight form, since both the flavin and the sulfur are highly labile in this form of the enzyme, in contrast to the high molecular weight form in which they are stably bound. As a result, the exact ratio depends on the preparative history and on how long after isolation the analysis was performed (Lusty *et al.,* 1965).

Loss of the nonheme iron and labile S on breakdown of the high molecular weight form to the low molecular weight type is accompanied by major changes in the absorption spectrum, as would be expected from the fact that most of the absorbance in the visible range is due to FeS groups (Ringler *et al.,* 1963; Singer and Gutman, 1971). Further, the substrate-induced EPR signal of the native enzyme, which manifests itself as an asymmetric signal at $g = 1.94$ at 77°K (Beinert *et al.,* 1965) but is resolved into at least four signals at 13°K (Orme-Johnson *et al.,* 1971; Gutman *et al.,* 1971c), also disappears during these transformations.

The transformation to the low molecular weight form is accompanied by profound changes in catalytic properties, among which the appearance of NADH-cytochrome c reductase and NADH-diaphorase (i.e., indophenol reductase) activities are best known (Cremona *et al.,* 1963; Kaniuga, 1963; Watari *et al.,* 1963). Under suitable conditions there is also an emergence of rotenone-insensitive CoQ_1 reductase and *partially* rotenone-sensitive CoQ_{10} reductase activities of considerable magnitude (Machinist and Singer, 1965a; Salach *et al.,* 1967). These activities account for earlier assumptions in the literature of the existence of NADH-cytochrome c reductase and of NADH-CoQ reductase (Mahler *et al.,* 1952; Pharo *et al.,* 1966). Along with the appearance of these artifactual activities, there is a disappearance of the NADH-ferricyanide activity which is normally used for assay of the high molecular weight form of the enzyme in membrane-bound or soluble form. The residual NADH-ferricyanide activity seen in the low molecular weight form is quite different from that seen in more intact preparations, since it is neither inhibited by P_i nor by high NADH concentrations and thus appears to represent another artifactual activity (Singer and Gutman, 1971).

Interestingly, transformation of the dehydrogenase to the low molecular weight form results only in a lowering but not in a complete loss of the ability to oxidize NADH by the FMN prosthetic group. The substrate binding site, however, does not appear to be identical in the high and low molecular weight forms. Not only are their substrate specificities different (V_{max} and K_m values) (Cremona *et al.,* 1963), but rhein, a powerful competitive inhibitor of the former, has no effect on the latter (Kean *et al.,* 1971).

There has been a considerable amount of discussion in the literature concerning the relation of the high and low molecular weight forms to each other. The problem has been reviewed recently (Singer and Gutman,

1970), and a detailed discussion is beyond the scope of this chapter. It has been proposed (King *et al.*, 1966; Mackler, 1966; Pharo *et al.*, 1968) that the low molecular weight form exists as such in mitochondria or is a subunit of the high molecular weight one, but this hypothesis fails to account for the extensive differences in catalytic properties between the two forms especially as regards the substrate site. Another weakness of the hypothesis is that certain methods of transforming the high molecular weight form to the low molecular weight one, such as prolonged incubation with substrate (Rossi *et al.*, 1965), result in the formation of numerous polypeptides of varying chain length, including several catalytically active forms, a behavior seemingly incompatible with subunit dissociation. A recent hypothesis (Hatefi, 1968) proposed that NADH dehydrogenase, in fact, consists of a separate flavoprotein and of a nonheme iron protein. The evidence for this view is that the two fragments obtained on dissociation with urea interact catalytically. It was subsequently shown, however, that the interaction is unspecific and extremely slow, so that at present there appears to be little basis for regarding the products of urea cleavage as catalytically active, preexisting proteins (Gutman *et al.*, 1970a).

The alternative hypothesis, which seems to fit best the available data (Singer and Gutman, 1971) is that the dehydrogenase in the native form consists of many polypeptide chains which are relatively easily dissociated, one of the metastable products of the dissociation being the low molecular weight form, and that the structure is stabilized in the lipid environment of the membrane.

Since there appears to be substantial agreement that the properties of the high molecular weight form resemble those of the enzyme as it occurs in membranes more closely than do the properties of the low molecular weight form, the present review concentrates on the former, particularly in regard to its iron–sulfur components.

B. Chemical and Molecular Properties

NADH dehydrogenase is released from the respiratory chain by the action of pure phospholipase A and is therefore probably bound by way of phospholipids to the membrane (Salach *et al.*, 1970). There remains some uncertainty regarding the molecular weight. The enzyme has a pronounced tendency to aggregate in highly purified preparations, so that physical methods of molecular weight measurement do not yield reliable data. Hence, available values are based on estimates of the FMN content, a method which is subject to considerable error if even small amounts of flavoprotein contaminants are present, since the flavin content of the dehydrogenase itself is low. The best value available is 5.5×10^5 (Cremona and Kearney, 1964).

There are 16 to 18 gm-atoms of nonheme iron per mole of FMN in both soluble and particulate preparations of the high molecular weight form (Ringler *et al.*, 1963; Lusty *et al.*, 1965). The labile S/FMN ratio has been reported to be 21 for complex I and 27 for the soluble enzyme (Lusty *et al.*, 1965). The >1 ratio of S/Fe is unusual, in that most proteins in this class show a ratio of unity; thus the analytical values merit reexamination. The contribution of these iron–sulfur moieties to the absorption spectrum and their EPR behavior are discussed in Sections III,D and III,E. As mentioned above, on conversion to the low molecular weight form most of the iron–sulfur content is lost and the remainder rendered catalytically nonfunctional, since the EPR signal at $g = 1.94$ which appears on reduction by substrate is lost.

So far five types of —SH groups have been identified in the enzyme with the aid of —SH inhibitors. These are distinguished by the particular effects on the different catalytic activities or on the stability of the enzyme, by the conditions under which they react with mercurials or other —SH reagents, and by the form of the enzyme in which they are detectable (Gutman *et al.*, 1970d). Type I —SH appears to be buried in the native enzyme and reacts only after conversion to the low molecular weight form (Cremona *et al.*, 1963). Type II has not been detected in either particulate preparations or in the soluble low molecular weight forms of the enzyme. This thiol group reacts with —SH reagents rapidly at 0°C and the resulting product has full catalytic activity. Combination of Type II —SH with mercurials is fully reversible and in the case of negatively charged mercurials (mersalyl, PCMB, PCMS) is prevented by inorganic phosphate. If the mercaptide or NEM derivative of the enzyme, freed from unreacted—SH reagents, is incubated at 15° to 30°C, temperatures at which the free enzyme is stable, gradual inactivation occurs, both the rate and final extent of which depends on the temperature. This inactivation is irreversible and appears to be the result of a conformation change (Cremona and Kearney, 1965). Since the inactivation affects the NADH–ferricyanide reaction, which is thought to involve an iron–sulfur component (Beinert *et al.*, 1965), but not transhydrogenase activity, which involves only the flavin at the active center, the conformation change may be either in the vicinity of the irons which donate electrons to ferricyanide (iron–sulfur center 1) or, possibly, in the region between flavin and this iron–sulfur center. Type III —SH group may be detected in soluble and particle-bound preparations of the high molecular weight form of the enzyme (Tyler *et al.*, 1965; Gutman *et al.*, 1970d). When mercaptide is formed with these thiols, all

TABLE II
INHIBITION OF NADH DEHYDROGENASE AND NADH OXIDASE ACTIVITIES BY MERSALYL IN ESP PARTICLES[a]

Treatment	Oxidase activity (μmoles NADH/ minute/mg)	DPNH dehydrogenase: Activity at V_{max} ferricyanide (μmoles/ minute/mg)	Activity at 0.33 mM ferricyanide (μmoles/ minute/mg)	K_m for ferricyanide (mM)
Untreated	0.84	34.0	4.5	2.2
NADH, then mersalyl	0.066	4.5	1.0	0.95
NADH, then $Fe(CN)_6^{3-}$, then mersalyl	0.47	25.8	3.6	2.2
$Fe(CN)_6^{3-}$, then mersalyl	0.55	30.0	4.1	2.2
Mersalyl	0.54	111.0	3.8	9.0
$Fe(CN)_6^{3-}$	0.80	30.0	4.1	2.2
NADH	0.77	30.0	4.1	2.2

[a] ESP particles were incubated at 0°C in 80 mM phosphate, pH 7.4, at a concentration of 0.66 mg protein/ml for 9 minutes with or without NADH, as indicated. NADH was added during this period in 3 aliquots at 3 minute intervals, bringing its concentration each time to 0.17 mM, since it took 3 minutes at 0°C to oxidize 0.17 mM NADH. At the end of 9 minutes, mersalyl or $Fe(CN)_6^{3-}$ were added, where indicated, to give concentrations of 0.03 and 0.5 mM, respectively, and the incubation at 0°C was continued for 6 minutes. Where both were present, mersalyl was added 2 minutes after $Fe(CN)_6^{-3}$. A small aliquot was then added to the complete assay mixture (less enzyme) which had been brought previously to the temperature of the assay (22°C for oxidase, 30°C for dehydrogenase). From Gutman *et al.* (1970d).

catalytic activity is lost and the inactivation is not reversed by dithiothreitol (Gutman *et al.*, 1971b). The characteristic feature of these thiols is that they do not react with mercurials unless the enzyme is first pretreated with NADH (Tyler *et al.*, 1965) (Table II). This "conditioning" is reversible, for on treatment with ferricyanide, the —SH groups involved become again unreactive towards mercurials (Tyler *et al.*, 1965; Gutman *et al.*, 1970d).

There is substantial reason to suppose that type III —SH groups are located in the initial segment of the intramolecular electron transport system in the dehydrogenase molecule and may, in fact, be near the substrate binding site, for on mercaptide formation with these thiols transhydrogenase and NADH-ferricyanide activities are lost to the same degree (Gutman *et al.*, 1970d). In the context of this chapter the following hypothesis (Tyler *et al.*, 1965) for the effect of NADH in rendering this type of thiol reactive is of considerable interest. It has been proposed that the untreated enzyme contains these groups in disulfide form adjacent to an Fe^{3+}-flavin group and that NADH reduces these moieties to an iron–flavin semiquinone complex containing iron–sulfur ligands, and that in the presence of O_2 this complex is converted to an oxidized form, which corresponds to the "preconditioned," mercurial-sensitive state.

$$\begin{matrix}-S \\ | \\ -S\end{matrix}\; Fe^{3+}\text{—Flavin} + 2\,NADH \longrightarrow 2\,NAD^{+} + \begin{matrix}-S\cdot. \\ -S\cdot\cdot\end{matrix}\!:Fe^{2+}\text{—Flavin}$$

$$\text{NADH} \uparrow\downarrow O_2$$

$$\begin{matrix}-S\cdot. \\ -S\cdot\cdot\end{matrix}\!:Fe^{3+}\text{—Flavin}$$

The scheme above calls for the participation of electron transport to O_2 in the conditioning process and implies that these —SH groups undergo oxidation reduction in the course of the normal catalytic cycle (Gutman *et al.*, 1970d). Two objections have been raised to this mechanism (Gutman *et al.*, 1970d; Singer and Gutman, 1971): first, the conditioning occurs normally in particles in which electron transport has been fully inhibited with piericidin and, second, both the conditioning by NADH and the deconditioning by ferricyanide are several orders of magnitude slower than the turnover number of the enzyme. Possibly, the appearance of type III —SH groups is an early sign of the conformation changes, fragmentation, and inactivation produced on prolonged incubation of the enzyme with NADH, which have been recognized for a long time (Rossi *et al.*, 1965). The properties of NADH dehydrogenase in the "conditioned state" would seem to be compatible with this explanation (Singer and Gutman, 1971). Type IV —SH groups are responsible for the direct effect (i.e., no preconditioning) of low concentrations of mercurials on the enzyme and are detected in particulate and soluble preparations of the high molecular weight form under conditions where none of the other types of thiols reacts. Mercaptide formation with type IV —SH is reflected in NADH-ferricyanide, transhydrogenase, and NADH oxidase assays: (in particles the partial inhibition of NADH oxidase activity in the range of 10 to 30 μM mersalyl is due to blocking of type IV —SH groups (Table II). The NADH-ferricyanide activity is altered in a characteristic manner (Table II): both the K_m for ferricyanide and the V_{max} (infinite ferricyanide concentration) are greatly increased, but at low ferricyanide concentrations major inhibition is observed (Gutman *et al.*, 1970d). These effects are completely reversed by treatment with thiols (Gutman *et al.*, 1971b).

Type V —SH has so far been studied only in particulate preparations, because the diagnostic tests used to recognize its reactions are not applicable to the soluble, purified enzyme. This type of —SH is titrated only are relatively high (30 to 80 μM) mercurial concentrations. On mercaptide formation with Type V —SH, electron transport from the dehydrogenase to the respiratory chain is lost, with consequent inactivation of NADH oxidase activity (Gutman *et al.*, 1970d) and one of the two specific binding sites of rotenone and piericidin (cf. below) is lost, resulting in a shift from sigmoidal titration curves of inhibition of NADH oxidase activity by piercidin (two binding sites) to a hyperbolic one (one site) Gutman *et al.*, 1970c,d).

Another consequence of mercaptide formation with type V —SH group is seen in Fig. 12. The effects illustrated here concern spectrophotometric observations on the redox cycle of the dehydrogenase, which will be discussed in greater detail later in this chapter. One of the parameters which may be monitored by dual wavelength spectrophotometry is the rate of reoxidation of the iron–sulfur components of the dehydrogenase by the respiratory chain, following reduction by NADH. The rate of this reoxidation ($\Delta A_{470-500}$ reox/second in Fig. 12) is inhibited in the

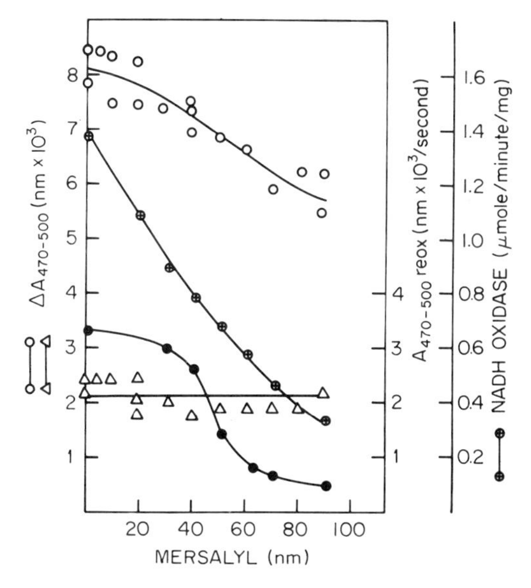

Fig. 12. The effect of mersalyl treatment of ETP on the redox cycle parameters. ETP (NADH-$K_3Fe(CN)_6$ reductase activity 35 μmoles NADH/minute/mg) in 0.1 *M* KP_i, pH 7.4, was incubated for 10 minutes at 0°C with the indicated mersalyl concentrations; the particles were then sedimented by centrifugation and resuspended in 0.1 *M* KP_i and assayed for NADH oxidase activity and redox cycle parameters. (○——○) ΔA_{red}; (⊗——⊗) NADH oxidase activity; (Δ——Δ) ΔΔ value; (●——●) $\Delta A_{reox} \times 10^3$/second. From Gutman and Singer (1970).

same range of mersalyl concentrations required to reduce the number of piericidin binding sites from 2 to 1. This is in accord with the interpretation that type V —SH group is located on the O_2 side of those iron–sulfur centers of the dehydrogenase which are thought to be involved in electron transport to endogenous CoQ. The inhibition of oxidase activity in Fig. 12 covers a broader range of mersalyl concentrations (0–80 μM), in accord with the idea that destruction of either type IV or type V —SH groups will restrict oxidase activity.

Table II illustrates one of the ways in which effects on type V —SH may be distinguished from those on type IV (or type III). The presence of ferricyanide during treatment with mercurial prevents both effects on V_{max} and K_m in the ferricyanide assay, but does not prevent inhibition of the oxidase activity. Under the conditions of Table II, inhibition of the oxidase is partial, for while the concentration of mersalyl (30 μM) is sufficient for maximal effects on type III and IV —SH groups, concentrations near 80 μM mercurial are required for nearly complete inhibition of oxidase activity. The effects of mercaptide formation with type V —SH group appear to be fully reversed by dithiothreitol, as is also true of effects associated with type IV —SH, but the regenerated enzyme, in the case of type V thiol, is not completely identical with the untreated one (Gutman *et al.*, 1971b).

As is apparent from the foregoing summary, sulfhydryl groups play varied and complex roles in NADH dehydrogenase. There is no clear-cut evidence to suggest that any of the five types of —SH groups participate directly in redox reactions catalyzed by the enzyme or that they form part of the substrate-binding site. On the contrary, there is evidence that the inactivations noted following mercaptide formation with some of these groups are the results of conformation changes, indicating that the —SH groups involved play a structural role (Singer and Gutman, 1971).

Another group of inhibitors of the enzyme which has been extensively studied is rotenone, piericidin A, and barbiturates. Inhibition by these agents, in the manner observed in mitochondria and membrane preparations, depends on the presence of lipids, since both lipid and protein are involved in their binding at the inhibitory site (Gutman *et al.*, 1970a; Horgan *et al.*, 1968). Inhibition is not observed with the soluble, purified high molecular weight form of the enzyme, which is lipid free. It should be mentioned that rotenone and piericidin A also cause *partial* inhibition of the low molecular weight form of the enzyme, but this effect appears to be different from the potent and complete inhibition observed with the membrane-bound form of the enzyme in all respects studied (Salach *et al.*, 1967; Horgan *et al.*, 1968a,b).

As is true of the effect of inhibitors on type V —SH groups of the enzyme, the action of rotenone and piericidin is of considerable interest in the context of the iron–sulfur components of the dehydrogenase and will, therefore, be summarized briefly. Rotenone, piericidin A, and Amytal inhibit electron transport from the dehydrogenase to the respiratory chain by reacting with the same two sites, which are located on the O_2 side of the enzyme.* Of this group of inhibitors Amytal is least tightly bound and, hence, the least potent inhibitor. Piericidin is most tightly bound, being effective at extremely low concentrations, although the binding even in this case is noncovalent.

When reacting with membranal preparations, rotenone and piericidin A are bound in quantities far in excess of the amount needed to cause complete inhibition of NADH oxidase activity. If the particles are washed with bovine serum albumin following treatment with these inhibitors, most of the inhibitor is released, leaving the enzyme completely inhibited (Horgan *et al.*, 1968a,b). To distinguish the two types of binding, the operational definitions "unspecific binding" and "specific binding" were adopted. The former denotes the fraction of inhibitor bound at loci from which serum albumin releases it; the latter refers to the fraction not dissociated by serum albumin. Although binding at unspecific sites may cause some inhibition of electron transport from NADH dehydrogenase to the respiratory chain under appropriate conditions, which is reversed by bovine serum albumin (Gutman *et al.*, 1970c), if piericidin is added in excess of the amount required to saturate the specific binding sites, all the inhibition of NADH oxidase is due to specific binding (Gutman *et al.*, 1970c).

As already mentioned, both protein and lipid play a role in the specific binding of these inhibitors. The protein component involved in this specific binding is NADH dehydrogenase (Gutman *et al.*, 1970a,c), and there are two such sites per FMN in the enzyme. The two sites contribute unequally to the inhibition of enzyme activity, with the result that titration curves of activity versus inhibitor concentration are sigmoidal, although the binding curve, measured with ^{14}C-labeled inhibitors, is hyperbolic, indicating lack of cooperativity in binding at the two sites (Gutman *et al.*, 1970c). The two specific sites are at or near the O_2 side of the enzyme, distal to the iron–sulfur center 1 (in terms of a

* For convenience of presentation, in the jargon of the field, reaction sites are indicated as if a linear electron transport occurred within enzyme and between the enzyme and the respiratory chain. Thus "substrate side" of the enzyme refers to the region where substrate is oxidized, and "O_2 side" as the region where electron transfer from the enzyme to CoQ and, therefore, via the respiratory chain to O_2 occurs.

linear sequence), since in preparations completely inhibited with rotenone and piericidin NADH elicits the EPR signal of center 1 at normal rates and to a normal extent (Gutman *et al.*, 1970a, 1971c). Spectrophotometric experiments and EPR studies at 13°K, to be discussed later, further localize these specific sites in the electron transport sequence of the enzyme.

Perhaps the most unusual molecular property of NADH dehydrogenase is the pronounced lability of its structure to a wide variety of agents and treatments. As mentioned above, the high molecular weight form of the enzyme is readily destroyed by exposure to temperatures above 30°C, organic solvents, acid pH, or a combination of these (the so-called heat-acid-ethanol treatment), to freeze-drying, relatively low concentrations of urea, thiourea, and Triton X-100, and to prolonged incubation with substrate. Many of these treatments, as well as proteolysis, lead to conversion to the low molecular weight form and to profound changes in properties (Singer, 1966; Singer and Gutman, 1971). This conversion or fragmentation is usually attended by loss of iron–sulfur moieties and to changes in the reactivity of the remaining ones, so that it has been possible to trace both the disappearance of certain catalytic and spectral properties and the emergence of new ones to changes in certain iron–sulfur components (Watari *et al.*, 1963; Cremona *et al.*, 1963; Beinert *et al.*, 1965; Singer and Gutman, 1971). Not unexpectedly, the membrane-bound form of the enzyme is more stable to such modifying action than the soluble, purified enzyme.

C. Catalytic Properties

The turnover number of the enzyme in the high molecular weight form is extremely high for a flavoprotein [about 800,000/minute at 30°C (Cremona and Kearney, 1964)]. Besides NADH, a number of reduced analogues are oxidized, but the only other rapidly oxidized substrate is deamino NADH. The K_m value is highest for NADH (0.108 mM at 30°C) and lowest for its thionicotinamide analogue (Minakami *et al.*, 1963). Since the action of the enzyme is readily reversible, it can act as a good transhydrogenase. This activity appears to require only the flavin, not the iron–sulfur moieties, since it survives treatments which detach or modify the reactivity of iron–sulfur components (Cremona *et al.*, 1963).

The specificity for electron acceptors is unusually high, ferricyanide being the only efficient oxidant known. In the membrane-bound form, electrons are thought to be passed on to endogenous CoQ_{10}, and in preparations capable of catalyzing the ATP-driven reduction of NAD^+ by succinate, electrons flow from reduced CoQ_{10} to NADH dehydrogenase.

Both of these activities are inhibited by rotenone and related compounds, while ferricyanide reduction is not.

It appears from kinetic studies (Minakami *et al.*, 1962) that there are two binding sites for NADH in the enzyme; the second one is observable only at relatively high NADH concentrations and expresses itself in a competitive inhibition toward ferricyanide.

A useful competitive inhibitor of the enzyme, rhein, has been recently described (Kean *et al.*, 1971). Although there is no obvious structural similarity between rhein and NADH, this inhibitor blocks the NADH–FMN reaction competitively at very low concentrations ($K_i = 2$ μM at 30°C). In submitochondrial particles a secondary, noncompetitive component is also apparent.

On conversion to the low molecular weight form most of these catalytic properties are changed. The turnover number per mole of flavin declines, the relative rates of oxidation of various NADH analogues and their K_m values change, inhibition by rhein disappears, and reactivities appear with a series of oxidants with which the native enzyme does not react significantly.

D. Spectrophotometric Studies on the Nonheme Iron Centers

1. Redox Cycle of the Enzyme

It was recognized during early studies on the isolation of the enzyme that the relatively featureless absorption spectrum (generalized absorption in the entire visible range with a maximum between 400 and 420 nm) is due to iron components (Ringler *et al.*, 1963). Although the difference spectrum obtained on reduction of the purified enzyme with NADH shows extensive bleaching in the 400 to 500 nm range, where flavins absorb light (Fig. 13), it was clear from both the nature of the spectrum and the very low flavin content per gram of protein that most of the color must be due to iron, not flavin (Ringler *et al.*, 1963). This fact, coupled with the very low concentration of the enzyme in mitochondria (Cremona and Kearney, 1964), contraindicated spectrophotometric studies of the flavin component in complex preparations.

Before these facts became known, however, Chance (1956) had proposed that absorbance changes monitored at 465 minus 510 nm might be used to study the redox state of the flavin in this enzyme in mitochondria. The method was adopted by many laboratories and led to erroneous assignments of the reaction site of Amytal (Chance, 1956) and rotenone (Oberg, 1961; Ernster and Lee, 1964) on the substrate side of NADH dehydrogenase. The theoretical basis of the method was questioned on

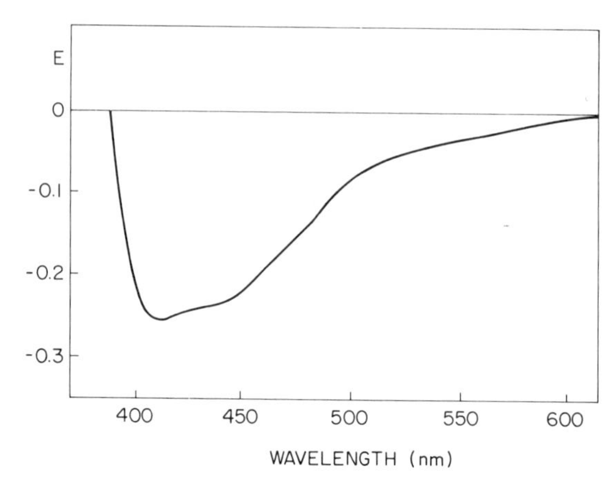

Fig. 13. Difference spectrum of the enzyme reduced by NADH. NADH dehydrogenase, specific activity = 350, 9.2 mg of protein per ml, in 0.05 *M* tris buffer, pH 7.4, was reduced with 0.8 m*M* NADH and the spectrum immediately recorded against a sample of oxidized enzyme. Downward deflection denotes bleaching.

the grounds that, since such studies were usually carried out in anaerobic conditions or in the absence of cytochrome oxidase activity, NAD-linked substrates reduced not only the flavin and the iron–sulfur components of NADH dehydrogenase (of which the latter contributes much more absorbance than the former), but also the various cytochromes and the iron–sulfur and flavin components of other enzymes, all of which tend to obscure the minor change due to reduction of the FMN component of NADH dehydrogenase (Singer, 1961; Minakami *et al.*, 1963; Nicholls and Malviya, 1968; Gutman *et al.*, 1970a).

More recently, Hatefi (1968), working with membrane fragments, ascribed the absorbance changes at 460 minus 510 nm to iron–sulfur components associated with NADH dehydrogenase. This assignment is subject to most of the criticisms which had been raised against the method to measure the redox state of the flavin. It was shown later, in fact, that a substantial part of the absorbance changes measured by Hatefi were due to cytochromes (Gutman *et al.*, 1970a), and Albracht and Slater (1969) pointed out that endogenous CoQ_{10} also contributes to the absorbance changes at this wavelength pair under the experimental conditions.

A spectrophotometric method for measuring oxidation–reduction of the iron–sulfur moieties of the dehydrogenase in membranal preparations emerged subsequently as a result of studies by Bois and Estabrook (1969) and Gutman and Singer (1970). These authors measured absorbance changes occurring on addition of substrate to inner membrane preparations in the *aerobic* state, i.e., under conditions where electron transport from CoQ_{10} to O_2 occurred unimpeded, minimizing thereby the likelihood that CoQ_{10} or cytochromes would become reduced. The extent of bleaching under these conditions is very small, compared with that obtained in anerobiosis, another indication that reduction of components other than the dehydrogenase has been eliminated.

Figure 14 illustrates the cyclic bleaching (reduction) and subsequent recolorization (reoxidation) of the chromophore measured at 470 minus 500 nm. As seen in Fig. 14A, NADH initiates essentially instantaneous bleaching, followed by a rapid but measurable rate of recolorization when the NADH is exhausted. Reoxidation, however, is not complete, so that residual bleaching (denoted as $\Delta\Delta$) remains. The extent of bleaching (ΔA_{red}) is not dependent on the amount of NADH added and is not influenced by the presence of piericidin or rotenone (cf. below) but is strictly dependent on the NADH dehydrogenase content of the particles (Gutman and Singer, 1970). Figure 14B shows that 0.5 m*M*

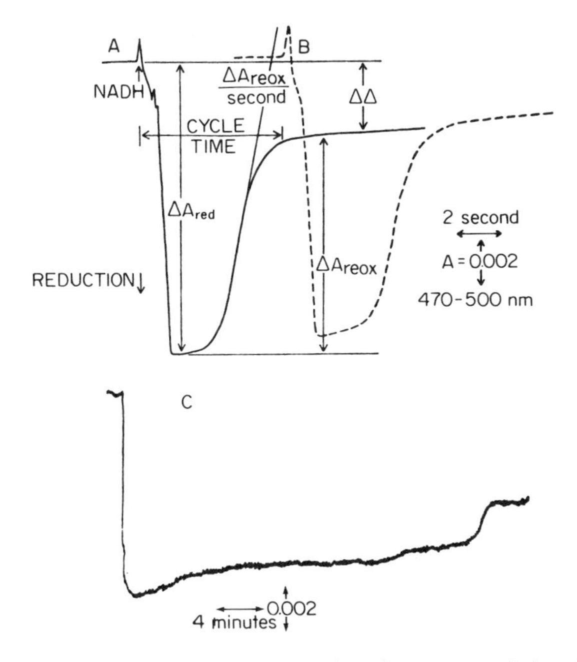

Fig. 14. Kinetics of absorbance changes at 470 minus 500 nm induced by NADH. ETP (NADH oxidase activity, 1.6 μmoles NADH oxidized/minute/mg; NADH-$K_3Fe(CH)_6$ reductase activity, 32 μmoles NADH/minute/mg) at 4 mg of protein/ml in 0.1 *M* KP_i, pH 7.4, 25°C. The reaction was followed at 470 minus 500 nm in the Aminco-Chance spectrophotometer. The reaction was started with the addition of 15 μl of 50 m*M* NADH. (A) Untreated ETP; (B) ETP treated with 0.5 m*M* TTF for 5 minutes before the addition of NADH; (C) ETP treated with 1.15 nmoles of rotenone per milligram of protein for 30 minutes at 0°C before assay. Note the difference in time scale for line C. From Gutman and Singer (1970).

thenoyltrifluoroacetone, which completely blocks succinoxidase activity, does not have a significant effect either on the extent of bleaching by NADH, the extent of reoxidation by O_2 (ΔA_{reox}), or on the cycle time. This shows that the reduction of succinate dehydrogenase by electrons originating from NADH does not contribute to the absorbance changes monitored under these experimental conditions. Figure 14C demonstrates the effect of rotenone on the redox cycle. Rotenone greatly increases the cycle time by lengthening the interval between the addition of NADH and its exhaustion by NADH oxidase activity, decreases the rate of reoxidation of the chromophore (ΔA_{reox}/second), and increases the amount of bleached chromophore which remains at the end of the cycle, but does not affect the rate or extent of initial reduction. Piericidin produces identical effects (Gutman and Singer, 1970). In general, the higher the concentration of rotenone or piericidin, the more pronounced these effects become. It may also be noted in Fig. 14C that the reoxidation does not follow a smooth time course but is characterized by a long, slow phase, followed by a faster, second phase. The phase transition corresponds to the time when NADH is exhausted; thus the slow phase is the result of two opposing processes: reduction of the chromophore by NADH and its slow reoxidation by the respiratory chain, since rotenone inhibition is incomplete and some oxidase activity always remains.

The effects of rotenone and piercidin on ΔA_{reox}/second and on the cycle time are understandable if one assumes that the absorbance changes reflect reduction and oxidation of chromophores associated with NADH dehydrogenase, since these inhibitors are known to block electron flux from the enzyme to the respiratory chain. The effect on $\Delta\Delta$ the residual bleaching, is more complex and will be dealt with later on.

The experiments cited were performed with both specifically and unspecifically bound rotenone or piericidin present. A somewhat different behavior is observed if unspecifically bound inhibitor is first removed by washing with bovine serum albumin. This procedure results in a remarkable shortening of the cycle time and a modest increase in the rate of the fast phase of reoxidation. This shows that most of the lag time is due to unspecifically bound inhibitor and that the latter also contributes to the inhibition of the observed rate of reoxidation (ΔA_{reox}/second). When only specifically bound rotenone or piericidin remains, the cycle time bears a strict inverse relationship to residual NADH oxidase activity (Fig. 15, curve A), suggesting that under these conditions the lag time represents a steady-state reduction of the enzyme and that its duration is determined only by the rate of electron flux from the reduced enzyme to the respiratory chain. If the bovine serum

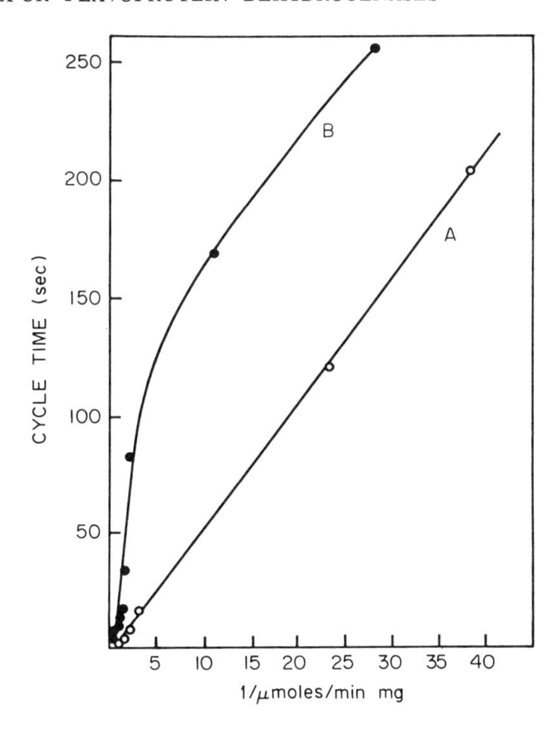

Fig. 15. Variation of the cycle time with the reciprocal value of NADH oxidase activity after inhibition with rotenone. ETP samples were treated for 30 minutes at 0°C with various concentrations of rotenone, and the redox cycle was measured as in Fig. 14. (A) Calculated for specific inhibition; (B) calculated for specific and unspecific inhibition (i.e., no BSA wash). From Gutman and Singer (1970).

albumin wash is omitted, secondary events are superimposed on the linear relation (Fig. 15, curve B), so that the cycle time is longer than predicted from the rate of NADH oxidation. As discussed below, the reason may be that unspecifically bound inhibitor appears to interfere with the redox equilibrium among the iron–sulfur components of the enzyme.

Figure 16 illustrates the relation of NADH oxidase activity in samples inhibited to varying extents with rotenone to the rate of reoxidation of the chromophore (ΔA_{reox}/second) after exhaustion of the NADH. A strictly linear relation is observed, as expected, both in samples inhibited only at the specific sites and in samples containing unspecifically as well as specifically bound rotenone. In order to explain the difference between the two curves, i.e., the effect of unspecifically bound rotenone (or piericidin) on ΔA_{reox}/second, as well as on the cycle time (Fig. 15), it has been postulated that unspecifically bound inhibitor constrains the redox equilibrium of the eighteen iron–sulfur moieties of the dehydrogenase molecule (Gutman and Singer, 1970).

Turning to the question of the identity of the chromophore bleached by NADH, as well as that part of it which remains "permanently bleached" (ΔΔ in Fig. 14), it has been proposed by Gutman and Singer (1970) and the ΔA_{red} value, after correction for a small contribution of a b-type cytochrome (ΔΔ on Fig. 14A), represents iron–sulfur and flavin components of NADH dehydrogenase. This interpretation is consistent with the facts that (a) rotenone and piericidin do not affect the extent of reduction by the substrate; (b) these inhibitors block electron flux from the enzyme to the CoQ pool and, since the conditions are aerobic, reduced cytochromes or $CoQH_2$ are not likely to persist, so that these pigments do not contribute to the absorbance changes at 470 minus 500 nm; (c) since the reduction of other mitochondrial flavoproteins proceeds by way of the CoQ pool, the same reasoning precludes the contribution of other iron flavoproteins to the observed color changes (Gutman and Singer, 1970). An exception is the "permanently bleached"

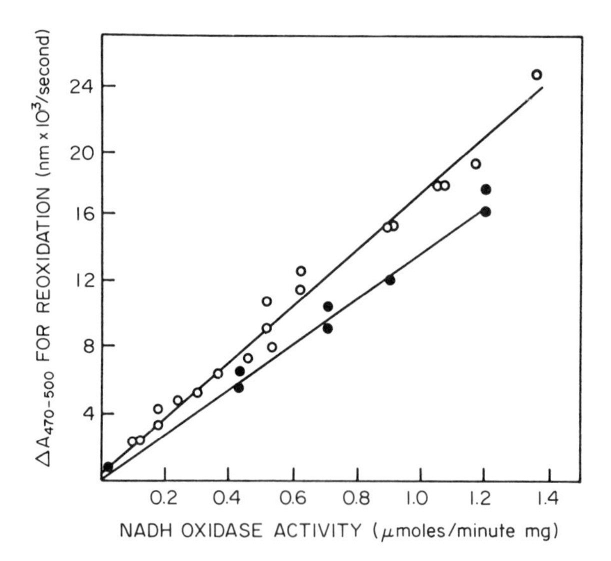

Fig. 16. Dependence of the rate of reoxidation of the chromophore measured at 470 minus 500 nm on the residual NADH oxidase activity after inhibition by piericidin A. ETP (NADH oxidase activity, 1.4 μmoles of NADH/minute per mg) were inhibited with various concentrations of piericidin A for 30 minutes at 0°C. The redox cycles and NADH oxidase activities were measured before (closed circles) and after (open circles) bovine serum albumin washings. The rate of reoxidation was estimated from the steep phase at the end of the cycle. From Gutman and Singer (1970).

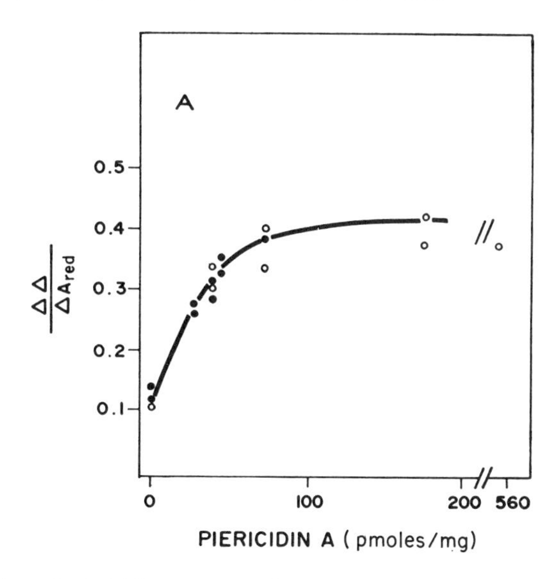

Fig. 17. Effect of piericidin concentration on the amunt of "irreversibly reduced" chromophore in ETP. Samples of ETP were treated with piericidin as in Fig. 16 and the redox cycle was measured on aliquots taken before and after bovine serum albumin (BSA) washings. The abscissa denotes the actual amount of piericidin in contact with the enzyme by using [^{14}C]piericidin A for the inhibition and determining the amount present before (open circles) and after (closed circles) BSA wash by scintillation counting. From Gutman and Singer (1970).

chromophore seen in normal (unihibited) samples, which will be referred to as $\Delta\Delta_0$, and which has been identified as a b-type cytochrome on the basis of its absorption spectrum. It represents a small fraction of the cytochrome b content of inner membrane preparations (Bois and Estabrook, 1969).

In the presence of rotenone or piericidin the fraction of chromophore which remains reduced after exhaustion of the NADH (denoted as $\Delta\Delta_i$) is much greater (Fig. 14C) but the contribution of the b-type cytochrome to the remaining absorbance is not changed, so that it can be corrected for (Gutman and Singer, 1970). The resulting value ($\Delta\Delta_i$ minus $\Delta\Delta_0$) increases with the concentration of inhibitor (Fig. 17) and is also a function of the degree of inhibition of NADH oxidase activity (Fig. 18). Bois and Estabrook (1969) suggested that the $\Delta\Delta_i - \Delta\Delta_0$ value represents iron–sulfur moieties of NADH dehydrogenase on the basis of its absorption spectrum and the fact that residual bleaching is accompanied by a residual EPR signal at $g = 1.94$ (measured at 77°K). This provisional identification is consistent with the later interpretation (Gutman and Singer, 1970) that all of the absorbance change

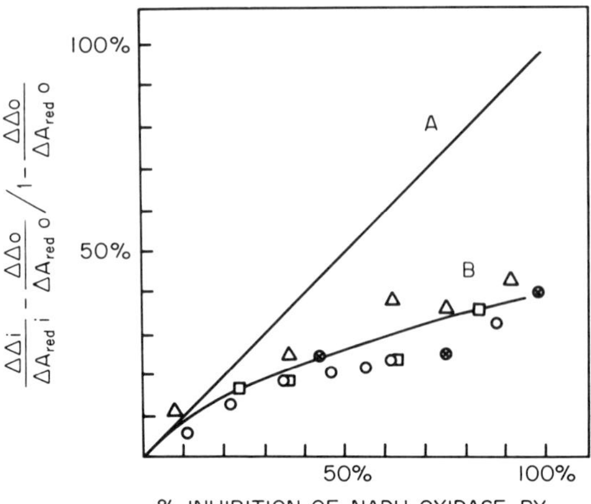

Fig. 18. Variation of the fraction of irreversibly reduced chromophore with the extent of inhibition of NADH oxidase activity. All points refer to specifically bound piericidin A (Δ, □, and ○) or rotenone (⊗). The data are derived from a series of experiments, representing different electron transport particle preparations: A, theoretical line corresponding to a situation where the irreversible bleaching is dependent only on the inhibition of oxidase activity; B, experimental curve. The symbols on the ordinate are $\Delta\Delta_i$ and $\Delta\Delta_0$, extent of irreversible bleaching in the presence and absence of inhibitor, respectively; ΔA^{i}_{red} and ΔA^{0}_{red}, extent of initial bleaching in the presence and absence of inhibitor, respectively. From Gutman and Singer (1970).

at 470 minus 500 nm (after correction for $\Delta\Delta_0$) may be due to reduction of components of the dehydrogenase.

Since only a part of the color bleached by NADH is reoxidized by the respiratory chain in inhibited samples on exhaustion of the substrate, it would seem likely that the iron–sulfur components which remain reduced in the presence of excess O_2 are different from those which are reoxidized. Recent EPR studies at 13°K provided experimental verification of this idea (Gutman *et al.*, 1971c, 1972). As detailed below, it was found that of the iron–sulfur centers of the dehydrogenase only center 2 remains "permanently" reduced at the end of the cycle, while the others are reoxidized. On the addition of ATP, however, center 2 iron sulfur is also reoxidized and the initial color returns (Gutman *et al.*, 1970b, 1971c, 1972).

The logical question arises why in the presence of >90% inhibition of NADH oxidase activity a major fraction (65% in Fig. 13) of the chromophore bleached by NADH is still reoxidized, considering the fact that this inhibitor blocks reoxidation of the enzyme by CoQ. The answer seems to be in the fact that inhibition by rotenone and piericidin is "leaky"; a small fraction of NADH oxidase activity—of the order of 1% or less—always remains. This small, residual electron flux from NADH dehydrogenase to the respiratory chain is thought to represent that fraction of NADH dehydrogenase molecules which are either inaccessible to or incapable of binding rotenone or piericidin and therefore remain functional. Reoxidation of the reduced chromophore in inhibited preparations on exhaustion of NADH may then be visualized in one of two ways. Electrons may flow back from reduced iron–sulfur centers 1, 3, and 4 (but not 2) and from reduced flavin to NAD, which is formed during the cycle. The latter would be thereby reduced and could be reoxidized via the remaining, uninhibited enzyme molecules. Alternatively, one may visualize lateral (intermolecular) electron transport between adjacent NADH dehydrogenase molecules at the level of flavin and iron–sulfur centers 1, 3, and 4 (but not 2), until an uninhibited enzyme molecule is reached, which provides an open path of electron flux to CoQ and thence to the cytochrome system and O_2 (Gutman and Singer, 1970). While either of these indirect routes may be expected to be too slow to contribute significantly to catalytic activity (e.g., NADH oxidase), they would be significant in achieving reoxidation of part of the chromophore reduced by NADH, since this requires a single turnover. Reoxidation of reduced iron–sulfur center 2 by either route would be prevented by the energy gap which appears to exist between center 2 and the other iron–sulfur centers, as indicated by the fact that energizing by ATP is required for reoxidation of center 2 (Gutman *et al.*, 1970b, 1971c, 1972). This is the mechanism currently visualized for the existence of the "permanently" reduced chromophore ($\Delta\Delta_i - \Delta\Delta_0$).

The fact that of the four hitherto identified iron–sulfur centers of the enzyme (Orme-Johnson *et al.*, 1971) only center 2 appears to remain reduced at the end of the cycle may also explain the curious behavior noted in Fig. 18. As shown in that figure, in plots of $\Delta\Delta_i - \Delta\Delta_0$ against inhibition of NADH oxidase obtained by titration with piericidin there is a tendency to reach a plateau, suggesting that only a fraction of ΔA_{red} can remain permanently bleached, no matter how high the concentration of piericidin. The plateau being approached in Fig. 18 may represent that fraction of the ΔA_{red} value which requires ATP for reoxidation, hence largely center 2 iron sulfur.

2. Effect of Mercurials on the Redox Cycle

With a method in hand for monitoring the iron–sulfur centers and flavin component of the enzyme during the events of oxidation–reduction, it is possible to verify the assignment of the different types of —SH groups in the sequence of intramolecular electron transport, given in Section III,B. It has been mentioned that Type IV —SH reacts at relatively low mercurial concentrations (0–30 μM) and is thought to occur relatively early in the electron transport sequence (at or prior to the iron–sulfurs which participate in ferricyanide reduction), since on titration of type IV —SH the NADH–ferricyanide activity is modified. In contrast, type V —SH requires >30 μM mercurial concentration for binding and is thought to belocalized on the O_2 side of the iron–sulfur centers in membrane preparations. The latter assignment is also in accord with the report of Estabrook (Estabrook *et al.*, 1968) that on titration of type V —SH with mersalyl, NADH can still reduce the flavin as well as the iron–sulfur moieties which are observed by means of EPR at 77°K.

As seen in Fig. 19, titration with mersalyl produces entirely different effects on the various parameters of the redox cycle than rotenone or piericidin. While the amount of "irreversibly bleached" chromophore

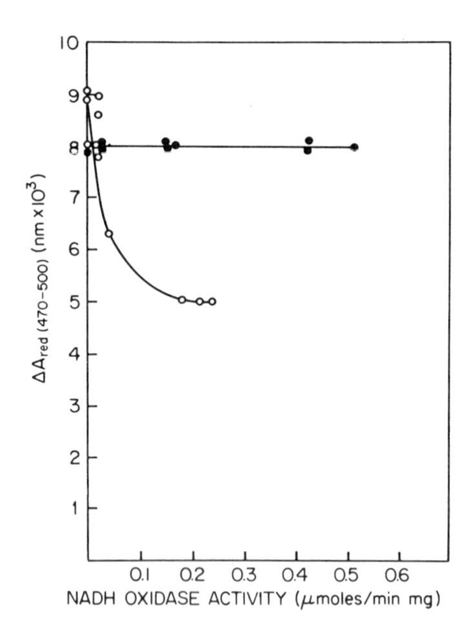

Fig. 19. The effect of piericidin A on the steady state reduction on the chromophore at 470 minus 500 nm in mersalyl-treated ETP preparations. ETP samples were treated with 80 μM mersalyl, as described in Fig. 12. Following centrifugation, NADH oxidase activity and the redox cycle were measured after incubation with various concentrations of piericidin A for 30 minute incubation at 0°C. Open circles, mersalyl-treated enzyme; shaded circles, untreated enzyme. From Gutman and Singer (1970).

is unaffected, the extent of bleaching by NADH (ΔA_{red}) declines to a moderate extent already at low mercurial concentration, where type V —SH does not react: this is thought to be the result of blocking type IV —SH group. A more pronounced effect is seen on monitoring the rate of reoxidation of the chromophore: this is greatly inhibited in the same range of mercurial concentration (30 to 80 μM) where type V —SH is titrated. This is in accord with the localization of this thiol on the O_2 side of the iron–sulfur cênters monitored at 470 minus 500 nm.

If ΔA_{red}, the maximal extent of bleaching, represents a balance between the rates of reduction and reoxidation of the chromophore, and if mersalyl inhibits both processes, then any inhibition of the reoxidation by another agent, superimposed on this balance, should lead to more extensive reduction. It may be seen in Fig. 19 that when piericidin inhibition is superimposed on the action of mersalyl, the greater the block of NADH oxidase activity, the more extensive is the bleaching of the chromophore during the steady state. When piericidin inhibition becomes maximal, the ΔA_{red} value in mersalyl-treated particles even exceeds the extent of bleaching observed in untreated particles, showing that in normal ETP the ΔA_{red} value does not represent complete bleaching of the fiavin and iron–sulfur components but is a function of the balance between their reduction and reoxidation (Gutman and Singer, 1970).

3. Effect of ATP on the Redox Cycle

It has been known for many years that energy coupling site I is located in the immediate vicinity of NADH dehydrogenase. Efforts at more exact localization have been hampered by the fact that in assays involving only part of the electron transport sequence of the dehydrogenase, such as NADH–ferricyanide activity (Schatz and Racker, 1966), ATP does not seem to be generated. As a result various authors have suggested that coupling site I is on the substrate side of the enzyme, others that it was on the O_2 side, according to some before the piericidin combining site, according to others after the piericidin site in the electron transport sequence. The first demonstration that iron–sulfur components of the dehydrogenase are involved in phosphorylation at this site and data permitting partial localization of the site within the electron transport sequence of the enzyme came from studies on the effect of ATP on the redox cycle of the enzyme (Gutman *et al.*, 1970a; Gutman and Singer, 1971). In the experiment of Fig. 20 absorbance changes due to iron–sulfur moieties of NADH dehydrogenase were followed at 470 minus 500 nm, so that a downward deflection denotes reduction of this chromophore (curve A), while absorbance changes attributable to cyto-

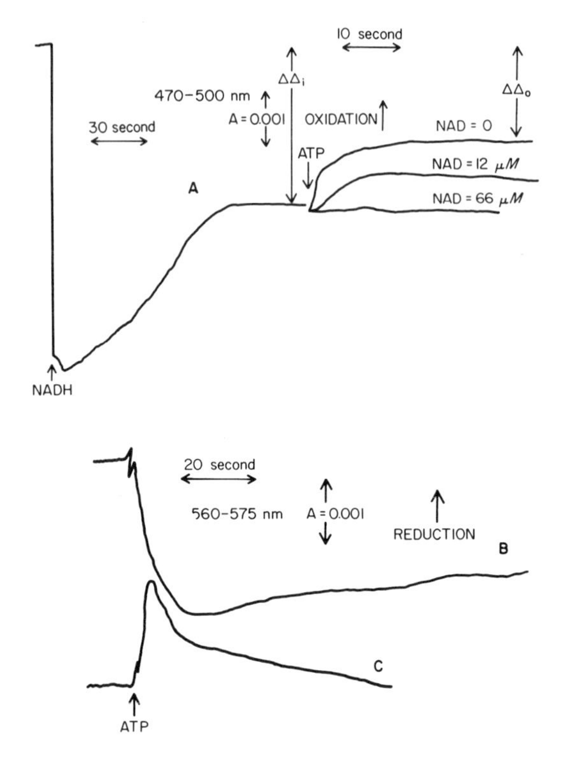

Fig. 20. The effect of ATP on the redox state of the NADH dehydrogenase and cytochrome b. (A) Redox cycle of NADH dehydrogenase in rotenone-inhibited ETP_H, followed by addition of 3 m*M* ATP in presence of various concentrations of dinitrophenol. Note the change in time scale after the addition of ATP. (B) The effect of ATP on the redox state of cytochrome b in uninhibited ETP_H. (C) Effect of ATP on the redox state of cytochrome b in rotenone-inhibited ETP. From Gutman *et al.* (1970b).

chrome b were monitored at 560 minus 575 nm (curves B and C), so that a downward deflection denotes oxidation of cytochrome b. It may be seen that the addition of ATP causes rapid reoxidation of the "permanently reduced" chromophore measured at 470 minus 500 nm. ATP also causes oxidation of the b-type cytochrome in uninhibited preparations (curve B), which results in reduced absorbance at both 560 minus 575 nm and 470 minus 500 nm but in the presence of rotenone (curve C) transient *reduction* of cytochrome b occurs. Since the absorption changes due to cytochrome b (the $\Delta\Delta_0$ value) may be easily corrected for, it can be calculated that the ATP-induced reoxidation of the iron–sulfur chromophore ($\Delta\Delta_i - \Delta\Delta_0$) is complete. In fact, in some experiments the final absorbance at 470 minus 500 nm was greater than at the start of the experiment, showing that this chromophore was partially reduced prior to adding NADH (Gutman *et al.*, 1971c).

Figure 20 further shows that the uncoupler dinitrophenol and oligomycin, an inhibitor of oxidative phosphorylation, abolish the effect of ATP at concentrations compatible with their actions on the oxidative phosphorylation system. Figure 21 demonstrates that at a series of concentrations dinitrophenol affects the ATP-driven reduction of NAD^+ and the ATP-induced oxidation of the iron–sulfur chromophore to identical extents. These observations and the fact that while all other features of the redox cycle are observed in phosphorylating as well as nonphosphorylating membrane preparations, the ATP effect is seen only in the former, suggest that the action of ATP on the reoxidation of the chromophore is intimately related to the energy conservation mechanism (Gutman and Singer, 1971; Gutman *et al.*, 1970b, 1972).

Understanding of the mechanism of the ATP-induced reoxidation of the iron–sulfur chromophore requires identification of the electron acceptor. An obvious possibility is that the oxidized NAD accumulated

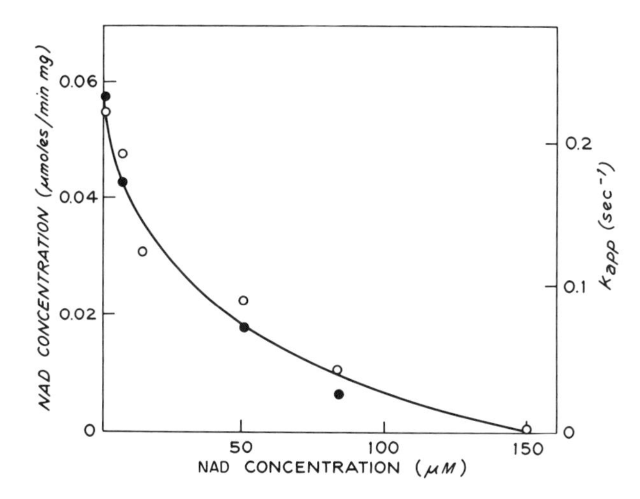

Fig. 21. Effect of dinitrophenol on the energy-linked reduction of NAD by succinate and on the rate of ATP-induced oxidation of the chromophore. ETP_H was washed with 0.25 *M* sucrose—50 m*M* tris—5 m*M* $MgSO_4$, pH 7.4, and resuspended in the same medium at a concentration of 4 mg of protein per ml. Aliquots were used to measure rates of NAD reduction at various concentrations of dinitrophenol (DNP). Another aliquot was treated with 0.5 nmole of piericidin A per mg of protein for 30 minutes at 0°C, which gave complete inhibition of energy-linked NAD reduction. The redox cycle was then measured at 470 minus 500 nm in the presence of varying concentrations of NAD. ATP was added when the absorbance leveled off, and the first order rate constant for oxdation of the chromophore (k_{app}) was determined. From Gutman *et al.* (1972).

during the first phase of the redox cycle fulfills this role. In accord with this postulate, on incubation of an ETP_H preparation with *Neurospora* NADase at the end of the redox cycle and then subsequent addition of ATP, it was found that reoxidation of the chromophore was 80% inhibited (Gutman *et al.*, 1970a). This experiment suggests that in piericidin-blocked membrane preparations NAD is the main electron sink for the ATP-triggered reoxidation of the "irreversibly reduced" chromophore.

If the identification of the "irreversibly bleached" chromophore as iron–sulfur components of NADH dehydrogenase (Bois and Estabrook, 1969; Gutman and Singer, 1970) is accepted [and this identification has received added support from recent EPR studies at 4° to 20°K (Gutman *et al.*, 1971c, 1972), as discussed below], these experiments provide direct evidence for the role of iron–sulfur centers of NADH dehydrogenase in phosphorylation at site 1. They also permit closer localizaton of this site than has been hitherto possible. The conclusion that site 1 is associated with the dehydrogenase and, assuming a linear arrangement of components, that it is located between iron–sulfur center 1 of the enzyme and both type V —SH group and iron–sulfur center 2 (Singer and Gutman, 1971; Gutman *et al.*, 1970b, 1971c, 1972) is based on the following considerations. Prior work by Hinkle *et al.* (1967) suggested that coupling site 1 is located on the substrate side of the rotenone block. The present experiments fully confirm this, since the chromophore reoxidized in the presence of ATP remains reduced in rotenone-blocked preparations until energy is supplied for its reoxidation. Since the specific binding site of rotenone and piericidin is in the vicinity of type V —SH group of the dehydrogenase (Gutman *et al.*, 1970d, 1971b), it also follows that site 1 is on the substrate side of Type V thiol. Second, recent EPR studies by Gutman *et al.*, (1971c, 1972), detailed in the next section, have shown that during the redox cycle monitored at 470 minus 500 nm both the low-potential iron–sulfur center 1 and the high-potential center 2 are reduced by NADH, but while the former is reoxidized by the respiratory chain in the absence of ATP once NADH is exhausted, center 2 is reoxidized only after the system is energized by ATP.

E. EPR Studies on the Iron–Sulfur Centers

Starting with the discovery by Beinert and Sands (1960; Sands and Beinert, 1960) that the addition of NADH to mitochondria and submitochondrial particles elicits an asymmetric, strongly temperature-dependent EPR signal at $g = 1.94$ (minor component at $g = 2.00$), EPR studies on NADH dehydrogenase have until recently been conducted at 77°K. At this temperature, this signal is readily distinguishable from those pertaining to iron–sulfur components of succinate dehydrogenase and an iron–sulfur protein in the cytochrome $b\text{–}c_1$ region. The signal, which was correctly ascribed to nonheme iron components of NADH dehydrogenase in these early studies (Beinert and Sands, 1960; Beinert *et al.*, 1963b), was seen in all membranal preparations and in the soluble, high molecular weight form of the enzyme (Beinert and Sands, 1960; Beinert *et al.*, 1965) but was absent in the low molecular weight form (Beinert and Sands, 1959; Beinert *et al.*, 1965). It was early recognized, in fact, that the absence of the EPR signal at $g = 1.94$ is one of the signs of modification in the low molecular weight form of the enzyme (Beinert *et al.*, 1963a, 1965).

At 77°K the soluble, purified enzyme gives qualitatively the same EPR spectra on reduction with NADH as membrane-bound preparations (Beinert *et al.*, 1965). Since the turnover number of the dehydrogenase is far too high to permit studies of the kinetics of the reduction with NADH as substrate, slowly oxidized substrate análogues and low temperature were used in comparing catalytic activity in the ferricyanide assay with the rate of appearance of the EPR signal. Under these conditions an excellent correlation was found between the two parameters compared: the half-time of the appearance of the $g = 1.94$ signal on mixing enzyme and substrate corresponded to one catalytic cycle in the ferricyanide assay (Beinert *et al.*, 1963a, 1965). These experiments provided solid support for the contention that the spectrophotometric ferricyanide assay, at V_{max}, measures the true activity of the enzyme. In other experiments it was shown that the rate-limiting step in the catalytic cycle, as monitored by ferricyanide, is the reduction of the iron–sulfur components by the substrate, not their reoxidation (Beinert *et al.*, 1965).

Table III shows that on thermal degradation of the enzyme to the low molecular weight form ferricyanide activity and substrate-induced EPR signal at $g = 1.94$ decay at the same rate. This is in agreement with the hypothesis that iron–sulfur moieties (those designated as center 1 in the discussion to follow) are involved in electron transport from NADH to ferricyanide, at least in the high molecular weight form.

Since anaerobic titration of the purified enzyme with NADH did not yield a sharp end point, the question of how many iron atoms of the enzyme correspond to the substrate-induced EPR signal at full development had to be approached indirectly. Since the time required for full appearance of the EPR signal at $g = 1.94$ corresponds to the oxidation of 2 moles of substrate per mole of enzyme or 4 e equivalents, no more than one flavin and a pair of iron atoms of four irons could be

TABLE III
TIME COURSE OF INACTIVATION OF NADH DEHYDROGENASE, EMERGENCE OF CYTOCHROME REDUCTASE, AND DECAY OF SUBSTRATE-INDUCED EPR SIGNAL AT $g = 1.94$[a]

	Ferricyanide reduction		Cytochrome c reduction		
Minutes at 35°C	Rate (μmoles NADH/ minute/ml)	Activity (% initial)	Rate (μmoles NADH/ minute/ml)	Activity (% maximal)	Substrate-reducible EPR signal[b] (% initial)
0	8800	100	2.2	1.7	100
5	7500	85			82
10	7150	81	19.0	15	73
13	5710	65	24.6	19	62
20	4650	53	46.2	36	51
40	3320	38	65.5	50	40
60	2000	23	92.5	71	32
120	1500	17	130	100	20
225	1000	11	65	50	16

[a] The enzyme (27.3 mg/ml; specific activity, 324) in 0.03 *M* phosphate, pH 7.8, was incubated aerobically at 34.9°C in the dark. Samples removed at the times indicated were rapidly cooled in ice. Aliquots were assayed after appropriate dilution with 0.06 *M* triethanolamine–1% serum albumin (w/v), pH 7.8, for ferricyanide activity (triethanolamine, pH 7.8), and for cytochrome reductase activity. Results were calculated for infinite concentration of electron acceptor. Ferricyanide activity is expressed as percentage of initial activity, and cytochrome reductase as percentage of maximal activity emerging. Aliquots of 0.20 ml were placed in anaerobic quartz tubes, 0.02 ml of 1 *M* glycine, pH 9, was added, and the tubes were repeatedly evacuated and filled with purified N_2. Under a flow of N_2, 0.01 ml of 0.1 *M* NADH was added; the samples were frozen 1 minute after mixing, and the EPR was measured at −176°C at 25-mW microwave power and 19 G modulation amplitude. The percentage of the initial peak-to-peak amplitude of the $g = 1.94$ signal (first derivative) is given in the table. From Beinert *et al.* (1965).

[b] EPR signal appears only on reduction. Signal size was, therefore, measured after addition of NADH anaerobically. The values were corrected for a small signal (4% maximal signal observed after addition of NADH) recorded with the samples before reduction.

accommodated out of the sixteen to eighteen present. If two electrons are taken up or reside in a cluster of two iron atoms, then a maximum of eight irons would be accounted for. More recent studies at 4° to 20°K (Orme-Johnson *et al.*, 1973) suggest that probably only eight and at the most sixteen iron–sulfurs are represented by the full EPR signal.

When the EPR spectrum of NADH dehydrogenase is examined at very low temperatures (4° to 20°K), it becomes apparent that, instead of a single asymmetric peak at $g = 1.94$ seen at 77°K, several components are detectable. Studies in Chance's laboratory (Ohnishi *et al.*, 1970) with particles from *C. utilis* resulted in the detection of at least two iron–sulfur components associated with NADH dehydrogenase, while Orme-Johnson *et al.* (1971, 1973), using a complex I preparation from beef heart, detected four iron–sulfur centers in the enzyme. The four centers seen in preparations of the heart enzyme at 4° to 20°K are of the ferredoxin type according to their average g values. Their positions on the g scale are at center 1 (2.022, 1.938, 1.923); center 2 (2.054, 1.922); center 3 (2.101, 1.886, 1.864); center 4 (2.103, 1.861). Although these are not true g values, they are likely to be close to them. These four centers have also been seen in ETP, ETP_H, as well as in the soluble high molecular weight enzyme (Gutman *et al.*, 1971c). In submitochondrial particles they are present at approximately equimolar concentrations. Resolution of the four iron centers is strongly temperature dependent: centers 2, 3, and 4 are not well resolved at temperatures exceeding 25°K, and the resonances of centers 3 and 4 overlap sufficiently, so that they are not always resolved even below 25°K (Orme-Johnson *et al.*, 1971). Center 1 is the only iron–sulfur component detectable at 77°K, where previous studies were carried out; it is reoxidized readily by ferricyanide, so that it may possibly be the point where electrons are donated from the enzyme to ferricyanide in routine assays. It is clear that in earlier kinetic studies at 77°K (Beinert *et al.*, 1963a, 1965), demonstrating agreement between catalytic activity in the ferricyanide assay and the rate of appearance of the $g = 1.94$ signal, it was the reduction of center 1 iron–sulfur which was being followed. The two resonances of center 1 seen at 4° to 20°K are not resolved, however, at liquid-nitrogen temperature. Reductive titrations have permitted assignment of the relative oxidation–reduction potentials as $2 > 3 > 4 > 1$ (Orme-Johnson *et al.*, 1971).

These studies laid the necessary foundation for correlating spectrophotometric observations on the oxidation reduction of the iron–sulfur components of the enzyme, described in the previous section, with their EPR behavior. The redox cycle of the enzyme was monitored spectrophotometrically at 470 minus 500 nm in piericidin-treated phosphorylating membrane preparations; samples were withdrawn and quickly frozen in liquid nitrogen and the EPR signal subsequently recorded at 13°K. In these studies the iron–sulfur components were found to be partially reduced at the outset of the experiment; in the experiments illustrated in Figs. 22–24 center 1 and 2 were 5 and 50% reduced, respectively, prior to the addition of NADH. It was found that NADH caused immediate full reduction of centers 1 and 2 and on exhaustion of the NADH, when

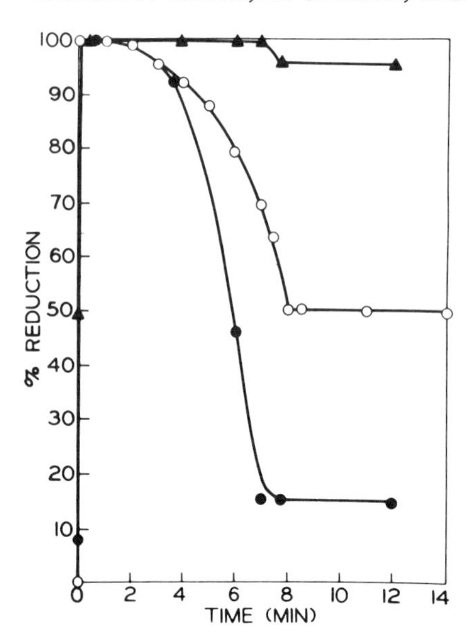

Fig. 22. Correlation of the redox state of centers 1 and 2 with the changes in absorbance at 470 minus 500 nm during oxidation of NADH. ETP_H (NADHD:36 μmoles/minute/mg at 30°C) was inhibited by 0.50 nmole of piericidin A/mg protein, suspended at 10 mg protein/ml in 0.18 *M* sucrose containing 50 m*M* tris chloride, 5 m*M* $MgCl_2$, and 2% (w/v) of BSA, and was reduced with 250 μ*M* NADH. The changes in absorbance were recorded and at intervals samples were removed and rapidly frozen in an isopentane bath. The sampling time was 10 seconds or less. The ordinate represents changes in reduction level (taking the reduction by NADH as 100%) according to ΔA 470 minus 500 nm (○) and EPR of centers 1 (●) and 2 (▲). The total content of center 1 was 60% of that of center 2. Estimates of absolute concentrations were derived by integration and comparison to a Cu-EDTA standard. From Gutman *et al.* (1971c).

the color began to return, center (1 and 3) iron–sulfurs became reoxidized by the respiratory chain at a rate paralleling the optical changes, but no significant reoxidation of center 2 was observed (Fig. 22) (Gutman *et al.*, 1971c). Thus, in accord with the previous assignment that the "permanently bleached" chromophore corresponds to iron–sulfur components, these experiments showed that center 2 iron–sulfur remains reduced at the end of the cycle. Figure 23 shows the effect of ATP on the reoxidation of the chromophore and of center 2 iron–sulfur, while Fig. 24 illustrates the EPR signals observed at various stages of the cycle. It may be seen that ATP indeed caused reoxidation of a large part of the center 2 iron–sulfur which had remained reduced at the end of the cycle. The rate of the ATP-induced return of color is faster than the decay of center 2 EPR signal, so that it remains possible that iron–sulfur components other than those which were monitored in this study (or other unidentified components) contribute to the "permanently reduced" chromophore to some extent (Gutman *et al.*, 1972). It is nevertheless clear that ATP causes reoxidation of iron–sulfur components of the enzyme and that energy coupling site 1 may be tentatively localized between centers 1 and 2 on the enzyme. This assignment of coupling site 1 was subsequently confirmed by Ohnishi *et al.* (1972).

One of the important advances which has resulted from recent EPR studies of the enzyme at 13°K is the assignment of the relative redox potential of the different iron–sulfur centers from reductive titrations (Orme-Johnson *et al.*, 1971). This may be the only basis we have at present for predicting the possible reaction sequence of these iron–sulfur

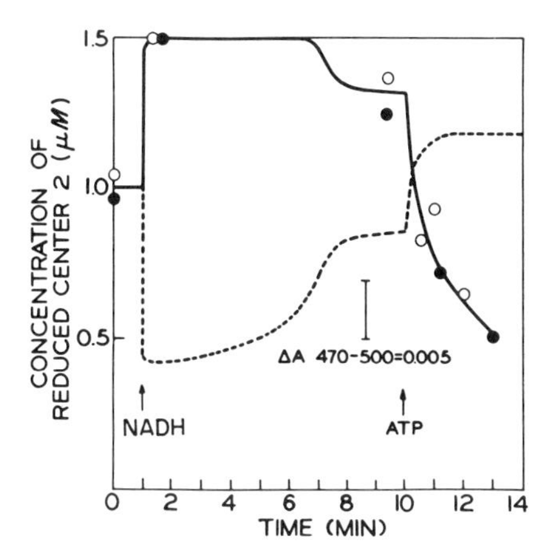

Fig. 23. Correlation between the redox state of iron–sulfur center 2 and absorbance changes at 470 minus 500 nm and the effect of ATP on these parameters. An ETP_H sample (NADH dehydrogenase activity, 41 μmoles NADH/minute/mg at 30°C), inhibited with 0.525 nmoles of piericidin A/mg of protein, was suspended in 0.18 *M* sucrose, 50 m*M* tris, 5 m*M* $MgCl_2$, 2% (w/v) bovine serum albumin to a concentration of 10 mg of protein/ml. The redox cycle was carried out as previously described (Gutman and Singer, 1970). NADH was added at 0 time, causing immediate reduction, the extent of which is shown by samples taken at 1 minute. Reoxidation ensuing on exhaustion of NADH is shown by the samples at 9 minutes. ATP (6 m*M*) was added at 10 minutes and samples were withdrawn at the times shown. The sampling time was less than 10 seconds; samples were frozen in isopentane. The dashed line represents absorbance changes at 470 minus 500 nm; the solid line, the micromolar concentration of reduced iron–sulfur center 2, calculated by comparison with Cu-EDTA standards. The solid and open circles represent samples from separate, parallel experiments. Note that the scale on the ordinate refers to the solid line only; the scale for absorbance is given separately in the figure. From Gutman *et al.* (1972).

components in the intramolecular electron transport sequence of the enzyme.

The fact that in experiments on the redox cycle of the type illustrated in Fig. 22 it was found that iron–sulfur centers 1 and 3 are reoxidized by the slow leak to the respiratory chain in piericidin-blocked prepara-

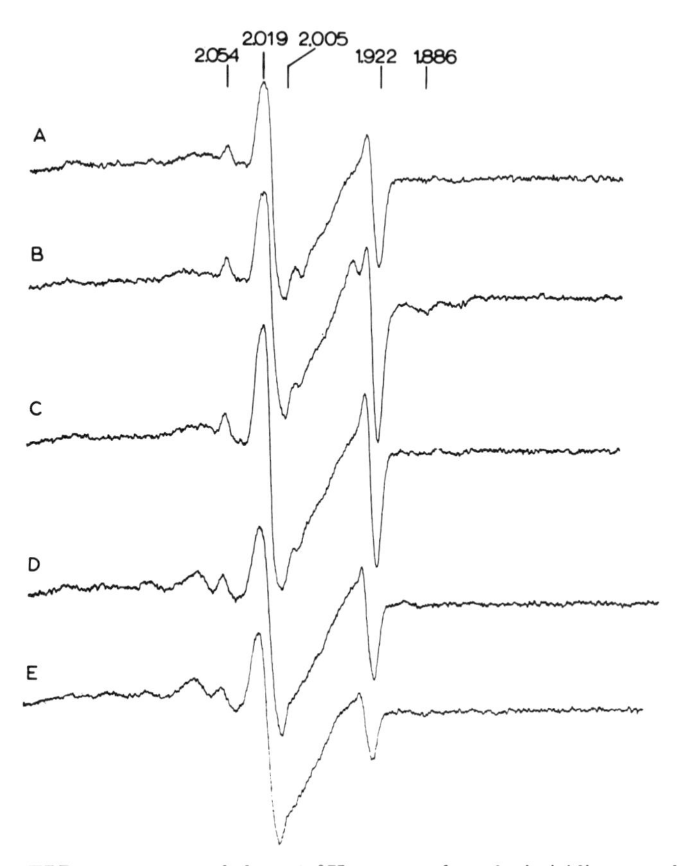

Fig. 24. EPR spectra recorded at 13°K on samples of piericidin-treated ETP_H before and after the oxidation–reduction cycle initiated by NADH and after addition of ATP. The experiment is that of Fig. 23. (A) Particles as prepared and piericidin treated; (B) after aerobic addition of 220 μM NADH; (C) after reoxidation; (D) 1 minute after addition of 6 mM ATP; (E) 5 minutes after ATP. The concentrations given are final concentrations. Note that after addition of ATP extra lines appear, particularly visible at low field. These lines are due to Mn^{2+}, which is a contamination of ATP, although the ATP used was purified by chromatography on Dowex 50 resin. There is also a large strongly temperature-dependent signal with derivative peaks at $g = 2.019$ and 2.005, which stems from the oxidized form of an unknown electron acceptor. However, neither this nor the Mn^{2+} resonances interfere with the evaluation of the signals from centers 1 and 2.

tions but center 2 is not until ATP is added (Gutman *et al.*, 1971c, 1972) is in accord with the sequence* of redox potentials: $2 > 3 > 4 > 1$. Thus center 2 appears to be nearest the oxygen side of the enzyme and the energy gap is on the substrate side of this center. (Center 4 is not distinguishable at protein concentrations compatible with such spectrophotometric studies.) The possible mechanism by which the low-potential iron–sulfur components on the substrate side of the energy gap are reoxidized was discussed in the preceding section.

The same sequence of four iron–sulfur centers appears to be operative in the soluble, purified enzyme. It is known from prior work (Beinert *et al.*, 1965) that when this preparation is reduced by NADH and acetylpyridine NAD is subsequently added, the ensuing reoxidation of the iron–sulfur species (center 1) detected at 77°K is incomplete. When the experiment with the soluble enzyme was recently repeated monitoring the EPR signals at 13°K (Gutman, *et al.*, 1971c), it was found that acetylpyridine NAD oxidized approximately 70% of reduced center 1, 60% of center 3, and only about 40% of center 2, in agreement with the order of redox potentials given above.

IV. DIHYDROOROTATE DEHYDROGENASE

A. Introduction

Dihydroorotate dehydrogenase was originally purified from the obligate anaerobic, *Zymobacterium oroticum*, by Lieberman and Kornberg (1953) who isolated the organism by enrichment culture on orotic acid media. Friedmann and Vennesland later developed a purification procedure which yielded a homogeneous crystalline enzyme (Friedmann and Vennesland, 1958, 1960). The enzyme was shown to be a flavoprotein, with the unique property of possessing equimolar amounts of FAD and FMN, and containing one atom of nonheme iron per mole of flavin (Friedmann and Vennesland, 1958, 1960). These findings have been confirmed by subsequent workers, who also determined that "labile" sulfide was present in equimolar quantities with the iron (Aleman and Handler, 1967; Aleman *et al.*, 1965; Brumby *et al.*, 1965). The *Zymobacterium* enzyme is specific for reaction with NAD: Udaka and Vennesland have also isolated a NADP specific enzyme of similar properties from an

* This denotes a temporal sequence of events, rather than a topographic one. In fact, the physical arrangement of the iron-sulfur centers in the enzyme is not necessarily linear.

unidentified aerobic bacterium (Udaka and Vennesland, 1962) as have Kerr and Miller from a pseudomonad (Kerr and Miller, 1967). The above enzymes all appear to have basically similar properties (cofactor content and reactivity either with NAD or NADP) and are considered metabolically to be degradative enzymes, concerned in the breakdown of orotate. In contrast, the biosynthesis of pyrimidines, proceeding from aspartate and carbamyl phosphate through orotate to uridylate, appears to utilize a dihydroorotate dehydrogenase of considerably different characteristics. Particulate biosynthetic enzymes have been reported from pseudomonad species (Miller and Kerr, 1967a), *E. coli* (Taylor and Taylor, 1964; Kerr and Miller, 1968) and beef liver (Miller and Curry, 1969; Miller *et al.*, 1968) and shown to link dihydroorotate oxidation to reduction of coenzyme Q. More recently the purification and properties of a soluble biosynthetic dihydroorotate dehydrogenase have been reported from the anaerobe, *Lactobacillus bulgaricus* (Taylor *et al.*, 1971). This enzyme has no activity with pyridine nucleotides either as electron donors or acceptors, contains only one molecule of flavin per molecule of protein (FMN), and no nonheme iron (Taylor *et al.*, 1971).

In the following sections, unless specified otherwise, the properties discussed apply to the iron-containing enzyme from *Zymobacterium oroticum.*

B. Physicochemical Properties

The minimum molecular weight per flavin was estimated by Friedmann and Vennesland as 31,000, i.e., 62,000 for FAD + FMN (Friedmann and Vennesland, 1960). By sedimentation equilibrium studies, Aleman and Handler (1967) determined the molecular weight to be 115,000 ± 7000. Similar values were reported by Nelson and Handler, who also determined that in concentrated guanidine solutions the subunit molecular weight was approximately 30,000 (Nelson and Handler, 1968). Peptide mapping revealed that these subunits were essentially identical in amino acid composition (Nelson and Handler, 1968). Thus it appears that this enzyme is a tetramer composed of four identical polypeptide chains, arranged in such a way as to bind two molecules of FMN, two molecules of FAD, and four atoms of iron and their associated "labile" sulfur atoms. The actual arrangement in the protein is presently unknown; it is commonly assumed that per molecule of enzyme there are two identical active sites, each comprising half the above composition. The unusually low iron–sulfur content, as well as the equimolar quantities of FMN and FAD, make this a rather atypical metalflavoprotein.

In keeping with the low iron content and the known spectral properties of simple iron–sulfur proteins the absorption spectrum is much closer to that of simple flavoproteins than is the case with other metalloflavoproteins (Rajogopalan and Handler, 1964). The ϵ_{454} value, expressed per molecule of flavin was estimated at 17.4×10^3 $moles^{-1}$ cm^{-1} by Friedmann and Vennesland (1960) and as 18.3×10^3 $moles^{-1}$ cm^{-1} by Miller and Massey (1965a).

C. Catalytic Reactions

Friedmann and Vennesland (1958, 1960) demonstrated that the crystalline enzyme catalyzes the following reactions:

$$\text{Orotate} + NADH + H^+ \underset{(b)}{\overset{(a)}{\rightleftharpoons}} \text{Dihydroorotate} + NAD^+ \quad (7)$$

$$NADH + H^+ + O_2 \longrightarrow NAD^+ + H_2O_2 \quad (8)$$

$$\text{Dihydroorotate} + O_2 \longrightarrow \text{Ororate} + H_2O_2 \quad (9)$$

Reactions (7a) and (7b) thus constitute dehydrogenase reactions; reactions (8) and (9) oxidase reactions. The redox potential of the dihydroorotate/orotate couple has been found to be —0.252 V (Krakow and Vennesland, 1961); hence reaction (7a) is much easier to follow experimentally than reaction (7b). Friedmann and Vennesland (1958) also determined that 5-fluoroorotate is an even better acceptor than orotate, and that acetylpyridine NAD will replace NAD in reaction (7b). It was also found that methylene blue was an effective electron acceptor in place of O_2 in reactions (8) and (9) (Friedmann and Vennesland, 1958). Aleman and Handler (1967) found that a number of 5-substituted orotates were substrates, although they were reduced at widely different rates; the compounds tested were the 5-fluoro, bromo, iodo, methyl, amino, and nitro derivatives, as well as the 2-thio derivative.

An unusual requirement for cysteine activation was also found by Friedmann and Vennesland (1958, 1960) who showed that reactions (7a), (7b), and (9) were very much activated in the presence of cysteine; on the other hand, the NADH oxidase activity was not dependant on cysteine. It was proposed that the cysteine effect was due to activation of a sulfhydryl group essential for reaction with the pyrimidine moiety (Friedmann and Vennesland, 1958, 1960); this concept was strengthened by the finding (Miller and Massey, 1965a) that pretreatment with cysteine followed by dialysis or passage through Sephadex had the same effect as including cysteine in the assay mixture.

In common with several other iron–sulfur flavoproteins, dihydroorotate dehydrogenase was found to catalyze the reduction of cytochrome c with NADH or dihydroorotate in an O_2-dependent reaction (Rajagopalan and Handler, 1964; Aleman *et al.*, 1965; Miller and Massey, 1965b; Aleman and Handler, 1967) owing to the univalent reduction of O_2 to O_2^- (McCord and Fridovich, 1969, 1970), which is the actual reductant of cytochrome c. This reduction was found to be strongly inhibited by the iron chelator, Tiron; this inhibition was interpreted as providing evidence for the reduced iron–sulfur chromophore being the site of reaction with O_2 (Rajagopalan and Handler, 1964; Aleman *et al.*, 1965; Aleman and Handler, 1967). However, it was found that Tiron was a good O_2^- trapping agent (Miller and Massey, 1965; Miller, 1970). The evidence with the similar reaction catalyzed by xanthine oxidase argues very strongly in favor of reduced flavin being the group on the enzyme responsible for the univalent reduction of O_2 (Komai *et al.*, 1969; Massey *et al.*, 1969; Ballou *et al.*, 1969).

In addition to methylene blue, *p*-nitroblue tetrazolium and ferricyanide were found to be efficient electron acceptors; the catalytic activity with ferricyanide is greater than with O_2 as acceptor (Miller and Kerr, 1966).

Miller and Massey (1965a) studied the kinetics of reactions (7)–(9) at pH 6.8, 25°C, varying the concentrations of both electron donor and acceptor. In each case a series of parallel Lineweaver–Burk plots was obtained, indicating that the reactions probably proceeded through binary complex mechanisms. These studies enabled the calculation of the kinetically significant V_{max} and K_m values, i.e., obtained by extrapolation to infinite concentrations. The values obtained under these conditions are shown in Table IV. Two sets of turnover values are given, depending on whether the turnover number is calculated per mole of total flavin, or per mole of FAD or FMN. As will be discussed later, the latter situation would appear to be the more likely (i.e., that the catalytic site involves both FAD and FMN).

Stopped-flow studies conducted under the same conditions of pH, but at 20°C, were also reported (Miller and Massey, 1965). It was found that anerobic reduction with NADH resulted in rapid decrease in absorption at 450 nm and increase in absorption at 600 nm (k, 1×10^4 min^{-1}). These rapid changes were followed by slower secondary decreases in absorption at both wavelengths (k, 35 min^{-1}). The extent of bleaching

TABLE IV

KINETIC CONSTANTS OF DIHYDROOROTATE DEHYDROGENASE

Reaction	Extrapolated maximum turnover number (minute/liter)		Michaelis constants
	per mole total flavin	per mole of FAD or FMN	
7a. NADH + H^+ + orotate → NAD^+ + dihydroorotate	2200	4400	$K_{NADH} < 5 \times 10^{-5}\ M$ $K_{orotate} = 1.8 \times 10^{-3}\ M$
7b. NAD^+ + dihydroorotate → NADH + H^+ + orotate	880	1760	$K_{dihydroorotate} = 1.7 \times 10^{-4}\ M$ $K_{NAD} = 3.6 \times 10^{-4}\ M$
8. NADH + H^+ + O_2 → NAD^+ + H_2O_2	2500	5000	$K_{NADH} = 3.6 \times 10^{-5}\ M$ $K_{O_2} = 5.4 \times 10^{-4}\ M$
9. Dihydroorotate + O_2 → orotate + H_2O_2	110	220	$K_{dihydroorotate} = 1.8 \times 10^{-5}$ $K_{O_2} < 10^{-5}\ M$

at 450 nm occurring in the rapid phase was so extensive ($\Delta\epsilon = 9.5 \times 10^3$ per mole of flavin) as to indicate reduction of both flavins at least to the semiquinoid form as well as reduction of the iron–sulfur. Reduction of flavin to the semiquinoid level, was, of course, evident from the increase in absorbance at 600 nm.

When anaerobic reduction of the enzyme by dihydroorotate was studied, very different results were obtained. Again there was a rapid decrease in absorbance at 450 nm ($k \sim 2.7 \times 10^3$ min^{-1}) but of small magnitude ($\Delta\epsilon_{450} = 2.4 \times 10^3$ per mole of flavin). There was no detectable rapid change in absorbance at 600 nm. However, in a much slower reaction ($k \sim 13$ $minute^{-1}$) increase in absorbance at 600 nm was observed, coincident with a large decrease in absorbance at 450 nm ($\Delta\epsilon_{450}$ slow 6.2×10^3 per mole of flavin). These results were interpreted as indicating no rapid reduction by dihydroorotate of the flavin, but reduction of the iron–sulfur linkage in such a way that the $g = 1.94$ signal would not be elicited.

Clarification of these results came from studies of the kinetics of reduction by EPR (Aleman *et al.*, 1968a,b). It was found that with either NADH or dihydroorotate as substrate, EPR signals due to flavin radical and iron were produced very rapidly. Further, it was found that different flavin radical signals were elicited with the two substrates, depending on the pH. At pH 8.2, dihydroorotate produced a radical signal with a peak to peak width of 15 G, while at pH 6.5 the linewidth was 19 G. At either pH, NADH produced a signal with a linewidth of 19 G (Aleman *et al.*, 1968a). Around this time it was also recognized that the absorption spectra of the neutral and anionic forms of flavin radicals differed greatly, the neutral radical being characterized by high extinction coefficient at 600 nm (~ 3000 M^{-1} cm^{-1}), the anion being characterized by negligible absorbance at this wavelength (Massey and Palmer, 1966). It was also found that the linewidth of the EPR spectra of the two forms differed significantly, that of the neutral radical being ~ 19 G, that of the anion radical being ~ 15 G (Palmer *et al.*, 1971). Thus it would appear that with NADH, the neutral radical is produced, while with dihydroorotate either form is produced depending on the pH. Accordingly it would appear likely that the failure to observe a rapid increase in absorbance at 600 nm by dihydroortate (Miller and Massey, 1965a) was due largely to the production of the anion radical, as well as possibly a canceling effect due to reduction of the iron–sulfur chromophore.

A potentially very informative experiment was also reported by Aleman *et al.* (1968a). It was observed that when enzyme was preincubated with small amounts of mercurial, production of the $g = 1.94$ signal by addition of NADH was largely eliminated without significantly affecting the development of flavin radical signal. In similar experiments where catalytic activity was followed (Aleman and Handler, 1967) mercurial treatment was found to inactivate NADH-orotate reductase activity much more readily than NADH-oxygen reductase activity. These results would suggest that whereas NADH-orotate reductase activity requires reduction of the iron–sulfur chromophore, NADH-oxygen reductase activity does not! These results have potentially great significance in the development of a reaction mechanism.

The $g = 1.94$ signal of the iron–sulfur chromophore is very similar to that found with plant ferredoxins (Palmer and Sands, 1966) and other iron–sulfur flavoproteins (see chapter on xanthine oxidase and aldehyde oxidase). The overwhelming evidence with these proteins is that the functional unit of such chromophores is a pair of iron atoms and a pair of inorganic sulfur atoms; each unit accepts only one electron on reduction to produce the typical $g = 1.94$ EPR signal (Mayhew *et al.*, 1969). Dihydroorotate dehydrogenase contains per molecule of protein of molecular weight $\sim$120,000, two molecules of FAD, two molecules of FMN, and four atoms of iron and four atoms of "labile" sulfide. Hence, it would appear likely that there are two identical active sites, each site comprising one molecule of FAD, one molecule of FMN and one iron–sulfur chromophore consisting of its pair of iron atoms and pair of inorganic sulfur atoms. It is tempting to speculate with Handler and his colleagues (Aleman *et al.*, 1965, 1966; Aleman and Handler, 1967; Rajogopalan and Handler, 1968) that the two different flavins are responsible for reacting separately with the pyridine nucleotide and pyrimidine substrates, and that the iron–sulfur chromophore acts as an electron transfer bridge between the two. This concept is illustrated below for the various activities of the enzyme. For convenience, FAD is taken as the flavin reacting with NADH and NAD and FMN as that reacting with orotate and dihydroorotate. It should be emphasized, that while no definitive information is available as to which flavin actually reacts with which substrate, there is a certain amount of circumstantial evidence that it is the FAD which is responsible for reaction with pyridine nucleotides and FMN for reaction with pyrimidines. Miller and Kerr (1967b) found that photoinactivation resulted in a more extensive destruction of enzyme-bound FAD, accompanied by a much greater loss in NADH-orotate reductase activity than in dihydroorotate-oxygen reductase activity. It has also been argued that the absence of FAD and absence of reactivity with pyridine nucleotides in the biosynthetic enzyme from *Lactobacillus bulgaricus* favors the hypothesis that FAD is the site of reaction with pyridine nucleotides (Taylor *et al.*, 1971).

However, as this enzyme also lacks iron, the argument is somewhat weakened.

Reaction (7a):

$$\begin{array}{c} NADH \rightarrow FAD \rightarrow Fe_2/S_2 \rightarrow FMN \rightarrow \text{orotate} \\ \uparrow \\ RHg^+ \end{array}$$

Reaction (8):

$$\begin{array}{c} O_2 \qquad\qquad O_2 \\ \uparrow \qquad\qquad \uparrow \\ NADH \rightarrow FAD \dashrightarrow Fe_2/S_2 \dashrightarrow FMN \\ \uparrow \\ RHg^+ \end{array}$$

Reaction (7b):

$$\begin{array}{c} NAD^+ \leftarrow FAD \leftarrow Fe_2/S_2 \leftarrow FMN \leftarrow \text{dihydrorotate} \\ \uparrow \\ RHg^+ \end{array}$$

Reaction (9):

$$\begin{array}{c} O_2 \qquad\qquad O_2 \\ \uparrow \qquad\qquad \uparrow \\ FAD \leftarrow Fe_2/S_2 \leftarrow FMN \leftarrow \text{dihydrorotate} \\ \uparrow \\ RHg^+ \end{array}$$

Such reactions, unfortunately do not appear to account satisfactorily for all the diverse information reported with the enzyme. From the extent of the extinction changes occurring rapidly with NADH as substrate (Miller and Massey, 1965), it would appear that each catalytic center is capable of rapid uptake of four electron equivalents, the rate limiting step presumably being the initial reduction of the FAD. Thus, at the end of the rapid phase of anaerobic reduction, these four electrons would presumably be distributed; FAD (1), Fe_2/S_2 (1), FMN (2). The slow secondary reduction observed ($\sim$35 minute^{-1}) may then represent slow full reduction of all chromophores. The $FADH^{\cdot}$ formed in the rapid reaction, by virtue of its observed absorbance at 600 nm and its EPR linewidth of 19 G would be the neutral radical. Reaction of this radical with O_2 would appear to account for the NADH–oxygen reductase activity [reaction (8)], and for the comparative lack of effect of mercurial on this reaction (Aleman and Handler, 1967) and the observation that mercurial treatment interferes with formation of the $g = 1.94$ signal without having much effect on the formation of flavin radical (Aleman *et al.*, 1968a). With the assumption of the rate limiting step being the initial reduction of FAD, the similar turnover number of the enzyme in reactions (7a) and (8) is also accounted for.

On reaction anaerobically with dihydroorotate there is a much smaller extent of rapid reduction of the enzyme, suggesting that the enzyme may only be able to accept one pair of electrons rapidly from dihydroorotate, to be distributed between FMN and the iron–sulfur chromophore. Thus the formation of the anion radical, typical of this substrate, would be accounted for. However, subsequent events appear to occur too slowly to account for catalysis. The major decrease in absorbance at 450 nm, and the accompanying increase in absorbance at 600 nm, occur with a first order rate constant of only ~ 13 min^{-1}. While this could readily be accomodated by a second pair of electrons entering the enzyme from substrate and being distributed; FAD (2), (Fe_2/S_2) (1), FMN (1), the observed rate is far too slow to account for catalysis (cf. Table IV). Hence if NAD is to react with $FADH_2$, the latter must be formed with a first order rate constant ≥ 1760 min^{-1}. From the stopped-flow and stopped-freeze data, this is feasible if one assumes that the small absorbance change is due to the unfavorable redox potential of the dihydroorotate/orotate couple. However, this leaves unexplained the subsequent slow and major changes in absorbance. It would also leave unexplained the observation that the V_{max} for the dihydroorotate-oxygen reductase activity is considerably smaller than both the dihydroorotate-NAD reductase activity and the NADH-oxygen reductase activity. The finding of different V_{max} values with O_2 as acceptor, in conjunction with the observed rate of reduction of the enzyme by NADH or by dihydroorotate, require that O_2 react at different sites in the two oxidase reactions. This requirement could be met by $FMNH_2$ or $FMNH^{\cdot}$ being the chief reactant with O_2 in the dihydroorotate-oxygen reductase activity, and $FADH_2$ or $FADH^{\cdot}$ being the reactant in the NADH-oxygen reductase activity. This would require both activities to be substantially unaffected by treatment with mercurial; the lack of effect of the latter (Aleman and Handler, 1967) on the NADH-oxygen reductase activity is consistent with this hypothesis. However, there appears to be some uncertainty about the effect of mercurial on the dihydroorotate-oxygen reductase activity (cf. Aleman *et al.*, 1965; Miller and Kerr, 1967b).

While it is clear that no single scheme so far advanced seems to provide an acceptable explanation of the electron transfer processes occurring, it would seem likely that a combined study employing conventional kinetic and rapid reaction techniques is the best hope for unraveling the catalytic events of this enzyme. The preferential removal of one of the two flavins, or the preparation of an undenatured deflavoenzyme which could be reactivated for different catalytic activities by addition of FMN, FAD, and FMN + FAD would also appear to be a goal of the utmost importance with this enzyme.

REFERENCES

Ackrell, B. A. C., and Kearney, E. B. (1973). *Fed. Proc.* **32,** 595.

Albracht, S. P. J., and Slater, E. C. (1969). *Biochem. Biophys. Acta* **189,** 308.

Aleman, V., and Handler, P. (1967). *J. Biol. Chem.* **242,** 4087.

Aleman, V., Rajagopalan, K. V., Handler, P., Beinert, H., and Palmer, G. (1965). *In* "Oxidases and Related Oxidation-Reduction Systems" (H. S. Mason, T. E. King, and M. Morrison, eds.), p. 327. Wiley, New York.

Aleman, V., Smith, S. T., Rajagopalan, K. V., and Handler, P. (1966). *In* "Flavins and Flavoproteins" (E. C. Slater, ed.) p. 99. Elsevier, Amsterdam.

Aleman, V., Handler, P., Palmer, G., and Beinert, H. (1968a). *J. Biol. Chem.* **243,** 2560.

Aleman, V., Handler, P., Palmer, G., and Beinert, H. (1968b). *J. Biol. Chem.* **243,** 2569.

Arrigoni, O., and Singer, T. P. (1962). *Nature (London)* **193,** 1256.

Baginsky, M. L., and Hatefi, Y. (1968). *Biochem. Biophys. Res. Commun.* **32,** 945.

Baginsky, M. L., and Hatefi, Y. (1969). *J. Biol. Chem.* **244,** 5313.

Ballou, D., Palmer, G., and Massey, V. (1969). *Biochem. Biophys. Res. Commun.* **36,** 899.

Beinert, H. (1965). *In* "Non-heme Iron Proteins" (A. San Pietro, ed.), p. 23. Antioch Press, Yellow Springs, Ohio.

Beinert, H. (1966). *In* "Flavins and Flavoproteins" (E. C. Slater, ed.), p. 49. Elsevier, Amsterdam.

Beinert, H., and Sands, R. H. (1959). *Biochem. Biophys. Res. Commun.* **1,** 171.

Beinert, H., and Sands, R. H. (1960). *Biochem. Biophys. Res. Commun.* **3,** 41.

Beinert, H., Palmer, G., Cremona, T., and Singer, T. P. (1963a). *Biochem. Biophys. Res. Commun.* **12,** 432.

Beinert, H., Heinen, W., and Palmer, G. (1963b). *Brookhaven Symp. Biol.* **15,** 229.

Beinert, H., Palmer, G., Cremona, T., and Singer, T. P. (1965). *J. Biol. Chem.* **240,** 475.

Bernath, P., and Singer, T. P. (1962). *In* "Methods in Enzymology" (S. Colowick and N. O. Kaplan, eds.), Vol. V, p. 597. Academic Press, New York.

Bois, R., and Estabrook, R. W. (1969). *Arch. Biochem. Biophys.* **129,** 362.

Bray, R. C. (1961). *Biochem. J.* **81,** 189.

Brumby, P. E., Miller, R. W., and Massey, V. (1965). *J. Biol. Chem.* **240,** 327.

Bruni, A., and Racker, E. (1968). *J. Biol. Chem.* **243,** 962.

Buchanan, B. B., and Arnon, D. I. (1970). *Adv. Enzymol.* **33,** 119.

Cerletti, P., Strom, R., Giordano, M., Balastero, F., and Giovenco, M. A. (1963). *Biochem. Biophys. Res. Commun.* **14,** 408.

Cerletti, P., Giovenco, M. A., Giordano, M. G., Giovenco, S., and Strom, R. (1967). *Biochem. Biophys. Acta* **146,** 380.

Cerletti, P., Zanetti, G., Testolin, G., Rossi, C., and Osenga, G. (1971). *In* "Flavins and Flavoproteins" (H. Kamin, ed.), p. 629. Univ. Park Press, Baltimore, Maryland.

Cerletti, P., Zanetti, G., and Righetti, P. G. (1972). *In* "Biochemistry and Biophysics of Mitochondrial Membranes" (G. F. Azzone, E. Carafoli, A. L. Lehninger, E. Quagliariello, and N. Siliprandi, eds.), p. 33. Academic Press, New York.

Chance, B. (1956). *In* "Enzymes: Units of Biological Structure and Function" (O. H. Gaebler, ed.), p. 447. Academic Press, New York.
Chapman, A. G., and Jagannathan, V. (1963). Unpublished data. Manuscript circulated by I. E. G. No. 1 of the Nat. Inst. of Health.
Coles, C. J., Tisdale, H., Kenney, W. C. and Singer, T. P. (1972). *Physiol. Chem. Phys.* **4,** 301.
Coles, C. J., Tisdale, H., Kenney, W. C., and Singer, T. P. (1973). *J. Biol. Chem.* (submitted).
Cremona, T., and Kearney, E. B. (1964). *J. Biol. Chem.* **239,** 2328.
Cremona, T., and Kearney, E. B. (1965). *J. Biol. Chem.* **240,** 3645.
Cremona, T., Kearney, E. B., Villavicencio, M., and Singer, T. P. (1963). *Biochem. Z.* **338,** 407.
Davis, K. A., and Hatefi, Y. (1971). *Biochemistry* **10,** 2509.
Dervartanian, D. V., and Veeger, C. (1964). *Biochim. Biophys. Acta* **92,** 233.
Dervartanian, D. V., and Veeger, C. (1965). *Biochim. Biophys. Acta* **105,** 424.
Dervartanian, D. D., Zeijlemaker, W. P., and Veeger, C. (1966). *In* "Flavins and Flavoproteins" (E. C. Slater, ed.), p. 183. Elsevier, Amsterdam.
Dervartanian, D. V., Veeger, C., Orme-Johnson, W. H., and Beinert, H. (1969). *Biochim. Biophys. Acta* **191,** 22.
Ernster, L., and Less, C. P. (1964) *Annu. Rev. Biochem.* **33,** 729.
Estabrook, R. W., Tyler, D. D., Gonze, J., and Peterson, P. A. (1968). *In* "Flavins and Flavoproteins" (K. Yagi, ed.), p. 268. Tokyo Univ. Press, Tokyo.
Felberg, N. T., and Hollocher, T. C. (1972). *J. Biol. Chem.* **247,** 4539.
Franke, W., and Siewerdt, D. (1944). *Z. Physiol. Chem.* **280,** 76.
Friedmann, H. C., and Vennesland, B. (1958). *J. Biol. Chem.* **233,** 1398.
Friedmann, H. C., and Vennesland, B. (1960). *J. Biol. Chem.* **235,** 1526.
Fry, K. T., and San Pietro, A. (1962). *Biochem. Biophys. Res. Commun.* **9,** 218.
Gawron, O., Glaid, A. J., Fondy, T. P., and Bechtold, M. M. (1961). *Nature (London)* **187,** 1004.
Gewitz, H. S., and Volker, W. (1962). *Z. Physiol. Chem.* **330,** 124.
Giuditta, A., and Singer, T. P. (1958). *J. Biol. Chem.* **234,** 666.
Gregolin, C., and Scalella, P. (1965). *Biochem. Biophys. Acta* **99,** 185.
Greville, G. D. (1966). *In* "Regulation of Metabolic Processes in Mitochondria" (J. M. Tager, S. Papa, E. Quagliariello, and E. C. Slater, eds.), p. 86. Elsevier, Amsterdam.
Griffin, J. B., Barnett, R. E., and Hollocher, T. C. (1967). *Arch. Biochem. Biophys.* **119,** 133.
Gutman, M., and Singer, T. P. (1970). *Biochemistry* **9,** 4750.
Gutman, M., and Singer, T. P. (1971). *In* "Energy Transduction in Respiration and Photosynthesis" (S. Papa, ed.), p. 486. Adriatica Editrice, Bari.
Gutman, M., Singer, T. P., Beinert, H., and Casida, J. E. (1970a). *Proc. Nat. Acad. Sci. U.S.* **65,** 763.
Gutman, M., Mayr, M., Oltzik, R., and Singer, T. P. (1970b). *Biochem. Biophys. Res. Commun.* **41,** 40.
Gutman, M., Singer, T. P., and Casida, J. E. (1970c). *J. Biol. Chem.* **245,** 1992.
Gutman, M., Mersmann H., Luthy, J., and Singer, T. P. (1970d). *Biochemistry* **9,** 2678.
Gutman, M., Kearney, E. B., and Singer, T. P. (1971a). *Biochemistry* **10,** 2726.
Gutman, M., Kearney, E. B., Mayr, M., and Singer, T. P. (1971b). *Physiol. Chem. Phys.* **3,** 319

Gutman, M., Singer, T. P., and Beinert, H. (1971c). *Biochem. Biophys. Res. Commun.* **44**, 1572.
Gutman, M., Kearney, E. B., and Singer, T. P. (1971d). *Biochem. Biophys. Res. Commun.* **44**, 526.
Gutman, M., Kearney, E. B., and Singer, T. P. (1971e). *Biochemistry* **10**, 4263.
Gutman, M., Singer, T. P., and Beinert, H. (1972). *Biochemistry* **11**, 556.
Hanstein, W. G., Davis, K. A., Ghalambor, M. A., and Hatefi, Y. (1971). *Biochemistry* **10**, 2517.
Hatefi, Y. (1968). *Proc. Nat. Acad. Sci. U.S.* **60**, 733.
Hatefi, Y., and Stempel, K. E. (1969). *J. Biol. Chem.* **244**, 2350.
Hatefi, Y., Haavik, A. G., and Griffiths, D. E. (1692). *J. Biol. Chem.* **237**, 1676.
Hauber, J., and Singer, T. P. (1967). *Eur. J. Biochem.* **3**, 107.
Hemmerich, P., Ehrenberg, A., Walker, W. H., Eriksson, L. E. G., Salach, J., Bader, P., and Singer, T. P. (1969). *FEBS Lett.* **3**, 37.
Hinkle, P. C., Butow, R. A., Racker, E., and Chance, B. (1967). *J. Biol. Chem.* **242**, 5169.
Hirsch, C. A., Raminsky, M., Davis, B. D., and Lin, E. C. C. (1963). *J. Biol. Chem.* **238**, 3370.
Hopkins, F. G., and Morgan, E. (1938). *Biochem. J.* **32**, 611.
Hopkins, F. G., Morgan, E., and Lutwak-Mann, C. (1938). *Biochem. J.* **32**, 1829.
Horgan, D. J., Ohno, H., Singer, T. P., and Casida, J. E. (1968a). *J. Biol. Chem.* **243**, 5967.
Horgan, D. J., Singer, T. P., and Casida, J. E. (1968b). *J. Biol. Chem.* **243**, 834.
Hucko, J. E., and King, T. E. (1964). *Fed. Proc.* **23**, 486.
Kaniuga, Z. (1963). *Biochim. Biophys. Acta* **73**, 550.
Kean, E. A., Gutman, M., and Singer, T. P. (1971). *J. Biol. Chem.* **246**, 2346.
Kearney, E. B. (1957). *J. Biol. Chem.* **229**, 363.
Kearney, E. B. (1960). *J. Biol. Chem.* **235**, 865.
Kearney, E. B., and Singer, T. P. (1955). *Biochim. Biophys. Acta* **17**, 596.
Kearney, E. B., Singer, T. P., and Zastrow, N. (1955). *Arch. Biochem. Biophys.* **55**, 579.
Kearney, E. B., Mayr, M., and Singer, T. P. (1972a). *Biochem. Biophys. Res. Commun.* **46**, 531.
Kearney, E. B., Ackrell, B. A. C., and Mayr, M. (1972b). *Biochem. Biophys. Res. Commun* **49**, 1115.
Kearney, E. B., Ackrell, B. A. C., and Mayr, M., and Singer, T. P. (1973). *Biochemistry* (submitted).
Keilin, D., and King, T. E. (1960). *Proc. Roy. Soc. (London) Ser. 13* **152**, 163.
Kenney, W. C., and Singer, T. P. (1971) (unpublished data).
Kenney, W. C. (1973). *Fed. Proc.* **32**, 595.
Kenney, W. C., Walker, W. H., and Singer, T. P. (1972). *J. Biol. Chem.* **247**, 4510.
Kerr, C. T., and Miller, R. W. (1967). *Can. J. Biochem.* **45**, 1295.
Kerr, C. T., and Miller, R. W. (1968). *J. Biol. Chem.* **243**, 2963.
Kimura, T., and Hauber, J. (1963). *Biochem. Biophys. Res. Commun.* **13**, 169.
Kimura, T., Hauber, J., and Singer, T. P. (1963). *Nature (London)* **198**, 362.
Kimura, T., Hauber, J., and Singer, T. P. (1967). *J. Biol. Chem.* **242**, 4987.
King. T. E. (1961). *Biochim. Biophys. Acta* **47**, 430.
King, T. E. (1962). *Biochim. Biophys. Acta* **59**, 492.
King, T. E. (1963). *J. Biol. Chem.* **238**, 4037.
King, T. E. (1964). *Biochem. Biophys. Res. Commun.* **16**, 511.

King, T. E. (1967a). *Adv. Enzymol.* **28**, 155.
King, T. E. (1967b). *In* "Methods in Enzymology" (R. W. Estabrook and M. E. Pullman, eds.), Vol. X, p. 322. Academic Press, New York.
King, T. E., and Howard, R. L. (1962). *J. Biol. Chem.* **237**, 1686.
King, T. E., Howard, R. L., and Mason, H. S. (1961). *Biochem. Biophys. Res. Commun.* **5**, 329.
King, T. E., Howard, R. L., Kettman, J., Jr., Hegdekar, B. M., Kaboyama, M., Nickel, K. S., and Possehl, A. E. (1966). *In* "Flavins and Flavoproteins" (E. C. Slater, ed.), p. 441. Elseiver, Amsterdam.
Komai, H., Massey, V., and Palmer, G. (1969). *J. Biol. Chem.* **244**, 1692.
Krakow, G., and Vennesland, B. (1961). *J. Biol. Chem.* **236**, 142.
Kröger, A., and Klingenberg, M. (1966). *Biochem. Z.* **344**, 317.
LaNoue, K., Nicklas, W. J., and Williamson, J. R. (1970). *J. Biol. Chem.* **245**, 102.
Lee, C. P., and King, T. E. (1962). *Biochim. Biophys. Acta* **59**, 716.
Lieberman, I., and Kornberg, A. (1953). *Biochim. Biophys. Acta* **12**, 223.
Lusty, C. J., Machinist, J. M., and Singer, T. P. (1965). *J. Biol. Chem.* **240**, 1804.
Machinist, J. M., and Singer, T. P. (1965a). *Proc. Nat. Acad. Sci. U.S.* **53**, 467.
Machinist, J. M., and Singer, T. P. (1965b). *J. Biol. Chem.* **240**, 3182.
Mackler, B. (1961). *Biochim. Biophys. Acta* **50**, 141.
Mackler, B. (1966). *In* "Flavins and Flavoproteins" (E. C. Slater, ed.) p. 427. Elsevier, Amsterdam.
Mahler, H. R., Sarkar, N. K., Vernon, L. P., and Alberty, R. A. (1952). *J. Biol. Chem.* **199**, 585.
Malkin, R., and Rabinowitz, J. C. (1966). *Biochem. Biophys. Res. Commun.* **23**, 822.
Massey, V. (1957). *J. Biol. Chem.* **229**, 763.
Massey, V. (1958). *Biochim. Biophys. Acta* **30**, 500.
Massey, V., and Palmer, G. (1966). *Biochemistry* **5**, 3181.
Massey, V., and Singer, T. P. (1957a). *J. Biol. Chem.* **228**, 263.
Massey, V., and Singer, T. P. (1957b). *J. Biol. Chem.* **229**, 755.
Massey, V., Strickland, S., Mayhew, S. G., Howell, L. G., Engel, P. C., Matthews, R. G., Schuman, M., and Sullivan, P. A. (1969). *Biochem. Biophys. Res. Commun.* **36**, 891.
Mayhew, S. G., Petering, D., Palmer, G., and Foust, G. P. (1969). *J. Biol. Chem.* **244**, 2830.
McCord, J. M., and Fridovich, I. (1969). *J. Biol. Chem.* **244**, 6049.
McCord, J. M., and Fridovich, I. (1970). *J. Biol. Chem.* **245**, 1374.
Miller, R. W. (1970). *Can. J. Biochem.* **48**, 935.
Miller, R. W., and Curry, J. R. (1969). *Can. J. Biochem.* **47**, 725.
Miller, R. W., and Massey V. (1965a). *J. Biol. Chem.* **240**, 1453.
Miller, R. W., and Massey, V. (1965b). *J. Biol. Chem.* **240**, 1466.
Miller, R. W., and Kerr, C. T. (1966). *J. Biol. Chem.* **241**, 5597.
Miller, R. W., and Kerr, C. T. (1967a). *Can. J. Biochem.* **45**, 1283.
Miller, R. W., and Kerr, C. T. (1967b). *Photochem. Photobiol.* **6**, 895.
Miller, R. W., Kerr, C. T., and Curry, J. R. (1968). *Can. J. Biochem.* **46**, 1099.
Minakami, S., Ringler, R. L., and Singer, T. P. (1962). *J. Biol. Chem.* **237**, 569.
Minakami, S., Cremona, T., Ringler, R. L., and Singer, T. P. (1963). *J. Biol. Chem.* **238**, 1529.
Nelson, B. D., Norling, B., Persson, B., and Ernster, L. (1971). *Biochem. Biophys. Res Commun.* **44**, 1312.

Nelson, C. A., and Handler, P. (1968). *J. Biol. Chem.* **243,** 5368.
Nicholls, P., and Malviya, A. N. (1968). *Biochemistry* **7,** 305.
Öberg, K. E. (1961). *Exp. Cell Res.* **24,** 163.
Ohnishi, T., Asakura, T., Wohlrab, H., Yonetani, T., and Chance, B. (1970). *J. Biol. Chem.* **245,** 901.
Ohnishi, T., Wilson, D. F., Asakura, T., and Chance, B. (1972). *Biochem. Biophys. Res. Commun.* **46,** 1631.
Orme-Johnson, N. R., and Beinert, H. (1972) (to be published).
Orme-Johnson, N. R., Orme-Johnson, W. H., Beinert, H., and Hatefi, Y. (1971). *Biochem. Biophys. Res. Commun.* **44,** 446.
Orme-Johnson, N. R., Orme-Johnson, W. H., Hansen, R. E., Beinert, H., and Hatefi, Y. (1973). *In* "Oxidases and Related Oxidation-Reduction Systems" (T. E. King, H. S. Mason, and M. Morrison, eds.). Univ. Park Press, Baltimore, Maryland (in press).
Palmer, G., and Sands, R. H. (1966). *J. Biol. Chem.* **241,** 253.
Palmer, G., Müller, F., and Massey, V. (1971). *In* "Flavins and Flavoproteins" (H. Kamin, ed.) p. 123. Univ. Park Press, Baltimore, Maryland.
Papa, S., Tager, J. M., and Quagliariello, E. (1968). *In* "Regulatory Functions of Biological Membranes" (J. Järnefelt, ed.), p. 264. Elsevier, Amsterdam.
Pharo, R. L., Andreoli, L. A., Vyas, S. R., and Sanadi, D. R. (1966). *J. Biol. Chem.* **241,** 4771.
Pharo, R. L., Sordahl, L. H., Edelhoch, H., and Sanadi, D. R. (1968). *Arch. Biochem. Biophys.* **125,** 416.
Quastel, J. H., and Wooldridge, W. R. (1928). *Biochem. J.* **22,** 689.
Rajagopalan, K. V., and Handler, P. (1964). *J. Biol. Chem.* **239,** 1509.
Rajagopalan, K. V., and Handler, P. (1968). *In* "Biological Oxidations" (T. P. Singer, ed.), p. 301. Wiley (Interscience), New York.
Righetti, P., and Cerletti, P. (1971). *FEBS Lett.* **13,** 181.
Ringler, R. L., Minakami, S., and Singer, T. P. (1963). *J. Biol. Chem.* **238,** 801.
Rossi, C., Cremona, T. Machinist, J. M., and Singer T. P. (1965). *J. Biol. Chem.* **240,** 2634.
Rossi, E., Norling, B., Persson, B., and Ernster, L. (1970). *Eur. J. Biochem.* **16,** 508.
Salach, J. I., Singer, T. P., and Bader, P. (1967). *J. Biol. Chem.* **242,** 4555.
Salach, J. I., Seng, R., Tisdale, H., and Singer, T. P. (1970). *J. Biol. Chem.* **246,** 340.
Salach, J., Walker, W. H., Singer, T. P., Ehrenberg, A., Hemmerich, P., Ghisla, S., and Hartmann, U. (1972). *Eur. J. Biochem.* **26,** 267.
Sanborn, B. M., Felberg, N. T., and Hollocher, T. C. (1971). *Biochim. Biophys. Acta* **227,** 219.
Sands, R. H., and Beinert, H. (1960). *Biochem. Biophys. Res. Commun.* **3,** 47.
Schäfer, G., Balde, P., and Lamprecht, W. (1967). *Nature (London)* **214,** 20.
Schatz, G., and Racker, E. (1966). *J. Biol. Chem.* **241,** 1429.
Singer, T. P. (1961). *In* "Biological Structure and Function" (T. W. Goodwin and O. Lindberg, eds.), Vol. II, p. 103. Academic Press, New York.
Singer, T. P. (1965). *In* "Oxidases and Related Redox Systems" (T. E. King, H. S. Mason, and M. Morrison, eds.), Vol. I, p. 448. Wiley, New York.
Singer, T. P. (1966). *In* "Comprehensive Biochemistry" (M. Florkin and E. H. Stotz, eds.), Vol. 14, p. 127. Elsevier, Amsterdam.

Singer, T. P. (1968). *In* "Biological Oxidations" (T. P. Singer, ed.), p. 339. Wiley, New York.
Singer, T. P., and Gutman, M. (1970). *In* "Pyridine Nucleotide-Dependent Dehydrogenases" (H. Sund, ed.), p. 375. Springer, Berlin.
Singer, T. P. (1971). *In* "Biochemical Evolution and the Origin of Life" (E. Schoffeniels, ed.), Vol. II, p. 203. North Holland Publ., Amsterdam.
Singer, T. P. (1973). *Methods Biochem. Anal.* **22,** (in press).
Singer, T. P., and Gutman, M. (1971). *Advan. Enzymol.* **34,** 79.
Singer, T. P., and Kearney, E. B. (1954). *Biochim. Biophys. Acta* **15,** 151.
Singer, T. P., and Kearney, E. B. (1963). *In* "The Enzymes" (P. D. Boyer, H. A. Lardy, and K. Myrbäck, eds.), Vol. VII, p. 383. Academic Press, New York.
Singer, T. P., and Massey, V. (1957). *Rec. Chem. Progr.* **18,** 201.
Singer, T. P., Kearney, E. B., and Bernath, P. (1956a). *J. Biol. Chem.* **223,** 599.
Singer, T. P., Kearney, E. B., and Massey, V. (1956b). *Arch. Biochem. Biophys.* **60,** 255.
Singer, T. P., Hauber, J., and Arrigoni, O. (1962). *Biochem. Biophys. Res. Commun.* **9,** 150.
Singer, T. P., Salach, J., Walker, W. H., Gutman, M., Hemmerich, P., and Ehrenberg, A. (1971a). *In* "Flavins and Flavoproteins" (H. Kamin, ed.), p. 607. Univ. Park Press, Baltimore, Maryland.
Singer, T. P., Salach, J., Hemmerich, P., and Ehrenberg, A. (1971b). *In* "Methods in Enzymology" (D. McCormick and L. Wright, eds.), Vol. XVIIIB, p. 416. Academic Press, New York.
Singer, T. P., Gutman, M., and Kearney, E. B. (1971c). *FEBS Lett.* **17,** 11.
Singer, T. P., Kearney, E. B., and Gutman, M. (1972a). *In* "Biochemical Control Mechanisms" (E. Kun and S. Grisolia, eds.), p. 271. Wiley, New York.
Singer, T. P., Gutman, M., and Kearney, E. B. (1972b). *In* "Biochemistry and Biophysics of Mitochondrial Membranes" (G. F. Azzone, E. Carafoli, A. L. Lehninger, E. Quagliariello, and N. Siliprandi, eds.), p. 41. Academic Press, New York.
Singer, T. P., Kearney, E. B., and Kenney, W. C. (1972c). *Adv. Enzymol.* **37,** 189.
Takemori, S., and King, T. E. (1964a). *Science* **144,** 852.
Takemori, S., and King, T. E. (1964b). *J. Biol. Chem.* **239,** 3546.
Taylor, W. H., and Taylor, M. L. (1964). *J. Bacteriology* **88,** 105.
Taylor, M. L., Taylor, W. H., Eames, D. F., and Taylor, C. D. (1971). *J. Bacteriol.* **105,** 1015.
Thorn, M. B. (1962). *Biochem. J.* **85,** 116.
Thunberg, T. (1933). *Biochem. Z.* **258,** 48.
Tisdale, H., Hauber, J., Prager, G., Turini, P., and Singer, T. P. (1968). *Eur. J. Biochem.* **4,** 472.
Tober, C. L., Nicholls, P., and Brodie, J. D. (1970). *Arch. Biochem. Biophys.* **138,** 506.
Tsou, C. C. (1951). *Biochem. J.* **49,** 512.
Tyler, D. D., Butow, R. A., Gonze, J., and Estabrook, R. W. (1965). *Biochem. Biophys. Res. Commun.* **19,** 551.
Udaka, S., and Vennesland, B. (1962). *J. Biol. Chem.* **237,** 2018.
Veeger, C., Dervartanian, D. V., and Zeijlemaker, W. P. (1963). *In* "Methods in Enzymology" (J. M. Lowenstein, ed.), Vol. XIII, p. 81. Academic Press, New York.

VonKorff, R. W. (1967). *Nature (London)* **214,** 23.

VanVoorst, J. D. W., Veeger, C., and Dervartanian, D. V. (1967). *Biochim. Biophys. Acta* **147,** 367.

Walker, W. H., and Singer, T. P. (1970). *J. Biol. Chem.* **245,** 4224.

Walker, W. H., Salach, J., Gutman, M., Singer, T. P., Hyde, J. S., and Ehrenberg, A. (1969). *FEBS Lett.* **5,** 237.

Wang, T. Y., Tsou, C. L., and Wang, Y. L. (1956). *Scientia Sinica* **5,** 73.

Warringa, M. G. P. J., and Giuditta, A. (1958). *J. Biol. Chem.* **230,** 111.

Warringa, M. G. P. J., Smith, O. H., Giuditta, A., and Singer, T. P. (1958). *J. Biol. Chem.* **230,** 97.

Watari, H., Kearney, E. B., and Singer, T. P. (1963). *J. Biol. Chem.* **238,** 4063.

Wu, J. T. Y., and King, T. E. (1967). *Fed. Proc.* **26,** 732.

Ziegler, D. M. (1961). *In* "Biological Structure and Function" (T. W. Goodwin and O. Lindberg, eds.), Vol. II, p. 253. Academic Press, New York.

Zeijlemaker, W. P., Dervartanian, D. V., and Veeger, C. (1965). *Biochim. Biophys. Acta* **99,** 183.

Zeijlemaker, W. P., Dervartanian, D. V., Veeger, C., and Slater, E. C. (1969). *Biochim. Biophys. Acta* **178,** 213.

Zeijlemaker, W. P., Klaase, A. D. M., Slater, E. C., and Veeger, C. (1970). *Biochim. Biophys. Acta* **198,** 415.

CHAPTER 10

Iron–Sulfur Flavoprotein Hydroxylases*

VINCENT MASSEY

I. GENERAL INTRODUCTION

Among the group of iron–sulfur containing flavoproteins, xanthine oxidase (and the related xanthine dehydrogenases) and aldehyde oxidase, stand out because of their marked similarities to each other and because of distinctive differences from other members of the group. While all the other enzymes of this class catalyze classical oxidation–reduction

* This review is dedicated to Professor Malcolmn Dixon, F. R. S., friend and teacher, whose love and understanding of enzymology has inspired several generations of workers in this discipline.

reactions, xanthine oxidase and aldehyde oxidase catalyze reactions which are formally hydroxylations of the substrate. They are also different in that they contain molybdenum in addition to iron–sulfur chromophores. Another characteristic aspect is their unusually wide specificity, which is almost unique in enzymology. However, despite this substrate promiscuity, all reactions (except NADH oxidation) which are catalyzed are hydroxylations. Even in this respect they are unusual enzymes; while most enzymic hydroxylation reactions utilize molecular oxygen as the source of the oxygen atom incorporated in the product, in xanthine oxidase and aldehyde oxidase catalyzed reactions the product-oxygen is derived from water, and molecular O_2 is used only as an electron acceptor to reoxidize reduced enzyme.

Finally, these enzymes are also unusual insofar that there is no clear-cut understanding of their physiological role!

II. XANTHINE OXIDASE

A. Introduction

Xanthine oxidase is an enzyme with a long and complicated history; its literature is perhaps the longest and most confused of any enzyme known. The existence in mammalian tissues of an enzyme catalyzing the oxidation of the purines hypoxanthine and xanthine to uric acid was recognized around the turn of the century. Milk was found to be a particularly rich source of the enzyme; owing to the pioneering studies of Dixon and his colleagues (Booth, 1935; Dixon, 1938–39; Dixon and Thurlow, 1924), it was gradually established that this enzyme was probably identical with the "Schardinger enzyme" of milk which had first been characterized by its ability to catalyze the oxidation of aldehydes to their corresponding carboxylic acids. These and many later studies revealed that xanthine oxidase was an enzyme of unusually broad specificity toward its electron donor and electron acceptor. Inevitable complications also arose through the recognition of a separate enzyme of similar broad specificity which has become known as aldehyde oxidase and which will be discussed separately in this chapter. This enzyme, like xanthine oxidase, is capable of oxidation of a wide variety of heterocyclic compounds and aldehydes; its main distinguishing feature is its lack of reactivity with xanthine.

Preparations of milk xanthine oxidase in a highly purified state were obtained as early as 1939 (Ball, 1939; Corran *et al.*, 1939) and revealed that FAD was an essential constituent. However, unlike the other flavoproteins which had been purified at that time (Old Yellow enzyme and the Haas flavoprotein) and free flavins, the isolated enzyme was reddish-brown rather than yellow. Corran *et al.* (1939) by quantitative estimation of the flavin, estimated that the bound flavin could contribute only about 35% of the absorption of the enzyme at 450 nm and postulated the presence of another chromophore whose absorption spectrum they calculated. Whether the unusual absorption spectrum was in fact due to another chromophore or to the effect of a spectral perturbation of the flavin on binding to the protein was a question which remained unsettled for many years. Then in the early 1950's, arising from nutritional studies, it was recognized that xanthine oxidase contained bound molybdenum; (Richert and Westerfield, 1953; Green and Beinert, 1953; Totter *et al.*, 1953) and later that it also contained nonheme iron (Richert and Westerfeld, 1954; Avis *et al.*, 1956). These findings lead to a great surge of work with xanthine oxidase which has continued unabated to the present day, and lead to the discovery that several other enzymes also contained iron (and in some cases molybdenum) as well as a flavin prosthetic group.

Studies by electron paramagnetic resonance (EPR) spectrometry have been invaluable in elucidating the role of the metals in enzyme catalysis, and will be considered in more detail later (as well as in other chapters of these two volumes). However, such studies have lead to a very confused picture, which is only now becoming somewhat clearer. Much of the confusion can probably be attributed to the existence in practically all preparations of enzyme so far studied of substantial amounts of inactive enzyme (sometimes more than half the total). While the existence of such inactive forms has long been postulated, it is only recently that definitive evidence has been obtained (Komai and Massey, 1971; Massey *et al.*, 1970; McGartoll *et al.*, 1970) and a method developed for the separation of active and inactive enzyme (Edmondson *et al.*, 1972). Studies on the differences between active and inactive enzyme have revealed the importance of yet another functional group in xanthine oxidase—a persulfide (protein —S—S^-) linkage (Edmondson *et al.*, 1972; Massey and Edmondson, 1970). This review will be devoted largely to the more recent studies with xanthine oxidase. Three previous reviews over the last 15 years (DeRenzo, 1956; Bray, 1963; Rajagopalan and Handler, 1968) have documented thoroughly previous studies and concepts existing at the time of their writing.

Note Added in Proof. Since this article was written, another review devoted essentially to xanthine oxidase and aldehyde oxidase has appeared (Bray and Swann, 1972).

B. Occurence and Purification

Enzymes capable of catalyzing xanthine oxidation have been found to have a wide distribution in nature, cow's milk and mammalian liver being particularly rich sources. A bacterial xanthine oxidase has also been described (Dikstein *et al.,* 1957). Xanthine dehydrogenases (i.e., enzymes capable of rapid oxidation of xanthine with various electron acceptors but not with molecular oxygen) have been found in the liver and kidneys of various birds (Richert and Westerfeld, 1951), in insects (Irzykiewicz, 1955), and in bacteria (Bradshaw and Barker, 1960; Smith *et al.,* 1967).

Purification procedures have been reported for the chicken liver (Remy *et al.,* 1955; Rajagopalan and Hander, 1967) and pigeon kidney dehydrogenases (Landon and Carter, 1960). Purification of the dehydrogenase from *Clostridium cylindrosporum* has also been described (Bradshaw and Barker, 1960). A highly purified xanthine dehydrogenase from *Micrococcus lactilyticus* has also been reported (Smith *et al.,* 1967); this enzyme is so far unique in its ability to use ferredoxin efficiently as an electron acceptor. The oxidase from calf liver has been partly purified (Kielley, 1955), and the oxidase from pig liver has been isolated in a homogeneous form (Brumby and Massey, 1963). The first extensive purification of xanthine oxidase from milk was obtained in 1939 by Ball (1939) and by Corran *et al.* (1939). The method of Ball relied on digestion with pancreatic lipase to remove troublesome lipid association. The procedure of Corran *et al.* used extraction with saturated NaCl to get rid of fatty material; their procedure resulted in a highly purified preparation (about 80% xanthine oxidase) which, however, on the basis of later work, would appear to be largely inactive. Horecker and Heppel (1949) described a purification procedure similar to that of Ball except that trypsin was used instead of pancreatic lipase. Later purification procedures were essentially variants of these original methods. Morell (1952) closely followed the procedure of Horecker and Heppel; the preparation of Mackler *et al.* (1954) uses only minor modifications of the method of Corran *et al.* Avis *et al.* (1955) described a purification procedure resulting in crystalline enzyme; this procedure followed Ball's in the initial stages and relied on column chromatography with calcium phosphate for most of the purification.

Despite the fact that most of the purification procedures led to material of similar spectroscopic properties, the enzymic activity varied widely, even within samples isolated by the same procedure. In 1949, the first evidence was put forward for the presence of inactive enzyme, i.e., of an enzyme having similar spectral properties to the active enzyme but lacking catalytic activity. Lowry *et al.* (1949) found that 6-pteridyl aldehyde was a potent competitive inhibitor (and weak substrate) of xanthine oxidase. They found by extrapolation that 0.6 mole pteridyl aldehyde per mole enzyme-bound flavin was sufficient for complete inhibition of xanthine oxidase activity and suggested that their preparation contained 40% of its flavin in an inactive form. Similar conclusions were reached in 1952 by Morell, who observed that when xanthine was mixed anaerobically with the enzyme, bleaching of the absorption at 450 nm occurred in a biphasic fashion. Part of the absorption change occurred rapidly (within 30 seconds), the remainder over a period of several hours. Morell observed that the extent of bleaching in the rapid phase was proportional to the activity per unit of total flavin content. He concluded that this phenomenon was due to the presence of inactive enzyme molecules; active enzyme being reduced rapidly by substrate, while the slow, secondary bleaching was ascribed to reduction of inactive enzyme by reduced active enzyme.

While in retrospect the combined evidence of Lowry *et al.* and Morell appears very clear for the existence of inactive species being responsible for slow, secondary changes in reduction by substrate, this interpretation has become widely accepted only recently. One of the reasons for this lack of acceptance of the hypothesis was the finding by several groups (Corran *et al.*, 1939; Mackler *et al.*, 1954; Lowry *et al.*, 1949) that the activity per total flavin remained constant during purification, and the isolation in recent years of enzyme still showing biphasic bleaching by substrate but having a higher catalytic activity than calculated by Morell for fully active enzyme (see below).

Bray and his colleagues (Avis *et al.*, 1956; McGartoll *et al.*, 1970; Bray *et al.*, 1961, 1966) also obtained considerable evidence for the existence of inactive forms of xanthine oxidase. Their evidence was indirect, based on the varying contents of iron, flavin, and molybdenum components found in different preparations, and on correlation of these components with the xanthine:oxygen reductase activity. Evidence was also obtained from the secondary slow changes of EPR signals with excess xanthine as substrate. They concluded that two inactive species of xanthine oxidase, i_1 and i_2, existed in their preparations; i_1 having a Mo:flavin:Fe ratio of 1:1:4 and i_2 having a ratio of 0:1:4. Until recently (McGartoll *et al.*, 1970), they did not correlate the existence of such inactive species with the degree of biphasic reduction of the enzyme by substrate.

Over the years, Bray and his colleagues have introduced several modifications of their initial purification procedure. The most significant of these was a fractionation with butanol following digestion with pancreatin (Gilbert and Bergel, 1964; Palmer *et al.*, 1964). More recently, selective denaturation of the demolybdoxanthine oxidase (i_2) has been achieved by prolonged incubation with high concentrations of sodium salicylate (Hart and Bray, 1967; Hart *et al.*, 1970). By this means, enzyme was obtained with a Mo:flavin:Fe ratio of 1:1:4, but still containing 20–30% of inactive xanthine oxidase (i_1).

Another recent purification procedure, very similar to that employed by Gilbert and Bergel (1964) but employing as starting material pasteurized buttermilk, has been reported (Massey *et al.*, 1969a). The enzyme so isolated had a Mo:flavin:Fe ratio of 1:1:4, i.e., it was free of demolybdoxanthine oxidase (i_2). The reason for the difference in result is not clear; it may be due to different nutritional conditions of Michigan and British cows or may be in some way connected with the pasteurization process. The turnover number of this enzyme was higher than any previously obtained; at pH 8.3 and 25°C the turnover number was 620–670 moles xanthine oxidized per minute per molecule of enzyme flavin and 370–400 at 19°C. This compares with the turnover number *calculated* by Morell (1952) for fully active enzyme of 330 per minute at 19°C. However, such an enzyme still shows a biphasic bleaching of absorption on the anaerobic addition of substrate; furthermore, the rate of bleaching in the slow phase varied by a factor of approximately 10 with different substrates (Massey *et al.*, 1969a). Hence it was concluded that the biphasic reduction was probably an intrinsic property of xanthine oxidase. While such quantitative discrepancies still await adequate explanation, it is now clear that the original interpretation of Morell is correct. This conclusion comes from recent work with inhibitors, which will be detailed in later sections. One of these, alloxanthine [2,4-dihydroxypyrazolo-(3,4-*d*)-pyrimidine], has been found to bind extremely tightly to reduced enzyme, resulting in stoichiometric inhibition (Massey *et al.*, 1970a,b; Spector and Johns, 1970). Titration of the most active preparation of xanthine oxidase revealed that complete inhibition was obtained with 0.73 mole alloxanthine per mole enzyme flavin, indicating that this preparation was only 73% functional. Smaller extents of binding of [^{14}C]alloxanthine for complete inactivation were found with lower activity enzyme, and good correlation found between alloxanthine binding, extent of rapid phase of bleaching by substrate and the activity/A_{450} ratio of the enzyme (AFR value). Such a correlation is shown in Fig. 1, and predicts that fully functional enzyme should have an $AFR^{25°C}$ value of 210. The finding of extremely tight association of pyrazolo-(3,4-*d*)-pyrimidines with reduced functional enzyme, but not with nonfunctional enzyme, has been used to isolate fully functional and nonfunctional enzyme (Edmondson *et al.*, 1972). This has been achieved by derivatizing agarose with a long side chain pyrazolo-(3,4-*d*)-pyrimidine and separating nonfunctional enzyme (which binds poorly) from functional enzyme (which binds strongly) by affinity column chromatography. Figure 1 also shows points for enzyme prepared in this way; values close to the predicted ones were obtained. Comparison of the properties of functional and nonfunctional enzyme reveal that they have very similar spectral properties and both possess molybdenum, flavin, iron, and labile sulfide in the ratio 1:1:4:4, The only difference found is the existence of a persulfide linkage in functional enzyme which is absent from nonfunctional enzyme (Edmondson *et al.*,1972; Massey and Edmondson, 1970). This subject will be pursued more fully in later sections.

Finally, in this section, we would like to comment on the use of pancreatin in the isolation of xanthine oxidase. Carey *et al.* (1961) have reported that this treatment results in digestion of xanthine oxidase

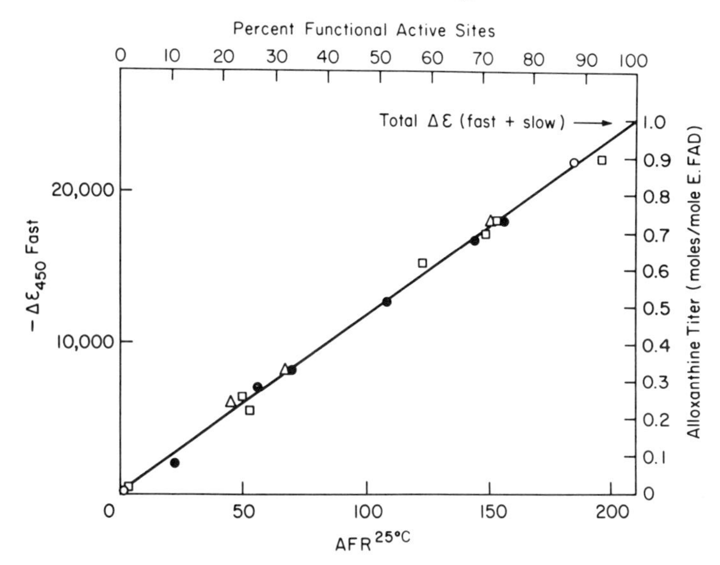

Fig. 1. Correlation of alloxanthine binding and the extent of rapid reduction of xanthine oxidase by xanthine. The extent of rapid reduction is expressed as $-\Delta\epsilon_{450}$ obtained 30 seconds after anaerobic addition of xanthine. The total $\Delta\epsilon_{450}$ (fast + slow phases) was found experimentally to be 24,500 M^{-1} cm^{-1}/mole E. FAD. (●), results obtained with enzyme as isolated (higher AFR values), and after partial inactivation by aging over many months (lower AFR values). (○), results obtained with fractions isolated by affinity chromatography. (Δ), moles [^{14}C]alloxanthine bound per mole E. FAD (□), moles alloxanthine per mole E. FAD required for complete inactivation of xanthine:oxygen reductase activity. These values are plotted against the observed AFR$^{25°C}$ value of the enzyme preparations. Conditions, 0.1 *M* pyrophosphate, pH 8.5, 25°C.

by proteolytic enzymes present in a sample of "steapsin" (of unstated origin) which they used. This treatment led to three peaks of enzymic activity when chromatographed with hydroxyapatite, whereas xanthine oxidase prepared without steapsin treatment gave only two peaks of active enzyme. The electrophoretic mobilities of "digested" and "undigested" enzyme were found to differ significantly. Later Nelson and Handler (1968) have reported a purification procedure without use of pancreatic enzymes, which by ultracentrifuge analysis appears to be 90–95% pure. No spectroscopic data were reported on this preparation, however, from its activity/A_{450} it would appear to be less than 50% functional. Massey *et al.* (1969a) have also prepared xanthine oxidase with and without digestion with pancreatin. In this work it was found that untreated enzyme adsorbed much less tightly to calcium phosphate gel than did enzyme treated with pancreatin. Apart from a dramatically lower yield when pancreatin treatment was omitted, this was the only difference found. Both methods lead to enzyme of similar AFR values, similar biphasic reduction properties and a single sharp band of the same mobility when electrophoresed on polyacrylamide gel.

C. Physicochemical Properties

1. Molecular Weight

The molecular weight of milk xanthine oxidase has been investigated in some detail by Bray and his colleagues. Avis *et al.* (1956) calculated a molecular weight of 290,000 for the crystalline enzyme from sedimentation and diffusion studies. Slightly lower values were reached by Andrews *et al.* (1964) who determined values of 274,000 ± 30,000 by sedimentation and diffusion, 265,000 ± 22,000 by the Archibald method, and 286,000 ± 35,000 by gel filtration. Hart and Bray (1967) estimated a molecular weight of approximately 310,000 by gel filtration of enzyme prepared by the method of selective denaturation with salicylate. While no direct molecular weight determinations have been made on the enzyme prepared by Massey *et al.* (1969a) (which gave a single symmetrical peak in the ultracentrifuge and a single band in polyacrylamide gel electrophoresis), the minimum molecular weight per atom of Mo and per molecule of flavin was found to be 181,000. Avis *et al.* (1956) report a value of 154,000 per FAD for the crystalline material; a value of 167,000 can be calculated from the data of Hart and Bray (1967) for their highest activity enzyme preparations. Nelson and Handler (1968) determined a molecular weight of 300,000 ± 5,000 for their preparation by sedimentation equilibrium. The xanthine dehydrogenase of chicken liver is also reported to have a molecular weight of 300,000 (Rajagopalan and Handler, 1967; Nelson and Handler, 1968), and the xanthine dehydrogenase of *Micrococcus lactilyticus* has a molecular weight of approximately 250,000 (Smith *et al.*, 1967). The xanthine oxidase of pig liver is reported to have a molecular weight of 385,000 by the Archibald method and 190,000 by gel filtration (Brumby and Massey, 1963) suggesting in this case reversible dissociation of a dimer.

2. Cofactor Composition

All preparations of xanthine oxidase and xanthine dehydrogenase have been found to contain FAD, molybdenum, nonheme iron, and inorganic ("labile") sulfide. The preparations of Bray and his group have been found routinely to possess nonintegral amounts of molybdenum, due to the presence of demolybdoxanthine oxidase. For example the crystalline material of Avis *et al.* (1956) was found to have flavin:Mo:Fe ratios of 1:0.56–0.87:4. Treatment of such preparations with high concentrations of salicylate has been found to cause selective denaturation of demolybdo enzyme and resulted in preparations with ratios of 1:1:4 (Hart *et al.*, 1970). Enzyme prepared from pasteurized buttermilk has the ratio of flavin:Mo:Fe:labile S of 1:1:4:4 (Massey *et al.*, 1969a). The same ratio was also found for pig liver xanthine oxidase (Brûmby *et al.*, 1965). Although earlier preparations of xanthine dehydrogenase from chicken liver (Remy *et al.*, 1955) and pigeon kidney (Landon and Carter, 1960) were reported to have flavin:Mo:Fe ratios of 1:1:8, the highly purified preparation of Rajagopalan and Handler (1967) was found to have the ratio 1:1:4, with 4 moles of labile sulfide. The same ratio has also been determined for the xanthine dehydrogenase of *Micrococcus lactilyticus* (Smith *et al.*, 1967).

3. Absorption Spectra

The unusual absorption spectrum of xanthine oxidase was evident from the preparations of Ball (1939) and Corran *et al.* (1939). The latter workers, by quantitative estimation of the FAD content, estimated that only 35% of the absorbance at 450 nm could be accounted for by the flavin and postulated the presence of another chromophore, whose spectrum they calculated. In the early 1950's, the recognition that Fe was a constituent of this enzyme, as well as of succinate dehydrogenase (Singer *et al.*, 1956) and NADH:cytochrome c reductase (Mahler and Elowe, 1954) coupled with the fact that these proteins had somewhat similar absorption spectra, led to the suspicion that the major part of the absorption spectrum was due to the nonheme iron. The spectrum of succinate dehydrogenase was simplified to that of free flavin by treatment with mercurial, suggesting that protein–S linkages were also involved (Massey, 1958). The discovery of ferredoxin in 1962, the similarity of its absorption spectrum to the difference spectrum calculated by Corran *et al.* (1939), and the abolition of its spectrum by treatment with mercurials (Lovenberg *et al.*, 1963), suggested that this group of metal flavoproteins contained the same type of iron–sulfur chromophore as existed with the ferredoxins. It remained for EPR studies to put this concept on a firm footing (see later section and other chapters in this volume). By treatment of xanthine oxidase with 40% methanol and 40% ethanol at 0°C, Rajagopalan and Handler (1964) succeeded in dissociating the flavin and precipitating the flavin-free protein in a denatured form. The reddish-brown precipitate was dissolved in 4 *M* urea and was found to have a spectrum quite similar to that of spinach ferredoxin. Aldehyde oxidase and dihydroorotate dehydrogenase were also treated in the same way with similar results. Rajagopalan and Handler found with these three denatured flavin-free proteins that per atom of Fe, E_{550} varied between 2750 and 2950 and E_{450} between 5000 and 5500. Thus with minimal assumptions it is possible to calculate the approximate ratio of flavin:Fe merely from a knowledge of the absorbance at 450 and 550 nm (Rajagopalan and Handler, 1964).

Per mole of FAD found, E_{450} for xanthine oxidase has been estimated as 35,000 by Bray and associates (Hart *et al.*, 1970) and as 37,800 by Massey *et al.* (1969a). The characteristic ratio A_{280}/A_{450} is given within the range 5.0–5.4 by many workers (Avis *et al.*, 1956; Hart *et al.*, 1970; Massey *et al.*, 1969a; Nelson and Handler, 1968). The absorption coefficient $E_{280}^{1\%}$ was estimated as 11.5 by Avis *et al.* (1956) and 11.26 by Massey *et al.* (1969a).

Recently, Komai *et al.* (1969) have succeeded in the reversible removal of flavin from milk xanthine oxidase. The method used was based on the observation of Morell (1952) that the addition of high concentrations of $CaCl_2$ appeared to result in dissociation of flavin. In the method of Komai *et al.*, xanthine oxidase is incubated at 20°C with 2 *M* $CaCl_2$ for 90 minutes and then dialyzed at 5°C to remove the $CaCl_2$ (and dissociated flavin). $MnCl_2$ was found to be even more effective than $CaCl_2$. The removal of the flavin appears to rely on at least two phenomena, a partial unfolding of the protein structure to release the FAD, and hydrolysis of the released FAD to FMN promoted by the $CaCl_2$. Rajagojalan and associates (Brady *et al.*, 1971; Rajagopalan *et al.*, 1970) have recently reported the similar removal of flavin from xanthine oxidase, chicken liver xanthine dehydrogenase, and aldehyde oxidase by high concentrations of $CaCl_2$, KI, and KSCN.

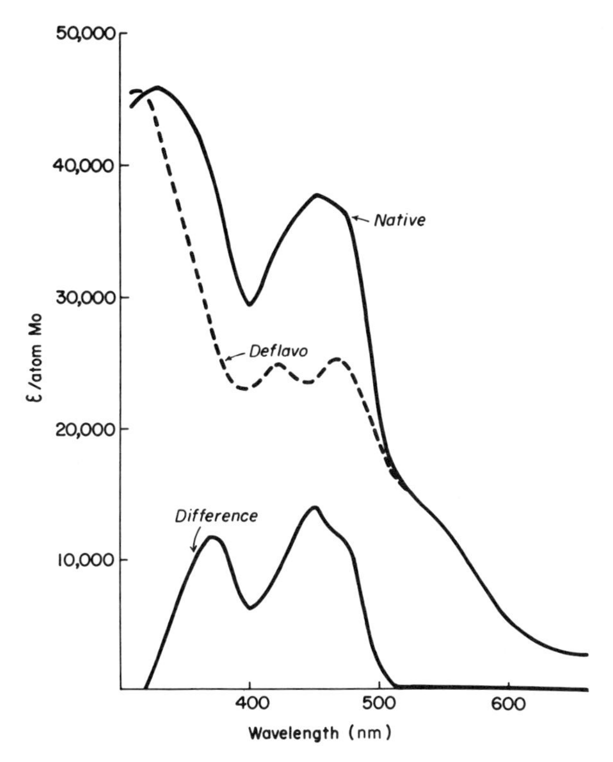

Fig. 2. Comparison of absorption spectra of native and deflavoxanthine oxidase, in 0.1 *M* pyrophosphate, pH 8.5. The spectra are expressed in terms of molar extinction coefficient (ϵ) (M^{-1} cm^{-1}) per atom of enzyme-bound molybdenum. The lower curve shows the calculated difference spectrum between the native and deflavo enzymes (from Komai *et al.*, 1969).

The deflavoxanthine oxidase obtained by $CaCl_2$ treatment still retains many of the catalytic properties of the native enzyme. The native enzyme can be regenerated by addition of FAD (Komai *et al.*, 1969). Such studies have been very helpful in elucidating the sites of electron transfer with this enzyme and will be considered in more detail later. Figure 2 shows the absorption spectra of native and deflavo enzyme, with the calculated difference spectrum of the contribution of the enzyme-bound FAD. The latter is typical of a simple flavoprotein, even to the extent of showing a shoulder around 480 nm, a phenomenon seen with many simple flavoproteins (see Palmer and Massey, 1968). The absorption spectrum of deflavoxanthine oxidase is very similar to that of several iron–sulfur proteins such as the plant ferredoxins (Tagawa and Arnon, 1968), adrenodoxin (Kimura and Suzuki, 1967), and putidaredoxin (Tsibris *et al.,* 1968). As these proteins do not contain molybdenum, it may be assumed that the spectral properties of deflavoxanthine oxidase are due in large part to the iron–sulfur chromophores. The observed E_{467} per enzyme-bound iron atom is 6300 cm^{-1} M^{-1}. This is slightly higher than that found with other iron–sulfur proteins which contain no molybdenum. For example, per equivalent of iron present, spinach ferredoxin has an E_{465} value of 4400 (Tagawa and Arnon, 1968) adrenodoxin an E_{445} value of 4200 (Kimura and Suzuki, 1967), putidaredoxin an E_{455} value of 4800 (Tsibris *et al.,* 1968), and two *Azotobacter* iron–sulfur proteins have E_{450} values of 4860 and 5300 (Shethna *et al.,* 1968). While it is not possible to draw any definite conclusions from such a comparison, because of the different wavelengths at which the absorption maxima occur, the possibility should certainly be left open that the molybdenum of xanthine oxidase does contribute somewhat to the visible absorption spectrum. Evidence that the molybdenum in complex form can have substantial absorption in the Mo^{IV} state has come from inhibition studies with pyrazolo-(3,4-*d*)-pyrimidines (Massey *et al.*. 1970). This evidence will be considered in a later section.

4. Circular Dichroism Spectra

In common with other iron–sulfur proteins, xanthine oxidase possesses strong and detailed optical activity in the visible regions of the spectrum. The circular dichroism (CD) spectra of oxidized and reduced xanthine oxidase have been reported by Garbett *et al.* (1969) and by Palmer and Massey (1969). When allowance is made for the different extinction coefficients used (E_{450} per flavin of 33,000 used by Garbett *et al.* and 37,800 used by Palmer and Massey), the results agree very well. The following values are taken from the latter authors. The spectrum of the oxidized enzyme resembles markedly that of spinach ferredoxin and adrenodoxin (Palmer *et al.,* 1967) with strong negative Cotton effects at approximately 350–400 nm and 500–550 nm and intense Cotton effects between 400 and 500 nm. The similarities are greater than is obvious from superficial examination of the spectra. Thus, from the various maxima and inflections of the spectrum, transitions can be identified at 370(—), 432(+), 470(+), 510(—), 640(—), and ~700(+) nm. (The latter band was initially ascribed as —ve by Palmer and Massey and +ve by Garbett *et al.;* later work showed that the ascription of Garbett *et al.* was correct.) The first four of these transitions are also found in spinach ferredoxin and in adrenodoxin. Although the rotational strengths are clearly different for the latter two proteins, the most conspicuous difference is in intensity rather than sign or energy of the transition. The intensity of the CD spectrum at 432 nm is 48 (liters per mole of flavin per cm) compared to 25 for ferredoxin and 33 for adrenodoxin. Thus on an iron basis the values are 12, 12.5, and 16.3, respectively.

On reduction of xanthine oxidase with substrate, the CD spectrum changes markedly, and, again, certain similarities can be seen with respect to ferredoxin and adrenodoxin. The decrease in CD intensity at 432 nm is biphasic and appears to be similar to the biphasic reduction of the visible absorption, i.e., is associated with the degree of functionality of the enzyme preparation (Komai and Massey, 1971; Massey *et al.*, 1970; McGartoll *et al.*, 1970; Edmondson *et al.*, 1972; Massey and Edmondson, 1970). The observed change in molar circular dichroism (fast and slow changes combined) was 56 liters per mole of flavin per cm. Thus the decrease was 14 liters per iron atom per cm. This compares favorably with the iron–sulfur proteins spinach ferredoxin and adrenodoxin, where the corresponding decreases on reduction were 14.3 and 22.2 liters per iron atom per cm at their respective maxima (Palmer *et al.*, 1967). As these proteins accept one electron for each pair of iron atoms, the results with xanthine oxidase suggest that, per molecule of flavin, there are two pairs of iron atoms, which accept a total of two electrons from reducing substrates.

5. EPR Spectra

The study of the EPR spectra of xanthine oxidase has had a profound influence on the understanding of this enzyme, particularly in the role of the molybdenum, which is not readily followed with other techniques. The mammoth contribution of Bray and his colleagues in this area cannot be emphasized too strongly. Bray has been responsible for the introduction of the rapid-freezing technique (Bray, 1961a) which has permitted a delineation of the time course of reduction of the separate redox components of the enzyme, and for the recognition and characterization of the complex molybdenum signals. While a definitive interpretation of all the results yet remains to be done, the findings made have contributed in a major way to the present concepts of the reaction mechanism of the enzyme.

a. Molybdenum Signals. A molybdenum EPR signal from the reduced enzyme was first observed by Bray *et al.* in 1959. More recent work has shown that with suitable conditions of reduction a considerable number of definable molybdenum signals may be obtained. Some of the signals develop within a catalytically significant time, whereas others develop much more slowly. Previously the signals were referred to by Greek letters (Palmer *et al.*, 1964), but a more systematic classification

TABLE I
MOLYBDENUM EPR SIGNALS FROM REDUCED XANTHINE OXIDASE

Signal	Signal-producing agents	Proton interaction	Comments
Very rapid ($\gamma\delta$)	Xanthine, 6-methyl purine	−	Signal so far unique to xanthine and 6-methyl purine
Rapid ($\alpha\beta$)	Xanthine, purine, salicylaldehyde	+	Proton interaction also seen from 8-deuteroxanthine in H_2O
Slow	Dithionite, purine, salicylaldehyde	+	Correlated with inactive enzyme (McGartoll *et al.*, 1970)
Inhibited	Methanol and formaldehyde	+	Correlated with functional enzyme (Edmondson *et al.*, 1972)

has now been made (Bray and Vänngård, 1969) and is shown in Table I. The unequivocal demonstration that Mo was responsible, at least for the "very rapid" and "rapid" signals, was shown by isolation of ^{95}Mo-enriched enzyme obtained from the milk of a cow injected with ^{95}Mo and study of the hyperfine interactions of this enzyme compared to normal enzyme (Bray and Meriwether, 1966). The "very rapid" signal has so far only been seen when xanthine or 6-methyl purine is used as substrate (Palmer *et al.,* 1964; Bray and Meriwether, 1966; Bray *et al.,* 1967). It exhibits no detectable interaction with protons from the medium or from xanthine (Bray and Vänngård, 1969; Bray *et al.,* 1969). The "rapid" signals actually are a class of signals, which show distinctive interaction from the substrate used (Pick and Bray, 1969). In all cases they show interaction with protons from the medium (Pick and Bray, 1969; Bray *et al.,* 1968). The great significance of the "rapid" signals to the catalytic mechanism of the enzyme is apparent from the demonstration with specifically deuterated substrates that one of the protons interacting with the metal is derived initially from the substrate. This has been interpreted in terms of direct hydrogen transfer from substrate to enzyme (Bray and Vänngård, 1969; Bray *et al.,* 1969; Bray and Knowles, 1968). Such a reaction had been postulated earlier by Rajagopalan and Handler (1964a, 1969); the EPR results provide compelling evidence for some such mechanism.

The "slow" signal has recently been shown (McGartoll *et al.,* 1970) to be associated with inactive xanthine oxidase. Although Bray and

Vänngård (1969) list it as being produced only by dithionite, purine, and salicylaldehyde, it has also been reported to occur on long term incubation with xanthine (Palmer and Massey, 1969). Its characteristic absorption probably reflects therefore a changed environment of the molybdenum atom in nonfunctional enzyme.

Electron paramagnetic resonance studies with deflavo enzyme reduced with xanthine for short times (30 seconds to 1 minute) showed the same "rapid" signals as native enzyme (Komai *et al.*, 1969). Thus it is clear that the proton interaction associated with the molybdenum signals is not due to interaction with a proton associated with the flavin radical.

b. FLAVIN RADICAL. A characteristic EPR absorption in the neighborhood of $g = 2.00$, ascribable to flavin radical, has been found by all workers under suitable conditions of reduction (Bray *et al.*, 1969; Palmer *et al.*, 1964; Palmer and Massey, 1969). Unambiguous ascription of this signal to flavin radical was obtained by its complete absence in reduced samples of deflavo enzyme (Komai *et al.*, 1969). When native enzyme is reduced with pterin as substrate, the EPR spectrum obtained at short times is predominantly that of flavin radical with only a small contribution from Mo. It has therefore been possible to make a fairly accurate measurement of linewidth; a value of 19.4 ± 1 G was obtained (Palmer and Massey, 1969). This value is characteristic for the blue or neutral semiquinone of flavins (Palmer *et al.*, 1971) and is unambiguously different from the linewidth of 15 G obtained with red anionic semiquinones (Palmer *et al.*, 1971). The same conclusion was reached by Bray and Vänngård (1969) who showed that one of the protons in the flavin that interact with the unpaired electron is exchangeable with the solvent.

c. EPR SIGNALS ASSOCIATED WITH IRON–SULFUR CHROMOPHORES. Like all iron–sulfur proteins which possess "labile" sulfide, xanthine oxidase displays on reduction the temperature-sensitive signal with g values of 1.90, 1.93, and 2.00, similar to those initially discovered by Beinert and Sands in mitochondrial particles (Beinert and Sands, 1960; Sands and Beinert, 1960). By measuring the EPR spectra at temperatures much lower than 77°K, the iron signal which is normally observed at 77°K as a broad unresolved band (Bray *et al.*, 1961; Palmer *et al.*, 1964) intensifies considerably and resembles that of spinach ferredoxin (Palmer and Sands, 1966). At temperatures between 39° and 18°K the integrated intensity increases and a new resonance at $g = 2.11$ is found (Palmer and Massey, 1969). With full reduction of the enzyme accomplished by dithionite, the integrated intensity over the temperature range

51°–60°K was found to be 1.16 electron equivalents per mole of flavin; over the temperature range 7°–18°K it was close to two electron equivalents (Palmer and Massey, 1969). It is thus concluded that xanthine oxidase contains (per mole of flavin) two pairs of iron–sulfur ligands, each pair of which can accept a single electron, but the pairs have somewhat different temperature sensitivity. These findings are in agreement with the results of circular dichroism studies (see previous section).

D. Substrate Specificity

Xanthine oxidase is capable of oxidizing a wide variety of compounds in addition to the conventional substrates hypoxanthine and xanthine. A large number of purines are oxidized, although at greatly differing rates, including hydroxylated (Bergmann and Dikstein, 1956; Bergmann *et al.*, 1960a, 1961), amino (Bergmann *et al.*, 1958, 1961; Wyngaarden, 1957; Wyngaarden and Dunn, 1957; Klenow, 1952; Booth, 1938; Bergmann *et al.*, 1960b; Scott and Brown, 1962), methyl (Bergmann and Dikstein, 1956; Bergmann *et al.*, 1960a, 1961), mercapto (Coombs, 1927; Silberman and Wyngaarden, 1961; Bergmann and Ungar, 1960), halogenated (Scott and Brown, 1962; Duggan and Titus, 1959), and *N*-oxide (Brown *et al.*, 1958) derivatives. Azapurines, with the additional N atom at position 2 or 8, are also very effective substrates (Shaw and Woolley, 1952; Bergmann *et al.*, 1959). Replacement of the imidazole ring of the purines by a pyrazolo ring also gives a series of compounds which can be very effective substrates (Feigelson *et al.*, 1957; Elion *et al.*, 1966; Spector and Johns, 1968), although they are perhaps more remarkable by the dramatic inhibitions to which they can lead (Elion, 1966; Massey *et al.*, 1970a,b; Spector and Johns, 1970). A wide variety of pteridines is also oxidized, some of them quite rapidly (Lowry *et al.*, 1949; Bergmann and Kwietny, 1958b, 1959; Valerino and McCormack, 1969). Pyrimidines are also reported to be good substrates, although no quantitative data on reaction rates have been reported (Lorz and Hitchings, 1950). Other heterocyclic compounds, including *O*-phenanthroline, have been reported as substrates (Bergmann and Kwietny, 1958a; Hassall and Greenberg, 1963; Greenlee and Handler, 1964; Massey *et al.*, 1969a). In the case of all the heterocyclic compounds listed above, the K_m values (generally measured in air-equilibrated solution) are quite low, in the region of micromolar. Many of these compounds inhibit when used at concentrations greater than 10^{-4} M; this interesting phenomenon will be pursued further in a later section. In Table II we have collected the data available from the literature on the substrate specificity of milk xanthine oxidase, employing heterocyclic substrates with oxygen as the electron acceptor. All rates are compared with xanthine taken arbitrarily as 100. As the turnover number (per mole of enzyme flavin) for fully active enzyme at 25°C can be estimated as approximately 1000 per minute, multiplication of the values quoted by 10 will give an approximate turnover number.

Almost all aldehydes tested can be oxidized to the corresponding carboxylic acids (Booth, 1938). In most cases the K_m values for aldehyde substrates are quite high, of the order of millimolar or greater. In fact, it is very difficult to "saturate" xanthine oxidase with aldehyde substrates; for this reason it is difficult to compare the reaction rates of aldehydes and heterocylic substrates (see Massey *et al.*, 1969a).

Formally all the above reactions are hydroxylations. This is illustrated below for the oxidation of xanthine to uric acid, although other tautomers than shown are probably involved (see Lichtenberg *et al.*, 1971).

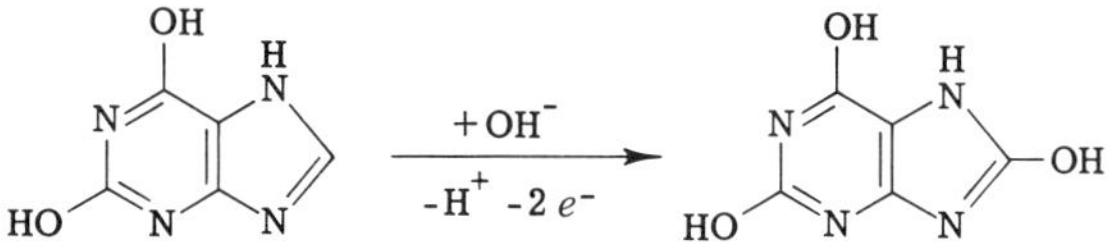

$$\text{Xanthine} \xrightarrow[-H^{+}\ -2e^{-}]{+OH^{-}} \text{Uric acid}$$

Similarly, in the oxidation of aldehydes

$$RCHO \xrightarrow[-H^{+}\ -2e^{-}]{+OH^{-}} RCOOH$$

The oxygen atom incorporated into product has been shown by ^{18}O studies to be derived from water, and not from molecular oxygen (Mason, 1957; Murray *et al.*, 1966). Hence the role of the latter is merely to reoxidize the reduced enzyme, as is commonly the case with flavoproteins. For many years, it was thought that the reaction proceeded by the enzyme reacting with the hydrated form of the substrate, followed by dehydrogenation with the enzyme becoming reduced (accepting two protons and two electrons). Recently, however, Fridovich (1966) showed with aldehydes that it is the unhydrated form which is the true substrate. The present evidence favors the view that the hydroxylation reaction proceeds through the abstraction from the substrate of a hydride ion, leaving a carbonium ion intermediate which then reacts with OH^- from the medium (Rajagopalan and Handler, 1964, 1968; Bray *et al.*, 1969; Edmondson *et al.*, 1972).

An intriguing finding from the specificity studies is that with compounds having several sites available for hydroxylation, distinct pathways of hydroxylation are followed rather than random attack. Such findings are also summarized in Table II. Thus hypoxanthine is con-

TABLE II

Substrate Specificity of Milk Xanthine Oxidase with Oxygen as Acceptor[a]

Substrate	Activity relative to xanthine (100)	Pathway of hydroxylation	Reference
Purines			
Purine	20	→ 6 → 2,6 → 2,6,8	*b*
7-Methyl	0.1		
9-Methyl	3.4		
2-Hydroxypurine	16	→ 2,8 → 2,6,8	*b*
1-Methyl	180		
3-Methyl	100		
9-Methyl	53		
6-Hydroxypurine (hypoxanthine)	70	→ 2,6 → 2,6,8	*b*
1-Methyl	0		
3-Methyl	0.15		
7-Methyl	0		
9-Methyl	0		
8-Hydroxypurine	1.5	→ 2,8 → 2,6,8	*b*
7-Methyl	2.2		
9-Methyl	0		
2,6-Dihydroxypurine (xanthine)	100	→ 2,6,8	*b*
1-Methyl	45		
3-Methyl	0		
7-Methyl	0		
9-Methyl	0		
2,8-Dihydroxypurine	0.2	→ 2,6,8	*b*
1-Methyl	0		
3-Methyl	2.1		
7-Methyl	0		

9-Methyl	0		
6,8-Dihydroxypurine	100	→ 2,6,8	*b*
1-Methyl	0		
9-Methyl	0		
2-Aminopurine	17.5	→ 6	*c*
6-Hydroxy (guanine)	0		
8-Hydroxy	1.3	→ 6,8	
2-Methylaminopurine	5.4		
2-Dimethylaminopurine	4.1		
8-Aminopurine	1.3	→ 6 → 2,6	*c*
6-Hydroxy	18	→ 2,6	
8-Methylaminopurine	0.3		*c*
6-Hydroxy	21		
6-Aminopurine (adenine)	0.6	→ 8 → 2,8	*d*
2-Hydroxy (isoguanine)	0.8	→ 8	*d*
2-Amino	0.16	→ 8	*d*
8-Hydroxy	2.1	→ 2	*e*
2-Fluoro	1.5		*f*
2-Chloro	1.2 (16)		*f*(*c*)
6-Mercaptopurine	3.8	{→ 8 → 2,8 {→ 2 → 2,8	*g*
8-Hydroxy	23	→ 2,8	*g*
2-Hydroxy (thioxanthine)	46.5	→ 2,8	*g*
8-Azapurines 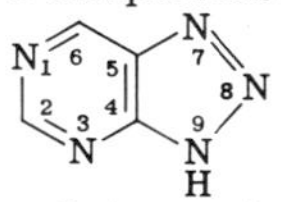			
8-Azapurine	14	→ 6 → 2,6 → 2,6,8	*h*
6-Hydroxy (8-azahypoxanthine)	41	→ 2,6 → 2,6,8	*h*
2-Hydroxy	6.5	→ 2,8 → 2,6,8	*h*
6-Amino	3.1	→ 6,8 → 2,6,8	*h*
2-Amino	21	→ 2,6 → 2,6,8	*h*
2-Methylamino	7.5	→ 2,8 → 2,6,8	*h*

TABLE II (*Continued*)

Substrate	Activity relative to xanthine (100)	Pathway of hydroxylation	Reference
2-Azapurines 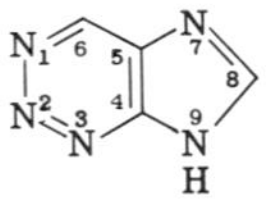			
6-Amino (azaadenine)	~100		*i*
6-Hydroxy (azahypoxanthine)	~60		*i*
Pteridines 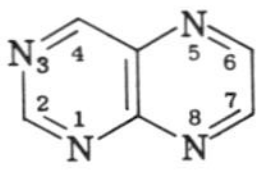			
Pteridine	Not readily measurable	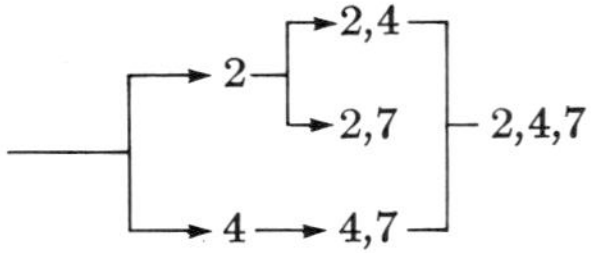 → 2 → 2,4 / → 2 → 2,7 / → 4 → 4,7; 2,4, 2,7, 4,7 → 2,4,7	*j*
2-Hydroxy	0.2	→ 2,4 → 2,4,7	*j*
	0.7	→ 2,7 → 2,4,7	*j*
4-Hydroxy	53	→ 4,7 → 2,4,7	*j*
6-Hydroxy	0		
7-Hydroxy	37	→ 2,7 → 2,4,7	*j*
2,4-Dihydroxy	7.7	→ 2,4,7	
2,7-Dihydroxy	21	→ 2,4,7	*j*
4,7-Dihydroxy	38	→ 2,4,7	
2,6-Dihydroxy	0		
4,6-Dihydroxy	~7	→ 2,4,6 → 2,4,6,7	*j*
6,7-Dihydroxy	~7	→ 2,6,7 → 2,4,6,7	

2,4,7-Trihydroxy	0		
2,4,6-Trihydroxy	1.2	→ 2,4,6,7	
4,6,7-Trihydroxy	25	→ 2,4,6,7	*j*
2,6,7-Trihydroxy	0.4	→ 2,4,6,2	*j*
2-Amino-4-hydroxy	20 (45)		*k*(*l*)
2-Amino-4,6-dihydroxy (xanthopterin)	5.5		*k*
2-Amino	20		*l*
2-Amino-7-hydroxy	46		*l*
2-Amino-4-methyl	0		*l*
2-Amino-7-methyl	13		*l*
2-Amino-6,7-dimethyl	16		*l*
4-Amino	2.2		*l*
4-Amino-2-hydroxy	29		*l*
4-Amino-7-hydroxy	146		*l*
4-Amino-6,7-dimethyl	0		*l*

[a] Most of the activities reported in the publications of Bergmann and co-workers are given directly relative to that of xanthine as 100, at pH 8, 28°C, using air saturated buffer. In other cases the activities were sometimes reported at different temperatures and pH values. In some cases no direct comparison of rate relative to that with xanthine is reported; in such cases rates have been calculated relative to substrates tested in common with Bergmann and co-workers, and so corrected back to xanthine as standard.

[b] Bergmann *et al.*, 1960a.

[c] Bergmann *et al.*, 1961.

[d] Wyngaarden, 1957.

[e] Wyngaarden and Dunn, 1957.

[f] Scott and Brown, 1962.

[g] Bergmann and Ungar, 1960.

[h] Bergmann *et al.*, 1959.

[i] Shaw and Woolley, 1952.

[j] Bergmann and Kwietny, 1959.

[k] Lowry *et al.*, 1949.

[l] Valerino and McCormack, 1969.

verted to uric acid almost exclusively via xanthine as an intermediate, with very little if any production of 6,8-dihydroxypurine (Bergmann and Dikstein, 1956). Furthermore, with such substrates, the accumulation of the intermediate is readily demonstrated in most cases. Thus, even though the enzyme molecule is capable of accepting at least six electrons from substrates (see later section), it is evident that the intermediate, even though it is a further substrate, is readily dissociated from the enzyme. Such behavior, conforming to the law of mass action, is of course not surprising. However, it does indicate that donation of electrons to the enzyme from substrate under anaerobic conditions probably occurs in stepwise pairs, with dissociation of product before attack by the next substrate molecule.

NOTE ADDED IN PROOF. The reader is also referred to a recent publication dealing with the substrate specificity of xanthine and aldehyde oxidases [Krenitsky *et al.* (1972)].

NADH is also oxidized by milk xanthine oxidase. In contrast to the reactions discussed above, this is a simple oxidation reaction not involving any hydroxylation; the product is NAD^+ (Mackler *et al.,* 1954). The nature of the enzymic reaction is poorly understood. The reaction has a pH optimum that is some 4 pH units lower than that for xanthine, and the reaction is considerably enhanced when ferricyanide is employed as electron acceptor rather than O_2 (Bergel and Bray, 1959). Under suitable conditions, turnover numbers comparable to that with xanthine can be measured (Massey *et al.,* 1969a). While NADH reduces the enzyme anaerobically, it does so only slowly, in fact, considerably more slowly than would be required to account for the catalytic turnover number (Massey *et al.,* 1969a). As the NADH:ferricyanide reductase reaction proceeds with a distinctive lag phase (which is not abolished by the addition of NAD^+ or ferrocyanide); these results suggest that the actual reductant may be a radical form of NADH. The flavin is considered to be the electorn acceptor in this reaction, since the deflavo enzyme is completely devoid of NADH:ferricyanide reductase activity (Komai *et al.,* 1969). This activity is also unaffected by the presence of enzyme which is nonfunctional in xanthine:oxygen reductase catalysis, i.e., it is unaffected by the AFR value of the enzyme (Bergel and Bray, 1959; Edmondson *et al.,* 1972) or by reaction of the enzyme with cyanide (Corran *et al.,* 1939), arsenite (Mackler *et al.,* 1954), or pyrazolo-(3,4-*d*)-pyrimidines (Massey *et al.,* 1970a). Reaction of the enzyme with these compounds has very dramatic effects on the reaction rates with purines and aldehydes.

In addition to its broad substrate specificity, xanthine oxidase utilizes a variety of oxidants as electron acceptors. The following compounds have been reported as effective acceptors: O_2 (Dixon, 1926), methylene blue (Dixon, 1926), dichlorophenol indophenol (Mackler *et al.,* 1954), ferricyanide (Mackler *et al.,* 1954), nitrate (Dixon, 1926), quinones (Dixon, 1926), cytochrome c (Horecker and Heppel, 1949), H_2O_2 (Dixon, 1926), dinitrobenzene (Dixon, 1926), picric acid (Dixon, 1926), permanganate (Dixon, 1926), iodine (Dixon, 1926), alloxan (Dixon 1926), trinitrotoluene (Bueding and Jolliffe, 1946), furacin (Taylor *et al.,* 1951), phenazine methosulfate (Fridovich and Handler, 1958), tetrazolium salts (Anderson and Patton, 1954), and NAD^+ (Morell, 1955).

In general, xanthine oxidizing enzymes from other sources (insofar as they have been tested) have a similar specificity pattern for the reducing substrate as does the milk enzyme, although distinctive differences do exist, particularly with the xanthine dehydrogenase of *Clostridium cylindrosporum* (Bradshaw and Barker, 1960). The chief difference between these enzymes which have been studied would appear to lie in their specificity for electron acceptors. As implied by their naming, the xanthine dehydrogenases of chicken liver (Rajagopalan and Handler, 1967) and *Clostridium cylindrosporum* (Bradshaw and Barker, 1960) react only slowly with O_2, but rapidly with acceptors such as ferricyanide or NAD^+. The so-called xanthine dehydrogenase of *Micrococcus lactilyticus* actually reacts reasonably efficiently with O_2; its unique property among this group of enzymes is its ability to use ferredoxin efficiently (Smith *et al.,* 1967). The latter enzyme is also unusual in its ability to catalyze rapidly the anaerobic dismutation of xanthine [uric acid, hypoxanthine, and 2,8-dihydroxypurine were identified as products (Smith *et al.,* 1967)]. A similar dismutation reaction with the milk enzyme had been reported in 1934 by Green with uric acid, as identified product, and hypoxanthine, a presumed product. This reaction is catalyzed extremely slowly by the milk enzyme and so it has tended to be regarded merely as a curiosity of little significance. However, the finding of rapid dismutation with the bacterial enzyme means that this reaction has to be explained adequately in any attempt at formulating a reaction mechanism for the enzyme.

Analysis of the mechanism of reduction of cytochrome c with xanthine or other substrates had led to very exciting discoveries within the last few years. In 1949, Horecker and Heppel (1949) made the interesting discovery that cytochrome c reduction was much faster in the presence of O_2 than it was under anaerobic conditions. Although this dependence was questioned by some workers (Morell, 1952), the studies of Fridovich and Handler and their colleagues adequately demonstrated its reality and finally its mechanism. These workers found that xanthine oxidase (and the related enzymes, aldehyde oxidase and dihydroorotate dehydrogenase) in the presence of substrate and O_2 initiated a chain reaction oxidation of sulfite whereby many more moles of O_2 were consumed than could be accounted for by the reducing substrate added, indicating the generation of free radicals (Fridovich and Handler, 1957, 1958a,b,c, 1961, 1962). Independently Totter *et al.* (1960a,b) demonstrated that the xanthine:oxygen reductase reaction, if carried out in the presence of dimethyl biacrydilium nitrate or of luminol, lead to the emission of chemiluminescence from these compounds. This was ascribed to the one electron reduction of O_2 to the superoxide anion, O_2^-, which was proposed as the reactive species both in the luminol reaction and in the sulfite oxidation. Greenlee *et al.* (1962) confirmed the chemiluminescence and showed that a similar phenomenon existed with aldehyde oxidase and dihydroorotate dehydrogenase (Greenlee *et al.*, 1962). Similarly they showed that the same enzymes exhibited the O_2-dependent reduction of cytochrome c. It was, therefore, proposed that the reduction of cytochrome c was due to its reaction with the O_2^- produced by reoxidation of these enzymes by O_2 (Fridovich and Handler, 1962; Greenlee *et al.*, 1962). The O_2-dependent reduction of cytochrome c was found to be surprisingly inhibited by the protein myoglobin (Fridovich, 1962) and carbonic anhydrase (Fridovich, 1967); this has been found recently to be due to a contamination with a newly recognized enzyme of very high catalytic efficiency, superoxide dismutase (McCord and Fridovich, 1968, 1969, 1970). This enzyme, which apparently catalyzes the reaction $2\ O_2^- \rightarrow O_2 + H_2O_2$, is effective in destroying O_2^- even at the level of $10^{-9}\ M$, and inhibits essentially completely the O_2-dependent reduction of cytochrome c by xanthine oxidase (McCord and Fridovich, 1969, 1970). It also inhibits in the same fashion the reduction of cytochrome c by O_2^- produced electrolytically (McCord and Fridovich, 1969). Unequivocal evidence for the univalent reduction of O_2 to O_2^- also came from EPR studies by Bray and colleagues (Knowles *et al.*, 1969; Nilsson *et al.*, 1969; Bray *et al.*, 1970), where the distinctive signal of O_2^- was found in rapid reaction studies with xanthine, xanthine oxidase, and O_2. More indirect evidence for the production of O_2^- was also obtained by Nakamura and Yamazaki (1969) who used lactoperoxidase as a radical trap. In their early studies Fridovich and Handler also tested a number of simple flavoproteins (glucose oxidase, D-amino acid oxidase, L-amino acid oxidase) reacting with their specific substrates and could not detect any initiation of the chain reaction oxidation of sulfite (Fridovich and Handler, 1961) or any chemiluminescence from luminol (Greenlee *et al.*, 1962). In keeping with the then current concepts of the role of the iron–sulfur chromophores, it was therefore proposed that it was due to reaction of O_2 with the reduced iron–sulfur centers that O_2^- was produced (Rajogopalan and Handler, 1968). This argument was also supported by the finding that the iron chelator, Tiron, was a strong inhibitor of the above reactions (Fridovich and Handler, 1962). However, it was pointed out that the effects of Tiron were complex, and that this substance might also act as a scavanger of O_2^- (Miller and Massey, 1965), a reaction since demonstrated directly (Miller, 1970). With the preparation of deflavoxanthine oxidase, the site of O_2 reaction in the enzyme became clarified (Komai *et al.*, 1969). The deflavo enzyme is rapidly reduced by purines and aldehydes (both molybdenum and iron–sulfur centers are rapidly reduced) and exhibits rapid catalysis of xanthine:ferricyanide reductase and xanthine:methylene reductase reactions. However it is almost completely inert in xanthine:oxygen reductase activity, and the xanthine:cytochrome c reductase activity it shows is completely independent of O_2 and unaffected by even high concentrations of superoxide dismutase (Massey *et al.*, 1969b). Thus it is clear that it is the reduced flavin of xanthine oxidase which reacts with O_2 to produce O_2^-. This conclusion is supported by EPR studies of Orme-Johnson and Beinert (1969) who found O_2^- signals on reaction of reduced xanthine oxidase with O_2, but not on reaction of reduced deflavo enzyme with O_2. It is also supported by the finding that free reduced flavins, on reaction with O_2, produce large yields of O_2^- (Ballou *et al.*, 1969). A number of metal-free flavoprotein enzymes have also been found to catalyze O_2-dependent reduction of cytochrome c, inhibited by superoxide dismutase (Massey *et al.*, 1969b).

E. Kinetic Studies

1. Results from Conventional Methods

Kinetic studies of milk xanthine oxidase are complicated by the phenomenon of inhibition by excess substrate, particularly with heterocyclic substrates. Dixon and Thurlow, using methylene blue as an electron acceptor, showed that the inhibition became less marked at higher acceptor concentrations (Dixon and Thurlow, 1924). Inhibition with excess substrate has also been observed with the acceptors: O_2 (Hofstee, 1955; Ackerman and Brill, 1962) and phenazine methosulfate (Fridovich and Handler, 1958a). It has been claimed that the phenomenon is not shown with cytochrome c as acceptor (Fridovich and Handler, 1958a); however, inhibition at high concentrations of xanthine in this system has been observed (V. Massey, unpublished results). Excess substrate inhibition is also observed in the pig liver oxidase (Brumby, 1963) and chicken liver dehydrogenase (Rajagopalan and Handler, 1967). It is not seen, however, with the bacterial dehydrogenases (Bradshaw and Barker, 1960; Smith *et al.*, 1967). On the contrary, with these enzymes, increased activity appears to occur at high substrate concentrations. The latter phenomenon appears to be associated with the dismutation reaction discussed previously (Green, 1934). Substrate inhibition of the milk enzyme appears to be greater the lower the pH (Fridovich and Handler, 1958a). Hofstee explained the inhibition in terms of a second molecule of xanthine inhibiting noncompetitively, with a dissociation constant of 10^{-3} M (Hofstee, 1955). An additional complication comes from the studies of Muraoka (1963a,b) who found that the presence of trace metals in reagent solutions could markedly affect the measurement of catalytic activity. In view of the complexities of the reaction, and the advances in knowledge and techniques since, the time appears ripe for a reinvestigation of this phenomenon.

Very little work involving the systematic variation of the concentrations of both the reducing substrate and the electron acceptor has been done. Fridovich (1964) reported a series of parallel Lineweaver–Burk plots on variation of both xanthine and O_2 concentrations. A similar pattern has been found with the chicken liver dehydrogenase (Rajagopalan and Handler, 1967). Thus the kinetics conform to that expected for a binary complex mechanism (Alberty, 1956; Dalziel, 1957) in which the product of the first half-reaction (uric acid) dissociates from the reduced enzyme prior to reaction of the latter in the second half-reaction with the electron acceptor (O_2). As pointed out previously (Palmer and Massey, 1968; Dalziel, 1962), under some conditions ternary complex mechanisms also simplify to this form. However, under these conditions further information may be obtained by use of inhibitors competitive to the substrate. In the case of a ternary complex mechanism, this would be expected to yield divergence from parallel plots in the Lineweaver–Burk plot of $1/v$ versus $1/O_2$. In the case of a binary complex mechanism the plots would be expected to remain parallel (Cleland, 1963). In the presence of a high concentration of salicylate, parallel plots were still obtained (Massey *et al.*, 1969a). Hence the available data, though not conclusive, point strongly to a binary complex mechanism.

The pH optimum of the milk enzyme is commonly quoted as being in the range 8–8.5. However, this would appear to be a consequence of a changing K_m at high pH values, rather than a changing V_{max}. Over the pH range of 9.4–11, Greenlee and Handler (1964) found no change in V_{max} but a large increase in K_m for xanthine. The effect of pH on K_m was extended to lower pH values by Fridovich (1965a); unfortunately, however, V_{max} values were not compared.

2. RESULTS FROM RAPID-FREEZING EPR STUDIES

When milk xanthine oxidase is reacted anaerobically with xanthine or other substrates, and the reaction is stopped at various times by the rapid-freezing technique of Bray (1961a), changes in EPR spectra characteristic of the reduction of the molybdenum, flavin, and iron–sulfur chromophores occur rapidly (Bray, 1961b; Bray *et al.*, 1964). With xanthine as substrate, the very rapid ($\gamma\delta$) Mo signal is the first to appear. It appears to be maximal at the shortest times studied (about 20 mseconds), and decays with a $t_{1/2}$ of about 100 mseconds at 22°C, pH 8.2. Under the same conditions the rapid Mo ($\alpha\beta$) and FADH$^{\bullet}$ signals appear at approximately the same rate, with a $t_{1/2}$ value in the neighborhood of 50 mseconds. The flavin radical then decays, with a $t_{1/2}$ of approximately 150 mseconds. In one preparation of the enzyme, the rapid Mo signal was also found to decay, while with another sample it remained essentially unchanged for 1–5 seconds (Bray *et al.*, 1964). The latter behavior appears to be the one consistently found with later preparations (Bray *et al.*, 1966). It should be noted, however, that the rapid Mo signal is very weak after reaction of the enzyme with xanthine for 1 minute, indicating further reduction over this longer time period (Palmer and Massey, 1969). Under the same conditions listed above, the $g = 1.94$ signal of the reduced iron–sulfur chromophore appears with a $t_{1/2}$ of 130 mseconds (Bray *et al.*, 1964). On the basis of these results, it was proposed that the sequence of electron transfer reactions within the enzyme was in the order →Mo→FAD→Fe/S, with the reduced species of the latter reacting with O_2 to generate O_2^- (Bray *et al.*, 1964). This latter conclusion has to be revised by the finding that deflavoxanthine oxidase, although reduced rapidly by xanthine, is oxidized extremely slowly by O_2 (Komai *et al.*, 1969).

The recognition that the occurrence of nonfunctional enzyme complicates the interpretation of many of the phenomena associated with this enzyme (Komai and Massey, 1971; Massey *et al.*, 1970a; McGartoll *et al.*, 1970; Edmondson *et al.*, 1972; Massey and Edmondson, 1970) warrants a reevaluation of the correlation of these changes in EPR signals with the overall catalytic velocity. Most of the rapid-freezing work was done with enzyme that has an $AFR^{23.5°C}$ value of around 100, i.e., with approximately 50% functionally active sites. The present evidence suggests that reduction of nonfunctional enzyme by reduced functional enzyme is a very slow process, so that the presence of a nonfunctional enzyme would not be expected to influence the *kinetics* of functional enzyme, but merely the *extent* of total changes which could be expected to be observed. Under the standard conditions of assay (pH 8.3 25°C,

$1-2 \times 10^{-4}$ *M* xanthine) the $AFR^{25°C}$ value of fully functional enzyme is estimated to be 210 (Komai and Massey, 1971; Massey *et al.*, 1970a; McGartoll *et al.*, 1970; Edmondson *et al.*, 1972; Massey and Edmondson, 1970). This corresponds to a turnover number (per molecule of flavin) of 815 per minute. From the observed effect of temperature on catalytic velocity (Massey *et al.*, 1969a), this would be equivalent to a turnover number of 645 per minute at 22°C. Thus a reduction of any group in the enzyme primarily concerned in the catalysis should proceed at a rate with a $t_{1/2}$ value of <64 mseconds. The appearance of the "very rapid" and "rapid" Mo signals, as well as the $FADH^{\bullet}$ signal occur at rates consistent with this requirement, whereas the Fe/S signal clearly does not.

With salicylaldehyde as substrate, further complications arise (Bray *et al.*, 1966). This substrate does not lead to the observable production of any "very rapid" Mo signal, and at 24°C leads to the production of the "rapid" signal with a $t_{1/2}$ value of approximately 1 second. Under the same conditions the $FADH^{\bullet}$ signal appears with a $t_{1/2}$ of less than 100 mseconds, reaches a maximum at 250 mseconds and decays with a $t_{1/2}$ of approximately 500 mseconds. No studies on the Fe/S signals were reported with this substrate, nor on rates of catalytic turnover.

It should be emphasized, of course, that drawing rational conclusions from such results is a difficult task, since it is not yet certain whether the reaction of oxygen with the flavin occurs with the fully reduced form, or the semiquinoid form, or both. Interpretation is also made difficult by the recognition that the enzyme may take up rapidly more than two electrons per active site (see later section). Thus the forms of the enzyme observed in rapid reduction studies may be different from those participating in catalysis, and their kinetics of appearance of only secondary significance. With respect to the rate of reduction of the iron–sulfur chromophores, it should be emphasized that at the temperature of observation, 77°K, the previous studies (Bray *et al.*, 1964) would fail to take into account the second species of iron–sulfur center, discovered subsequently (Palmer and Massey, 1969).

NOTE ADDED IN PROOF. Recent work has indeed shown that both iron–sulfur centers of xanthine oxidase are reduced at rates consistent with overall catalysis (Edmondson *et al.*, 1973).

3. RESULTS FROM STOPPED-FLOW SPECTROPHOTOMETRY

With an enzyme as complex as xanthine oxidase, containing in addition to its FAD prosthetic group, molybdenum and iron–sulfur chromophores, it might be expected that the kinetics of anaerobic reduction by substrates (followed spectrophotometrically) could be very complex. Indeed this has been found to be the case, although the results often appear to be disarmingly (and perhaps misleadingly) simple. The following section describes results obtained with approximately 70% functional enzyme at 25°C, pH 8.5 (Massey *et al.*, 1969a), where rates some 50% higher than previously reported (Gutfreund and Sturtevant, 1959) were obtained. When the anaerobic reduction of the enzyme with xanthine was followed, reduction to the extent expected for functional enzyme was observed to be complete in less than 500 mseconds. When the reduction was followed at 450 nm, apart from a small deviation at very short times, the reduction is represented accurately by a first-order plot with a rate constant of 950–1000 per minute. When the reaction was followed at wavelengths greater than 520 nm, more complex results were found. For example, at 610 nm, the decrease in absorbancy proceeds only after a distinct lag, probably due to the appearance of the neutral flavin radical, which has pronounced absorbance in this wavelength range. However, once this lag period was overcome, the subsequent decay was also found to be first order with a rate constant of 900–1000 per minute. As the wavelength region above 520 nm is considered to be due largely to absorption of the iron–sulfur chromophores, these results would suggest that these chromophores are also reduced sufficiently rapidly to be involved directly in catalysis. From kinetic analysis of varying xanthine and oxygen concentrations (Massey *et al.*, 1969a) it can be calculated that for fully functional enzyme the turnover number (extrapolated to infinite O_2 concentration) at 10^{-4} M xanthine, 25°C, pH 8.5 would be 815×1.20, or 975 per minute. Thus it would appear that the reoxidation of reduced enzyme with O_2 must be very much faster than this, and that all the chromophores of xanthine oxidase are reduced sufficiently fast to be involved in catalysis, reduction, indeed, being the rate limiting step in catalysis.

This conclusion, while apparently holding for xanthine, may not apply to other substrates. For example, when pterin (2-amino-4-hydroxypteridin) was used as substrate, a large increase in absorbance at 610 nm preceded the final reduction of the enzyme. The increase in absorbance, due to extensive flavin radical formation, proceeded with a rate constant of approximately 500 per minute; the subsequent decrease in absorbancy occurred with a rate constant of 50 per minute. The turnover number of the enzyme under these conditions was found to be 55 per minute, which was considered to be in reasonable agreement with the rate of full reduction of the enzyme. However, at that time the existence of nonfunctional enzyme was not appreciated; as the enzyme used was approximately 70% functional, the true turnover number would be approximately 80 per minute.

F. Anaerobic Titration of Xanthine Oxidase

In order to interpret the results from EPR and optical spectroscopy, it is necessary to know the number of reducing equivalents which can be accepted by the enzyme. Early attempts at measuring this were complicated by the presence of nonfunctional enzyme. In 1961 Bray *et al.* measured the Mo and FADH$^\bullet$ signals on anaerobic titration with xanthine, purine, and dithionite. With xanthine and purine as substrates, measuring signals after a 2 minute reaction, (i.e., measuring only functional enzyme) full reduction was obtained at approximately 3 moles xanthine per mole xanthine oxidase. As the enzyme used was probably only about 50% functional this corresponds to 6 moles xanthine per mole of functional enzyme or six electron equivalents per active center flavin. A similar end point was obtained by dithionite titration, between 6–8 moles dithionite per mole of enzyme, i.e., 6–8 electron equivalents per mole of flavin (dithionite reduces both functional and nonfunctional enzyme; hence no correction is necessary for titrations with dithionite).

In a later study, Bray *et al.* (1964) using enzyme of $AFR^{23.5°C}$ of 100 (i.e., $AFR^{25°C}$ of 112 = 54% functional enzyme) obtained end point titrations for Mo and FADH$^\bullet$ signals after 1 minute reaction of 6 moles xanthine per mole enzyme and an end point of approximately 8 moles xanthine for the reduced iron–sulfur signal. Per equivalent of functional active site these results would imply a total of 11–15 electron equivalents accepted by the chromophores of xanthine oxidase!

More recently, Massey *et al.* (1969a) have also reported the results of a large number of anaerobic titrations of xanthine oxidase with a variety of electron donors, using spectrophotometric measurements and waiting until no further absorbancy changes occurred with time (i.e., measuring both functional and nonfunctional enzyme). The results are shown in Fig. 3 and are expressed as moles of the two electron reducing agent per mole of enzyme flavin. It can be seen that all substrates can donate a total of seven electrons per mole of flavin, while dithionite can donate a total of eight electrons. These results were confirmed by titrations with dithionite and by measuring the appearance and disappearance of the EPR signals of Mo and FADH$^\bullet$ and the appearance of the two reduced iron–sulfur chromophores (Palmer and Massey, 1969). Similar anaerobic titrations of the deflavo enzyme indicated a total of six electron equivalents per atom of molybdenum (Komai *et al.*, 1969), in good agreement with theory. An independent confirmation of the uptake of seven electrons per mole enzyme flavin from substrates under anaerobic conditions has also been shown by analysis of products, when the reduced enzyme is inactivated by complexing with alloxanthine (Massey *et al.*, 1970b). Disagreement still exists on this point; Beinert

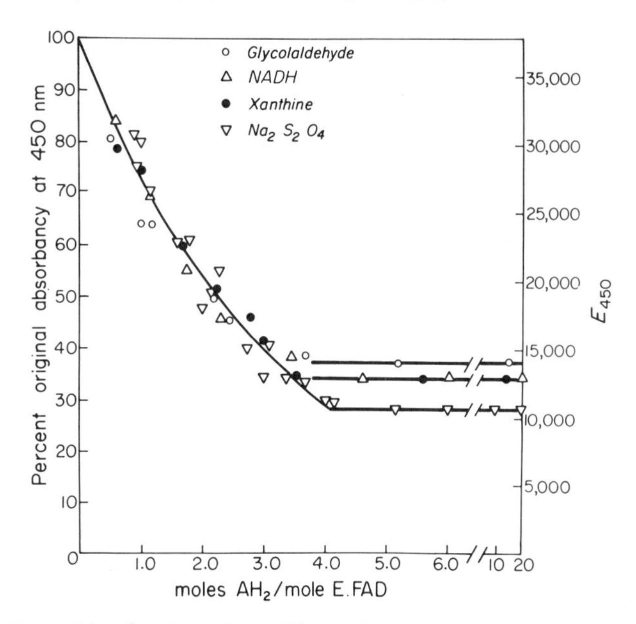

Fig. 3. Anaerobic titration of xanthine oxidase with substrates and dithionite. In most cases the results shown were obtained in separate experiments in which different amounts of substrate or dithionite were allowed to react with the enzyme until no further spectrophotometric changes occurred. Conditions, 0.1 *M* pyrophosphate, pH 8.5, 25°C (from Massey *et al.*, 1969a).

(1971) has recently reported that with dithionite as reductant, full titration of the $g = 1.94$ signal was obtained with six electron equivalents per mole flavin; however, full titration of the "slow" molybdenum signal required eight electron equivalents.

These surprising results have to be accomodated in any interpretation of the EPR signals of the enzyme. Bray and his colleagues (Pick and Bray, 1969; Bray *et al.*, 1964) have consistently interpreted all Mo signals as being due to Mo(IV) as the lowest likely oxidation state of the metal. In the absence of any other known electron acceptor in the enzyme, these titrations are difficult to explain unless one invokes reduction of Mo to lower valence states. Massey *et al.* (1969a) and Palmer and Massey (1969) have argued that of the observed eight-electron uptake with dithionite, the flavin can accept two electrons (FAD → $FADH^{\cdot}$ → $FADH_2$) and the two iron–sulfur chromophores can accept another two electrons, leaving Mo with the task of being the electron acceptor for the remaining four electrons. Thus it was suggested that with substrates, Mo may undergo the following valence changes; Mo(VI) → Mo(V) → Mo(IV) → Mo(III). With dithionite, it was argued that the lowest valence state reached could be Mo(II). Theoretically, the only valence state of Mo which could not show paramagnetism is Mo(VI). Surprisingly, such conclusions have been rejected rather vigorously by other workers (Pick and Bray, 1969), although it seems to us on no good grounds.

The importance of a correct interpretation of these results cannot be overemphasized. Rapid reaction studies with xanthine as substrate showed that all of the absorbance decreases associated with functional enzyme occur sufficiently rapidly to be catalytically significant (Massey *et al.*, 1969a). In other words, each enzyme molecule is capable of reacting sequentially with at least three substrate molecules (to accept at least six electrons) in a catalytically significant time. While it would seem likely that the catalytic reaction proceeds with an enzyme which has accepted only one pair of electrons from substrates, the possible involvement of further reduction states cannot be ruled out.

G. Action of Inhibitors and Conclusions Regarding the Difference between Functional and Nonfunctional Enzyme

1. Competitive Inhibitors

A wide variety of purines (Coombs, 1927; Wyngaarden, 1957; Bergmann *et al.*, 1960b), pteridines (Lowry *et al.*, 1949; Hofstee, 1949; Krebs and Norris, 1949), and other heterocyclic substances (Landon and Carter, 1960; Shaw and Woolley, 1952; Beiler and Martin, 1951; Fridovich, 1965b) which either are not oxidized by the enzyme, or are oxidized slowly, act as competitive inhibitors. One of the most powerful inhibitors known is 2-amino-4-hydroxy-6-formylpteridine; its K_i value is so low that by extrapolation it was concluded to inhibit completely at concentrations lower than stoichiometric with total enzyme flavin (Lowry *et al.*, 1949). Thus inhibition studies with this compound provided the first evidence for the existence of nonfunctional enzyme. Salicylate is also known to be a competitive inhibitor; it is commonly added to preparations of the enzyme to stabilize it from spontanteous loss of activity (Bergel and Bray, 1959). Dichlorophenol indophenol, as well as being an electron acceptor, has also been found to be a potent competitive inhibitor (competitive with substrate) in oxygen reductase assays (Fridovich, 1966b; Gurtoo and Johns, 1971).

2. Evidence for Importance of Protein Sulfhydryl Groups

Incubation of xanthine oxidase with thiol group reagents in the absence of substrates shows that the enzyme is relatively resistant to inhibition by such reagents. However, when incubation is carried out in the presence of substrate, relatively rapid inactivation occurs that is only partially reversible by incubation with excess sulfhydryl compounds (Fridovich and Handler, 1958b; Harris and Hellerman, 1956). Several possible explanations can be put forward to account for these results. Fridovich and Handler postulated that the sensitive thiol was exposed on reaction with substrate, owing to reduction of an iron–mercaptide linkage (Fridovich and Handler, 1958b). However, this explanation is ruled out in the light of later evidence, particularly by the finding that the iron–sulfur chromophores of the enzyme were not affected by this treatment (Massey *et al.*, 1969a). Bray *et al.* (1959) suggested that a molybdenum–mercaptide linkage might better fit the data. This remains an interesting hypothesis; no definitive evidence for or against it has since been obtained. Massey *et al.* (1969a) considered the possibility of reduction of a disulfide to a dithiol, but concluded that the body of evidence was against this. Unpublished experiments (Komai and Massey, unpublished data) reveal that substrate–mercurial inactivated enzyme has identical behavior with reducing agents as has native enzyme (Massey *et al.*, 1969a) providing further evidence against the possibility of disulfide reduction by substrate. Yet another explanation of the observed results is simply that a thiol group is exposed as a result of a substrate or reduction-induced conformational change. This area clearly needs more investigation.

In a recent paper, Gutoo and Johns (1971) have reported the interesting observation that 2,6-dichlorophenol indophenol (DCIP) is bound stoichiometrically (one molecule per molecule enzyme flavin) to milk xanthine oxidase with a characteristic spectral shift. They found that arsenite or *p*-chloromercuribenzoate interfered with binding but that prior reaction with cyanide did not. They interpreted these results as indicating that DCIP was bound either through a protein thiol or to molybdenum. The significance of these observations is not clear, however. We have found that arsenite does not interfere with complex formation, and that *S*-carboxymethyl serum albumin, a protein containing neither a free thiol nor molybdenum, binds DCIP with a similar shift in absorption spectrum to that observed with xanthine oxidase (Massey and Edmondson, unpublished data).

3. Substrate-Induced Inactivation by Iodoacetamide

It was shown by Bray and colleagues (Bray *et al.*, 1966; Bray and Watts, 1966; McGartoll and Bray, 1969; Bray, 1971) that inactivation of milk xanthine oxidase by iodoacetamide was greatly accelerated by the presence of xanthine. Similar results were also obtained by Green and O'Brien (1967). It was assumed that this inactivation was due to reaction with a thiol group exposed by reduction of the enzyme. It was shown that reaction of 1 mole of iodoacetamide per mole of enzyme (i.e., 1 mole per 2 moles FAD) is sufficient to inactivate the enzyme (McGartoll and Bray, 1969; Bray, 1971), and so it was proposed that there was but one active center per mole of xanthine oxidase, comprised in some way of the two molybdenum atoms, two molecules of FAD, and eight atoms of nonheme iron with their associated eight labile sulfur atoms (Bray *et al.*, 1969). In later papers Bray *et al.* observed that inactivation was also accompanied by a decrease in the visible absorption, and proposed that FAD was dissociated from the enzyme as a consequence of alkylation of the supposed sulfhydryl group (Bray, 1971; Bray *et al.*, 1966). In 1969 Komai and Massey reported at the Third International Flavin Symposium their results which showed that this reaction was due, in fact, to alkylation of $FADH_2$, and not to reaction with a thiol group (Komai and Massey, 1971). They showed not only the reductive alkylation of the flavin, but provided the first definitive evidence for the presence of nonfunctional active centers. It was shown that using 70% functional enzyme, 70% of the flavin was alkylated at the point of complete inactivation, whereas with dithionite, a nonspecific reducing agent, 100% inactivation corresponded to 100% alkylation of the flavin. Similarly, alkylation of the flavin by a nonspecific reaction (photobenzylation using phenylacetate, cf. Hemmerich *et al.*, 1967; Walker *et al.*, 1967, 1970) resulted in inactivation stoichiometric with the total flavin content. It was found that xanthine–iodoacetamide inactivated enzyme could be converted to a deflavo enzyme with similar properties to that prepared from native enzyme (Komai *et al.*, 1969) and reactivated in its xanthine:oxygen reductase merely by adding unmodified FAD. Many of these results have been confirmed in later work of McGartoll *et al.* (1970).

4. Inactivation by Pyrazolo-(3,4-*d*)-Pyrimidines

Compounds of this class have long been known to be both substrates and inhibitors of xanthine oxidase. In 1957 Feigelson *et al.* showed that 4-aminopyrazolo-(3,4-*d*)-pyrimidine was oxidized by xanthine oxidase to 4-amino-6-hydroxypyrazolo-(3,4-*d*)-pyrimidine, and that the latter was a more potent inhibitor than the former. Similarly, Elion *et al.* (1966) showed that allopurinol [4-hydroxypyrazolo-(3,4-*d*)-pyrimidine] is a potent inhibitor of xanthine:oxygen reductase activity and that alloxanthine [4,6-dihydroxyprazolo-(3,4-*d*)-pyrimidine] is the product of xanthine oxidase reaction with allopurinol.

$$\text{Allopurinol} + O_2 + H_2O \longrightarrow \text{Alloxanthine} + H_2O_2$$

It was also observed (Elion, 1966) that low concentrations of allopurinol inactivated xanthine oxidase when preincubated with the enzyme for several minutes before addition of xanthine. With low concentrations of alloxanthine, on the other hand, preincubation with enzyme caused no decrease in activity, but there was progressive inactivation after the substrate, xanthine, was added.

Recent work has elucidated the mechanism of the potent inhibitory action of this class of compounds (Massey *et al.*, 1970a,b; Spector and Johns, 1970). It has been found that inactivation is due to the extremely tight complexing of alloxanthine, and similar fully substituted pyrazolo-(3,4-*d*)-pyrimidines, with Mo(IV), or possibly still further reduced states of Mo, produced by reduction of the enzyme by substrates. Thus the inhibitory action of substrates such as allopurinol is due to a "suicide" reaction, in which the oxidized product, alloxanthine, interacts strongly with a reduced form of molybdenum produced by the reducing equivalents donated to the enzyme by the substrate. The complex formation is so tight that compounds such as alloxanthine may be used to titrate the number of functional active sites in xanthine oxidase preparations. Such studies (Massey *et al.*, 1970a,b) were very instrumental in establishing the $AFR^{25°C}$ value of fully functional enzyme to be 210 (see earlier section).

Inactivated enzyme complexed in this way has a modified absorption spectrum, which is characteristically different for each pyrazolo-(3,4-*d*)-pyrimidine used. Difference spectra between inactivated and native enzyme are shown in Fig. 4 and provide the first evidence that the molybdenum, at least in complexed form, may have significant absorption in the visible region of the spectrum. Nonfunctional enzyme, if reduced by dithionite, also complexes with pyrazolo-(3,4-*d*)-pyrimidines. In such complexes, which are formed slowly (over a period of hours, as opposed to seconds for functional enzyme) the difference spectrum for each complex is shifted 30–40 nm to longer wavelengths (cf. Section II,G,5).

Functional enzyme inactivated by complexing with pyrazolo-(3,4-*d*)-pyrimidines retains completely its NADH:ferricyanide reductase activity. All xanthine-acceptor reductase activities are essentially zero, but the enzyme gradually becomes reactivated in the course of the assays

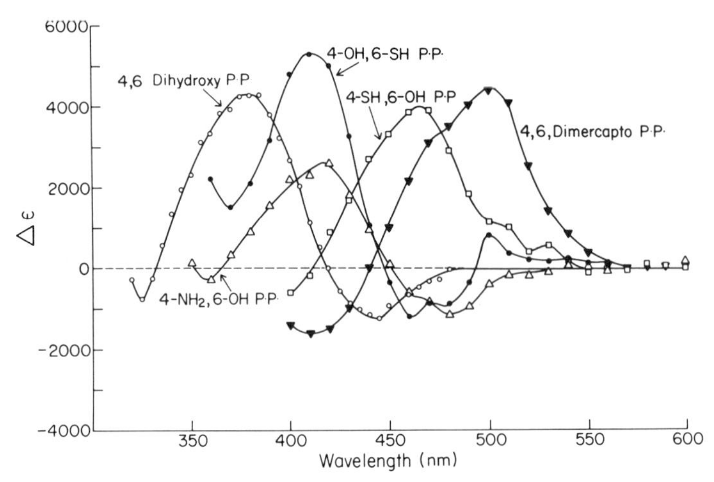

Fig. 4. Difference spectra between xanthine oxidase inactivated with various pyrazolo-(3,4-*d*)-pyrimidines and native enzyme. As the native enzyme used was approximately 70% functional, the extinction changes for fully functional enzyme would be expected to be some 40% greater than shown (from Massey *et al.*, 1970a).

because of reoxidation of the molybdenum and the concomitant loss of complexing with the pyrazolo-(3,4-*d*)-pyrimidine. Such reactivation is slow with O_2 (particularly at low temperatures) and much faster with acceptors such as ferricyanide or phenazine methosulfate. Reactivation is accompanied by complete return to the spectrum of native enzyme (Massey *et al.*, 1970a,b).

The very strong complexing of functional enzyme to pyrazolo-(3,4-*d*)-pyrimidines has been used as the basis of an affinity chromatography method to isolate fully functional and nonfunctional enzyme (Edmondson *et al.*, 1972).

The structural requirements for such tight binding seem to be quite rigorous; the analogous pyrazolo-(4,3-*d*)-pyrimidines, triazolo-(*d*)-pyrimidines and pyrrolopyrimidines are without effect (Massey *et al.*, 1970b).

5. Inactivation by Cyanide and Arsenite; Evidence for the Catalytic Importance of a Persulfide Group

The apparently irreversible inactivation of xanthine oxidase by incubation with cyanide was discovered by Dixon and Keilin (1936) and has been the focus of considerable attention since. Fridovich and Handler (1958b) reported the tight binding of ^{14}CN to cyanide-inactivated enzyme with a stoichiometry of 1 mole ^{14}CN per mole of enzyme flavin. These workers also reported that cyanide-treated enzyme though devoid of any activity with xanthine or hypoxanthine as substrate, could still bind one molecule of [^{14}C]hypoxanthine per molecule of enzyme flavin. This finding has been confirmed by using xanthopterin, where a pronounced shift in spectrum is observed on complexing with the inactive enzyme (Massey and Edmondson, unpublished data). Rajagopalan and Handler (1967) and Coughlan *et al.* (1969) have reported stoichiometric ^{14}CN binding with xanthine oxidase and two other molybdenum containing iron–sulfur flavoproteins, aldehyde oxidase, and chicken liver xanthine dehydrogenase. These workers concluded that cyanide inactivation was due to cyanide forming a complex with the enzyme-bound molybdenum, thus preventing substrate binding and electron transfer through the sequence of molybdenum, flavin, and iron–sulfur chromophores. Coughlan *et al.* showed that cyanide inactivation was accomplished by a change in the visible spectrum, with a decreased absorption maximal at 320 nm. These workers also studied the effect on the spectrum caused by inactivation due to reaction of the enzyme with arsenite. In this case the complexing resulted largely in an increased absorption, maximal at 380 nm. In contrast to cyanide inactivation, which was apparently irreversible, it was found that arsenite inhibition was reversible by dilution or dialysis (Coughlan *et al.*, 1969). [Inhibition by arsenite had first been reported by Mackler *et al.* (1954); it was also studied by Peters and Sanadi (1961).] Coughlan *et al.* found that cyanide and arsenite were mutually competitive in their action, and concluded that both agents reacted at the same site in the enzyme, viz., the molybdenum atom (Coughlan *et al.*, 1969).

In more recent studies, Massey and Edmondson (1970) showed that cyanide reaction was specific for functional enzyme, and that the inactivation was accompanied by the release from the enzyme of 1 mole of thiocyanite per functional active site, suggesting that the inactivation was due to cyanolysis of a persulfide linkage essential for catalysis

$$\text{Protein—S—S}^- + \text{CN}^- \rightarrow \text{protein—S}^- + \text{SCN}^-$$

It was also observed that when cyanide inactivation was carried out under anaerobic conditions, there resulted reduction of the enzyme equivalent to a two-electron uptake per functional active site (Massey and Edmondson, 1970). It was found that reactivation of cyanide-inactivated enzyme could be obtained by incubation with Na_2S, with reincorporation of sulfur into the enzyme. When such reactivated enzyme (labeled with ^{35}S) was resubjected to cyanide, inactivation again occurred, with the release of $^{35}SCN^-$. The following scheme (scheme I, below) was therefore proposed to explain the effects of cyanide (Massey and Edmondson, 1970).

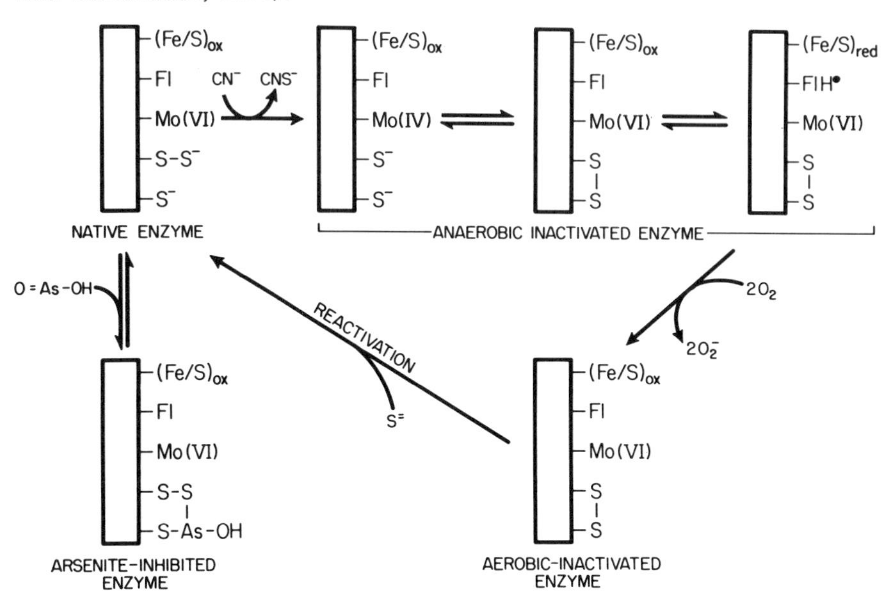

Scheme I

These studies also suggested an explanation for the competitive effect of arsenite on cyanide inactivation; it was proposed that arsenite could bridge the active center persulfide and an adjacent thiol group of the enzyme (Massey and Edmondson, 1970). The possibility also became evident that the difference between functional and nonfunctional enzyme might be due simply to the presence of the persulfide in the former and its lack in the latter.

Considerable evidence is now available to support these conclusions (Edmondson *et al.*, 1972). Figure 5 shows that the distinctive spectral changes accompanying cyanide and arsenite action are strictly proportional to the number of functional active sites, showing that arsenite, like cyanide, only reacts with functional enzyme. Secondly, nonfunctional enzyme, isolated by affinity chromatography can be activated by incubation with Na_2S under the same conditions as employed to reactivate cyanide-inactivated enzyme. Further evidence for the similarity between cyanide-inactivated and nonfunctional enzyme comes from their binding studies with pyrazolo-(3,4-*d*)-pyrimidines. In contrast to results with functional enzyme, where reaction of these compounds with reduced en-

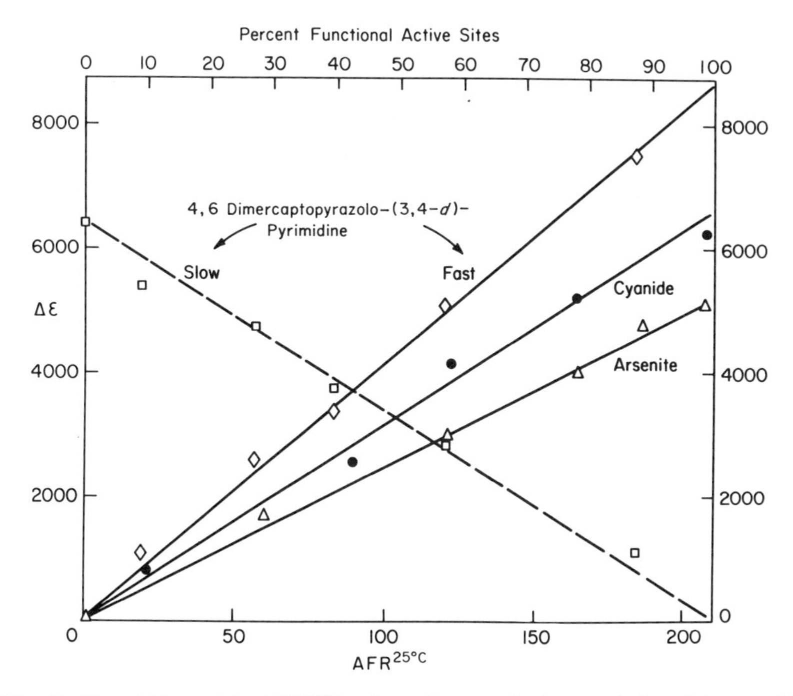

Fig. 5. Correlation with $AFR^{25°C}$ value of spectral changes induced in xanthine oxidase by complexing or reaction with inhibitors. The observed changes at characteristic wavelengths are expressed in terms of molar extinction coefficient (M^{-1} cm^{-1}) per mole E.FAD. (Δ) $+\Delta\epsilon_{380}$ on reaction with arsenite, (●) $-\Delta\epsilon_{320}$ on reaction with cyanide. The changes observed with 4,6-dimercaptopyrazolo-(3,4-*d*)-pyrimidine were obtained with enzyme previously reduced by dithionite; (◇), immediate increase in ϵ_{500}, (□) slow increase in ϵ_{545} (i.e., any immediate increase in ϵ_{545} subtracted from the finally obtained value). Conditions, 0.1 *M* pyrophosphate, pH 8.5, 25°C (from Edmondson *et al.*, 1972).

zyme is very rapid, and produces the spectral changes shown in Fig. 4, reaction with reduced cyanide-inactivated enzyme is slow ($t_{1/2}$ of the order of hours) and results in difference spectra shifted some 30–40 nm to longer wavelengths (Fig. 6). Nonfunctional enzyme, isolated by affinity chromatography, is found to show identical behavior to cyanide-inactivated enzyme in this reaction. Thus it is concluded that nonfunctional enzyme must have a structure similar to that shown for cyanide-inactivated enzyme in scheme I. Whether nonfunctional enzyme is produced entirely as an artifact during isolation or is secreted as such by the mammary gland, remains a fascinating question.

It should be emphasized that the inhibitory or inactivation effects described above apply only to the hydroxylation reactions catalyzed by xanthine oxidase. The NADH: ferricyanide reductase activity of the

	Wavelength Maxima (nm)	
	Functional	Nonfunctional
① 4, 6-Dihydroxypyrazolo-(3, 4-*d*)-pyrimidine	380	425
② 4-Hydroxy-6-mercaptopyrazolo-(3, 4-*d*)-pyrimidine	410	455
③ 4-Mercapto-6-Hydroxypyrazolo-(3, 4-*d*)-pyrimidine	465	515
④ 4, 6-Dimercaptopyrazolo-(3, 4-*d*)-pyrimidine	500	545

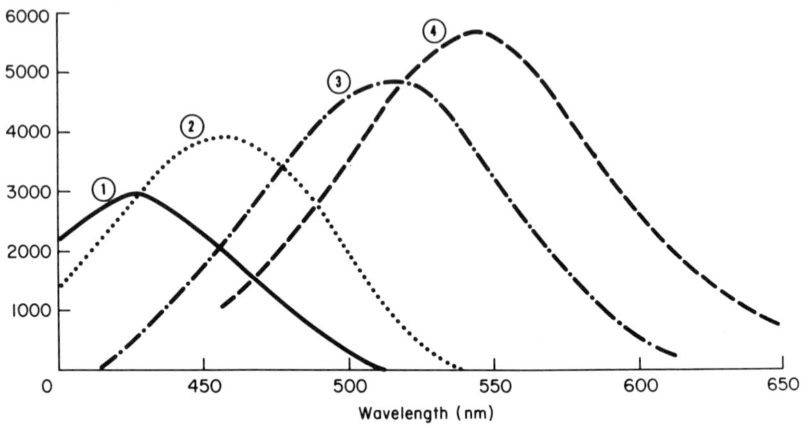

Fig. 6. Difference spectra obtained on adding various pyrazolo-(3,4-*d*)-pyrimidines to reduced, cyanide-inactivated xanthine oxidase. The latter was reduced by $Na_2S_2O_4$ under anaerobic conditions, spectra recorded, and the pyrazolo-(3,4-*d*)-pyrimidine added from a side arm of the anerobic cuvette. In all cases, the development of the characteristic difference spectrum was very slow; it is likely that incomplete changes are shown here. The figure shows, for comparison, the wavelength maxima of the difference spectra (produced immediately) obtained with native enzyme.

enzyme is completely insensitive to cyanide or arsenite treatment, or to the AFR value of the enzyme (Edmondson *et al.*, 1972). Thus it is evident that this activity depends on a completely different mechanism than that of the hydroxylation reactions, and does not rely on the presence of the active center persulfide.

6. Inactivation by Methanol and Formaldehyde

Handler and co-workers (Rajagopalan and Handler, 1968; Coughlan *et al.*, 1969) have studied in detail a progressive inactivation of xanthine oxidase by methanol which occurs when the enzyme is turning over in the presence of substrate. They observed that enzyme inactivated in this fashion, and freed of substrate, showed a characteristic EPR absorption ascribed to Mo(V) having methanol in its coordination sphere, and very resistant to oxidation by O_2. In further studies, Bray and his colleagues showed that a similar inactivation, and similar EPR signal (denoted by them as the "inhibited" molybdenum signal) could be produced by incubation with formaldehyde (Bray, 1971; Pick *et al.*, 1971). It was shown by isotope studies that a single nonexchangeable proton of the formaldehyde (or methanol) was associated with this signal (Pick *et al.*, 1971). Pick *et al.* proposed that a formyl residue, —CHO, becomes attached to a group in the active center and that interaction of molybdenum with this formylated residue stabilizes the pentavalent state of the metal, thus preventing enzyme turnover (Pick *et al.*, 1971). Methanol was presumed to be oxidized to formaldehyde at the active center as a prerequisite to inactivation.

Coughlan *et al.* (1969) reported that methanol-inactivated enzyme has a modified absorption spectrum, very similar to that of cyanide-inactivated enzyme. They showed that partial recovery of enzyme activity could be obtained by prolonged heating at 50°C; treatment with cyanide after methanol inactivation did not affect this reactivation, whereas cyanide-inactivated enzyme showed no reactivation under the same conditions. While more experiments remain to be done, it is tempting to speculate that the inactivation by methanol or formaldehyde is due in some way to formylation of the active center persulfide or that the latter is required for reduction of the enzyme by formaldehyde. In keeping with this idea, we have found that the production of the "inhibited" molybdenum EPR signal is proportional to the degree of functionality of the enzyme preparation; no signal was elicited from cyanide-inactivated enzyme (Edmondson *et al.*, 1972).

7. Reduction of Xanthine Oxidase by Borohydride

When milk xanthine oxidase is reacted under anaerobic conditions with $NaBH_4$, there is a partial bleaching of the visible absorption spectrum, correlating well with the percentage of functional active sites in the enzyme preparation. The extent of reduction found is the same as that produced rapidly by xanthine or other substrates, whether the reduction is followed by decrease in absorbancy at 450 nm, by the measurement of the $g = 1.94$ EPR signal of reduced iron–sulfur chromophores, or by measurement of the "rapid" molybdenum EPR signal. This correlation is shown in Fig. 7 (Edmondson *et al.*, 1972). The behavior of the chemical reductant, borohydride, is thus similar to that of substrates, and should be contrasted with that of dithionite, which reduces nonfunctional as well as functional enzyme. On admitting O_2, full return of the initial spectrum of oxidized enzyme is obtained; the reoxidized enzyme possesses its original catalytic activity. As borohydride is a good

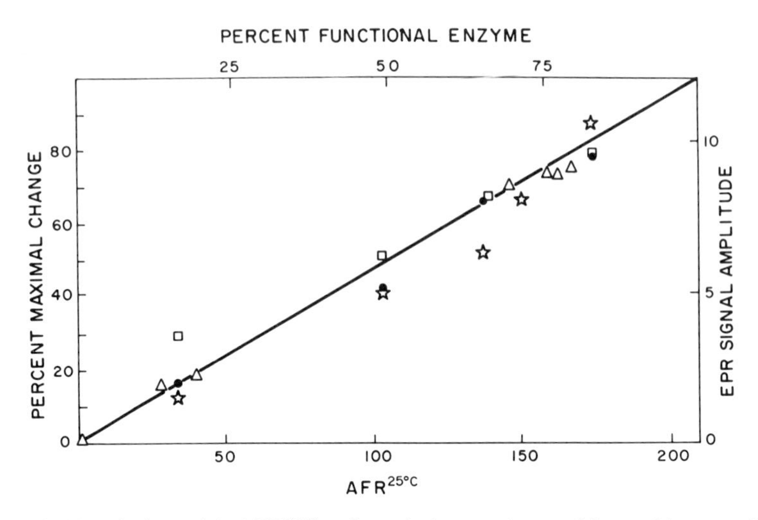

Fig. 7. Correlation with AFR$^{25°C}$ value of changes in xanthine oxidase produced by $NaBH_4$. (Δ) Bleaching of the visible absorption at 450 n*M*, expressed as a percentage of the total change (fast plus slow changes) induced anaerobically by xanthine ($\Delta\epsilon_{450}$ = 24,500 M^{-1} cm^{-1} per mole E.FAD for 100%); (●) and (☆), g = 2.11 EPR signals of reduced iron–sulfur chromophores, produced by incubation for 1 minute with $NaBH_4$ and xanthine, respectively, and expressed as a percentage of the same signal found after the addition of $Na_2S_2O_4$; (□), amplitude of the "rapid" molybdenum EPR signal, produced by reaction for 1 minute with $NaBH_4$. Conditions of reduction, 0.1 *M* pyrophosphate, pH 8.5, room temperature. (Data from Edmondson *et al.*, 1972.)

donor of hydride ions, this finding suggests strongly that the initial step in reduction of functional enzyme, by borohydride or by substrates, is the transfer to the enzyme of a hydride ion. The possible significance of this finding will be considered further in the following section.

H. Toward a Reaction Mechanism

The results summarized in preceding sections require that any reaction mechanism to be proposed for xanthine oxidase should account adequately for the following.

1. The rapid reduction of the molybdenum and flavin constituents, and possibly of the iron–sulfur chromophores.
2. The rapid appearance of two different kinds of reduced molybdenum EPR signals, the "very rapid" characteristic of xanthine and 6-methylpurine and the "rapid" signals produced by all heterocyclic and aldehyde substrates.

3. The finding that the "very rapid" molybdenum signal does not involve proton interaction, whereas the "rapid" signal does.

4. The finding that when 8-deuteroxanthine is employed as substrate, the "rapid" molybdenum signal observed indicates that the proton interacting with the molybdenum is derived from the 8 position, i.e., the position to be hydroxylated.

5. A functional role for the active center persulfide.

6. The observation that borohydride reduction is similar to substrate reduction.

7. A distinctly different reaction pathway is required for NADH oxidation compared to the hydroxylation reaction catalyzed for all other substrates.

8. The extremely probable situation (for which no contrary evidence presently exists) that per molecule of protein there are two active centers, each comprising one atom of molybdenum, one molecule of FAD, two somewhat different iron–sulfur centers, and a protein–persulfide group.

9. The finding that under anaerobic conditions each functional active site is capable of the rapid uptake of six to seven electrons from substrates and eight electrons from dithionite.

10. The observation that substrates capable of donating more than two electrons do not appear to do this without first dissociating from the enzyme.

11. The oxygen atom incorporated in the product is derived from water, not from molecular O_2.

12. Cyanide-inactivate enzyme can still bind strongly one molecule of substrate per molecule of enzyme flavin.

The studies of Bray and his colleagues on molybdenum–proton interactions on reaction of xanthine with the enzyme (Bray and Vänngård, 1969; Bray *et al.*, 1969; Bray and Knowles, 1968) suggest very strongly that a hydrogen atom is transferred from the substrate to the enzyme as one of the primary steps in catalysis. Such a possibility was in fact first suggested by Rajagopalan and Handler (1964, 1968). The recent finding of a persulfide group in the active center (Massey and Edmondson, 1970) and the similar behavior of borohydride and substrates (Edmondson *et al.*, 1972) suggests that this initial step may be a hydride ion transfer induced by the persulfide; which then acts as an "acceptor" for the resulting carbonium ion product. The stage would then be set for a separation of function; the reducing equivalents of the hydride ion being distributed among the various redox components of the enzyme and the hydroxylation reaction proceeding through attack by hydroxyl ion to liberate product and regenerate the persulfide group. In this scheme (scheme II) an intact persulfide is postulated for reaction with substrate;

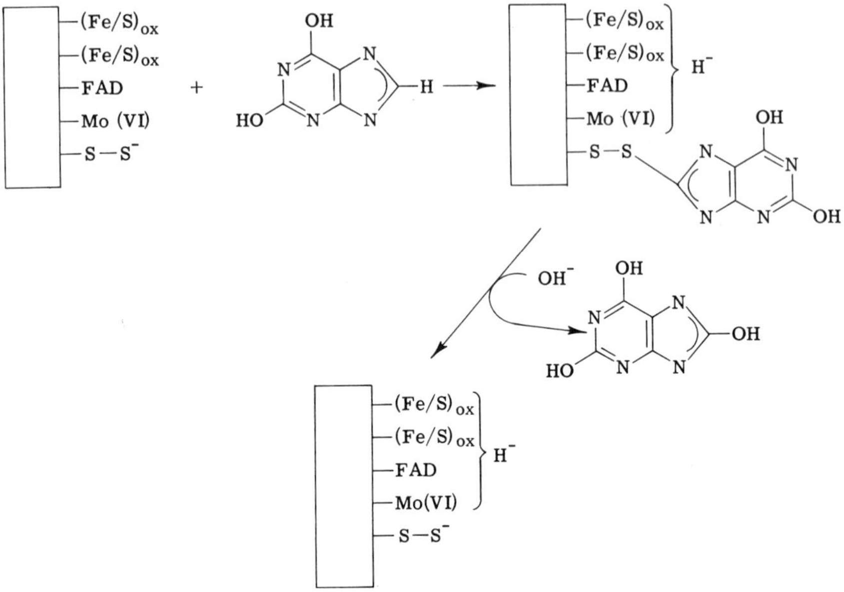

Scheme II

thus the hydroxylation reaction and liberation of the product would have to precede further reduction of the enzyme by subsequent molecules of substrate. Thus, with xanthine as substrate, the rate-limiting step in catalysis may be the regeneration of the persulfide with liberation of uric acid; the reducing equivalent donated to the enzyme being distributed among the other redox components with a given time sequence. If it is assumed that the molybdenum atom is the primary recipient of the hydride ion, it would be expected that this form of the enzyme would not show any paramagnetism [Mo(VI)$\bar{H}$]. If the next step were H atom transfer to the flavin, then one would expect simultaneous appearance of Mo(V) and FADH$^{\bullet}$ EPR signals, without any proton interaction with the molybdenum. If the secondary event were simply an electron transfer, the flavin abstracting a proton from the medium, then the resulting Mo(V)H signal would of course show proton interaction. In the former case, a Mo signal without proton interaction (the "very rapid" signal) would be accounted for, but at the expense of coincident flavin radical appearance. In the latter case, a proton-interacting Mo signal would be observed, accompanying the appearance of FADH$^{\bullet}$. Neither situation adequately accounts for the observed kinetics. However, as discussed previously, there is a discrepancy between the freeze-stopped

EPR studies (Bray *et al.*, 1964) and the stopped-flow spectrophotometric studies (Massey *et al.*, 1969a). If O_2 were not rigorously excluded from the former, and rapid equilibration of electrons between reduced iron–sulfur chromophores and flavin were to occur, then the time course of appearance and disappearance of FADH•, and the Fe/S chromophores, would be slower than really occurs.

Further experimentation, and extension of work with other substrates, is clearly desirable. The difficulty in interpretation of such results should, of course, be emphasized. In enzyme which has received a total of two electrons, these could be distributed as shown below.

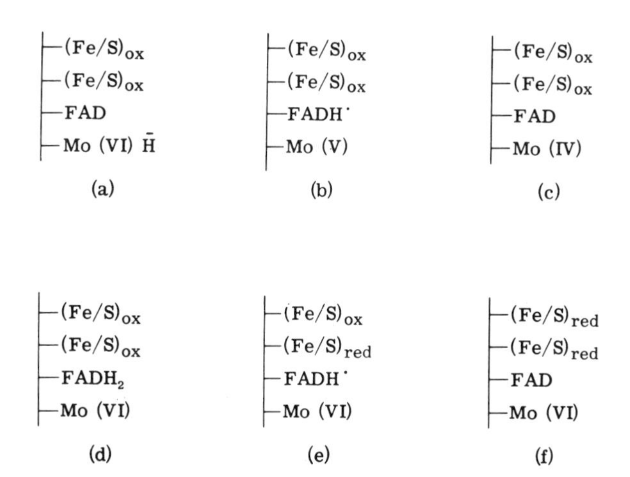

For enzyme which has received a total of four electrons, the forms shown below are possible.

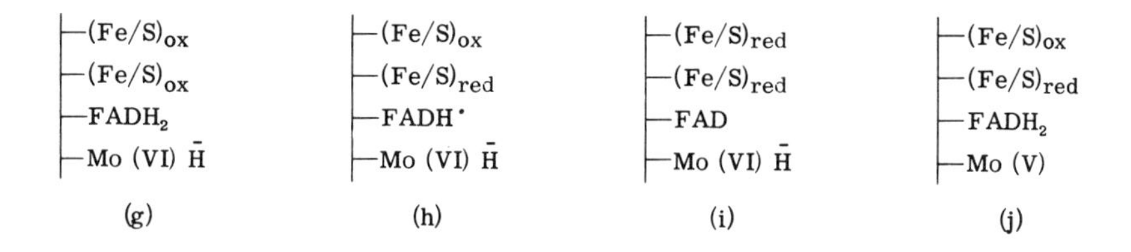

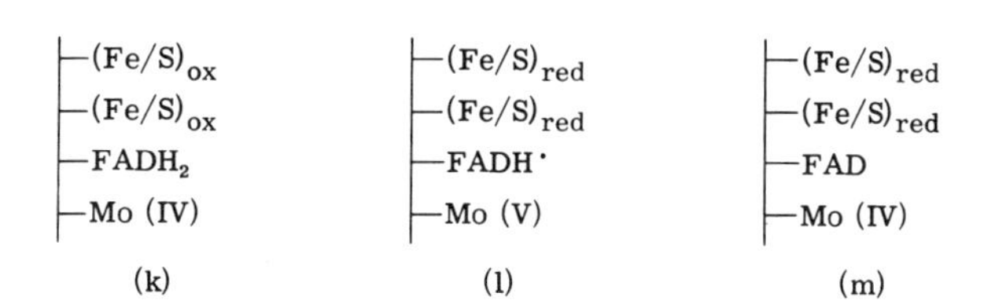

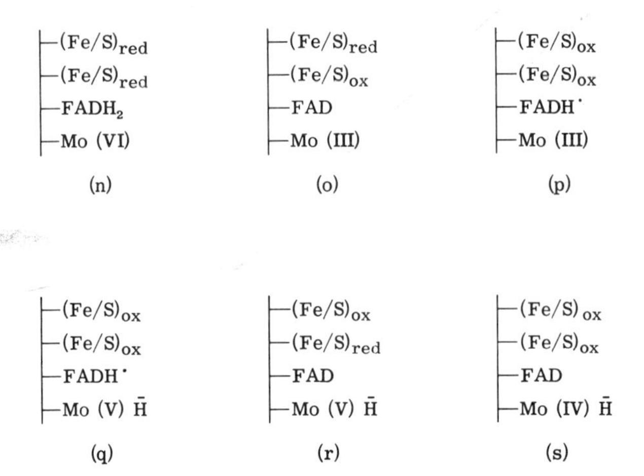

For an enzyme with a total uptake of six electrons forms (j)–(p) with a hydride ion associated with the Mo, as well as the forms shown below.

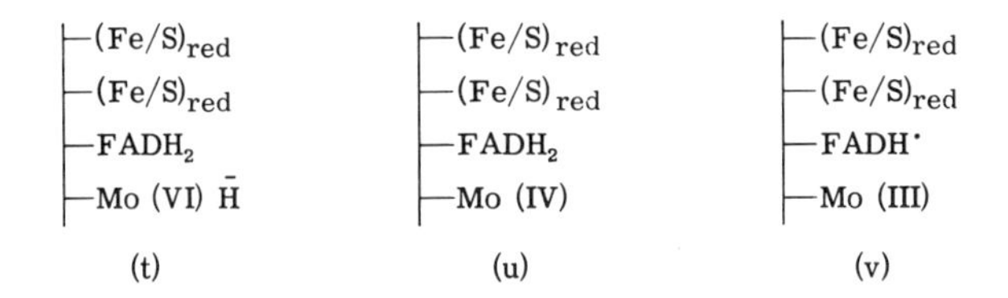

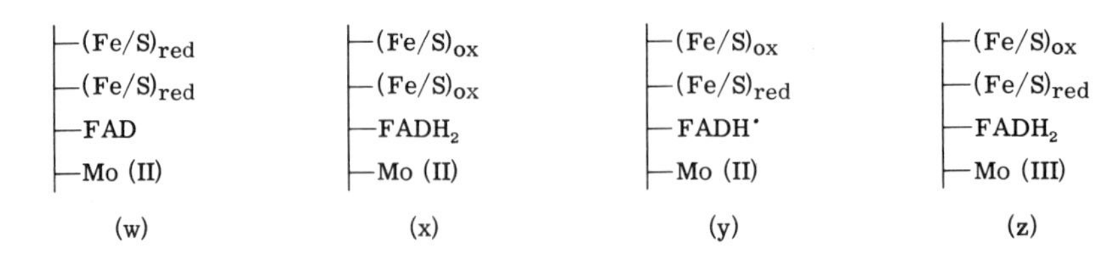

Each of the above forms could also occur with the carbonium ion of the substrate in linkage with the active center persulfide. It should be emphasized that the proposed role of the latter is not primarily as the binding site of the substrate, but as the acceptor of the carbonium ion product. It was shown by Fridovich and Handler that cyanide-inactivated enzyme still bound one molecule of hypoxanthine per molecule of enzyme flavin (Fridovich and Handler, 1958b); we have shown a similar stoichiometric binding with xanthopterin (Massey and Edmondson, unpublished data) as well as with pyrazolo-(3,4-*d*)-pyrimidines (see Section II,G,4). Hence it is likely that the initial attachment of substrate to the enzyme is by ligand formation with Mo, perhaps displacing in the case of functional enzyme, a preexisting ligand from the persulfide. With cyanide-inactivated enzyme (or nonfunctional enzyme), while substrate may still ligand with the molybdenum, it is proposed that hydride transfer does not occur because this requires the inductive effect of the persulfide.

Whatever the mechanism of the hydroxylation reactions may be, it is clear that the NADH-acceptor reductase activity of the enzyme proceeds through an entirely different pathway. In these reactions, it is clear that the persulfide grouping is not concerned, since factors which affect this grouping (cyanide inactivation, arsenite inhibition, and AFR value) have no effect. On the other hand, removal of the flavin (Komai *et al.*, 1969) or its alkylation (Komai and Massey, 1971) abolish this activity. It therefore appears likely that the uptake of electrons from NADH proceeds via the flavin. Whether the other redox components of the enzyme are at all involved in the catalysis of such reactions cannot be answered at the moment. As ferricyanide and phenazine methosulfate react rapidly with reduced deflavo enzyme, it is quite possible that either or both the molybdenum and iron–sulfur chromophores are involved; however, as with many metal-free flavoproteins which catalyze similar "diaphorase" reactions, the flavin alone may be utilized.

While much more work of a correlative nature remains to be done between rapid reaction spectroscopic and EPR studies, the present data suggest the electron flow scheme shown below.

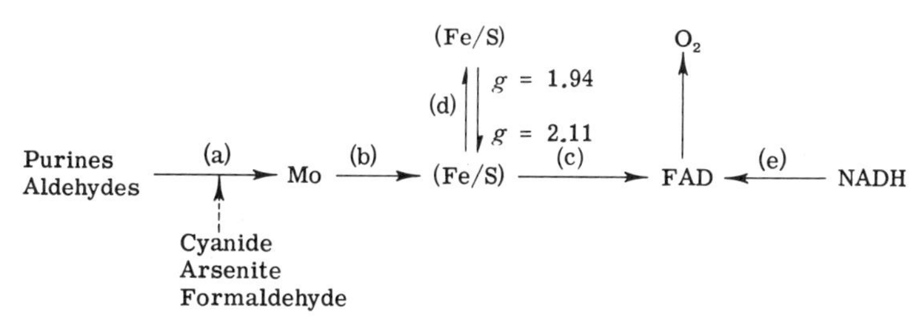

Such a scheme appears to be consistent with all the data currently available. While the previously reported rapid-freeze EPR data indicate a slower formation of the $g = 1.94$ signal than of FADH$^{\bullet}$, more recent work shows that the $g = 2.11$ signal is formed somewhat faster than the $g = 1.94$ signal (Edmondson *et al.*, 1973). For this reason the equilibrium reaction (d) is postulated between the two iron–sulfur centers. If reaction (c) were faster than reaction (b), the EPR signals of the reduced iron–sulfur centers would in fact appear to be formed more slowly than FADH•, since the FAD would have to receive its full complement of electrons before the iron–sulfur centers could become saturated. A further complication in the interpretation of results comes from uncertainty about anaerobiosis, particularly in the EPR studies. The kinetic evidence, outlined previously, suggests that O_2 reacts very rapidly with the reduced enzyme, presumably with $FADH_2$ or FADH•. Thus, unless O_2 is excluded rigorously, FADH• formation might appear to be considerably slower than is really the case. The role of an electron transfer bridge for the iron–sulfur centers is also suggested by studies with deflavo xanthine oxidase. While no rapid-freeze EPR data have been reported with this form of the enzyme, stopped-flow spectrophotometric studies (Komai *et al.*, 1969) showed that the iron–sulfur centers were reduced rapidly when xanthine or glycolaldehyde was used as substrate, in fact, some 30% more rapidly than with the intact enzyme. Hence a direct transfer of electrons between Mo and iron–sulfur centers is possible with the deflavo enzyme; the simplest hypothesis is that such a reaction can also occur efficiently with native enzyme.

III. ALDEHYDE OXIDASE

A. Introduction

In 1940 Gordon *et al.* purified a flavoprotein from pig liver that catalyzed the oxidation of aldehydes, but that was inactive with hypoxanthine. This specificity ruled out the possibility of identity with xanthine oxidase. In 1946 Knox, using a partially purified preparation from rabbit liver, presented strong evidence for the identity of aldehyde oxidase and quinine oxidase activities. Again, hypoxanthine was found not to be a substrate (Knox, 1946). In this study quinine and a large number of other quinolines were found to be effective substrates, as was *N′*-methylnicotinamide (Knox, 1946; Knox and Grossman, 1946, 1947). Knox commented that enzyme prepared from pig liver by the same procedure, while displaying aldehyde oxidase activity, was not capable of oxidation of quinolines. Later, Mahler *et al.* (1954) reported a modified preparation procedure of the enzyme from pig liver and found that in addition to FAD it contained molybdenum and heme iron, in the ratio 2:1:1. In subsequent work, Igo and Mackler (1960) and Igo *et al.* (1961) concluded that the heme was not catalytically active; they showed that the heme was due to catalase copurifying with the enzyme. A similar conclusion was reached by Palmer (1962a).

The first preparations of aldehyde oxidase free from heme contamination were reported by Rajagopalan *et al.* (1962) who isolated the enzyme from rabbit liver in a nearly homogeneous form. The availability of this purified enzyme has permitted a detailed study of the enzyme and revealed many properties which are similar to those of xanthine oxidase. Unless specified to the contrary, the properties reported below were obtained with the rabbit liver enzyme.

B. Physicochemical Properties

Purified aldehyde oxidase appears to have less than 10% contamination with other proteins, as judged by ultracentrifuge analysis (Rajagopalan *et al.*, 1962). The $s_{20°,w}$ value was determined as 11.5 S, and the molecular weight was determined by sedimentation equilibrium studies to be in the range 270,000–280,000 (Nelson and Handler, 1968). The enzyme was found to be free of heme iron, but to contain nonheme iron in addition to the previously recognized FAD and molybdenum components (Rajagopalan *et al.*, 1962). The Mo:FAD:Fe ratio was given as 2.0:1.8:8.2 per 300,000 molecular weight unit; in this estimation a value of E_{450} (oxidized–reduced) for liberated FAD of 1.13×10^4 was employed. If the more commonly accepted value of 1.05×10^4 (Beinert, 1960) is used, the above ratio becomes 2.0:1.94:8.2.

The absorption spectrum initially reported for the $\geq 90\%$ homogeneous enzyme showed a peak at 275 nm and no distinct peak at 450 nm. The ratio E_{280}/E_{450} can be estimated at 9.5–10 (Rajagopalan *et al.*, 1962). [*Note added in proof:* In a recent paper, E_{280}/E_{450} ratios of 5.2–5.8 have been reported for aldehyde oxidase preparations from rabbit and pig livers (Felsted *et al.* (1973).] On reduction with acetaldehyde, a difference spectrum was found with a large contribution at 275 nm. Analysis revealed the presence of coenzyme Q_{10} to the extent of 0.5–1.0 mole per mole of enzyme flavin and which is presumably responsible for the 275 nm difference peak. The significance of the CoQ content appears obscure at the moment. In later publications the E_{280}/E_{450} ratio has not been quoted, but a clearly defined absorption maximum at 450 nm was found (Rajagopalan and Handler, 1968; Coughlan *et al.*, 1969). The E_{450} value, per molecule of FAD, has been estimated as 31,500 (Rajagopalan and Handler, 1964), considerably less than the values found with xanthine oxidase and xanthine dehydrogenase. It is claimed (Rajagopalan and Handler, 1964c) that anaerobic addition of acetaldehyde results in rapid bleaching to the same extent as given by dithionite, although no experimental data are given. At face value, this would imply that aldehyde oxidase is free of nonfunctional active sites (Section II,G,7). However, other published experiments seem to be at variance with this claim. For example, from Figs. 2 and 3 of the paper of Rajagopalan *et al.* (1962), it can be calculated that 22.5% of the absorbance at 450 nm is bleached on addition of acetaldehyde. In later publications it may be seen that approximately 60% of the absorbance at 450 nm is bleached by substrate or dithionite (Rajagopalan and Handler, 1968).

As the interpretation of results with such enzymes is so dependent on whether all the active sites are in a functional form, the question of whether aldehyde oxidase preparations so far used contain nonfunctional sites is of the utmost importance.

C. Substrate Specificity

Like xanthine oxidase, aldehyde oxidase has a wide specificity for both electron donors and acceptors. Also the oxidation reaction is formally a hydroxylation, an oxygen atom derived from H_2O being incorporated into the product. This is illustrated below for the oxidation of *N′*-methylnicotinamide to *N′*-methyl-6-pyridone-3-carboxamide (Knox and Grossman, 1946, 1947).

$$\text{[}N'\text{-methylnicotinamide: } CONH_2,\ \overset{+}{N}-CH_3\text{]} + H_2O + O_2 \longrightarrow \text{[}N'\text{-methyl-6-pyridone-3-carboxamide: } CONH_2,\ O{=},\ N-CH_3\text{]} + H_2O_2$$

Recent evidence suggests that variable amount of the 4-pyridone may also be produced depending on the mammalian liver source (Felsted and Chaykin, 167). Similarly, indirect evidence has been obtained for the oxidation of pyridoxal by aldehyde oxidase (Stanulovic and Chaykin, 1971).

It is very difficult to extract from the literature estimates of the turnover number of the enzyme with any substrate, or relative activities between aldehyde and heterocyclic substrates. This difficulty arises from the fact that K_m values for aldehydes can be very high (of the order of 0.01–0.1 *M*) and that different electron acceptors have been used by different workers. Knox (1946) estimated that the turnover number (per molecule of enzyme flavin) at 37°C, pH 7.5, for cinchonidine as a substrate and methylene blue as an electron acceptor, was 150 per minute. Under the same conditions he lists the relative rates of a large number of other quinoline derivatives, *N′*-methylnicotinamide (55 per minute), crotonaldehyde (250 per minute), and benzaldehyde (135 per minute). With the pig liver enzyme as isolated by Gordon *et al.* (1940) with acetaldehyde as the substrate and methylene blue as the electron acceptor, the turnover number at 38°C, pH 7.2 was estimated at 520 per minute. The other substrates tested can therefore be calculated to have the following turnover numbers; crotonaldehyde, 340 per minute; benzaldehyde, 130 per minute; propionaldehyde, 80 per minute; butyraldehyde, 68 per minute; glycolaldehyde 40 per minute, and salicylaldehyde 10 per minute. It should be pointed out that these numbers must be less than V_{max} values, since they were estimated at a single, finite concentration of methylene blue. Considerably higher turnover numbers can be calculated from the data of Palmer, using the enzyme he prepared from pig liver (Palmer, 1962a). In this work, with $K_3Fe(CN)_6$ as the acceptor, the V_{max} turnover numbers at 25°C, pH 8.5, can be estimated as follows; formaldehyde, 340 per minute; butyraldehyde, 4100 per minute; valeraldehyde, 820 per minute; heptaldehyde 110 per minute; 2-methylpropionaldehyde, 290 per minute; 2-hydroxybutyraldehyde, 820 per minute; 2-methylbutyraldehyde 4100 per minute, 2-ethylbutyraldehyde, 2000 per minute; and crotonaldehyde 2600 per minute.

Rajagopalan *et al.* (1962) employed *N′*-methylnicotinamide as the substrate and O_2 as the electron acceptor in their standard assay. Although no estimate of turnover number is given, one can be calculated from their data and the published value of the extinction change at 300 nm when this substrate is oxidized (Knox and Grossman, 1947; Pullman and Colowick, 1954), assuming that the volume in their assay mixture was 3 ml. With these assumptions, a turnover number of 545 per minute per mole enzyme flavin at 25°C, pH 7.8 can be calculated. These workers (Rajagopalan and Handler, 1964c) confirmed and extended the studies of Knox on the substrate specificity. Even in concentrated solution, the inability of the enzyme to oxidize xanthine was confirmed. However, some purines were found to be oxidized, albiet slowly. Hypoxanthine was found to be hydroxylated to yield xanthine; purine was converted solely to 8-hydroxypurine. The latter reaction is in interesting contrast to the reaction of purine with xanthine oxidase, where the sequence of reactions, purine → hypoxanthine → xanthine → uric acid, was found (Bergmann and Dikstein, 1956). Rajagopalan and Handler (1962) also report that phenazine methosulfate is a good substrate, being oxidized to 10-methylphenazin-3-one. Thus phenazine methosulfate occupies an unusual position, as it is also a good electron acceptor (Rajagopalan and Handler, 1964b). Unfortunately, Rajagopalan and co-workers do not report data which would permit a comparison of the turnover efficiency of the enzyme with the various substrates and acceptors they have studied. The following acceptors were found to be utilized by the enzyme: O_2, $K_3Fe(CN)_6$, cytochrome c, methylene blue, dichlorophenol indophenol, phenazine methosulfate, silicomolybdate, nitro blue tetrazolium, and trinitrobenzene sulfonate. Quinones could not be used as electron acceptors, since in agreement with previous studies (Mahler *et al.*, 1954) they were found to be potent inhibitors (Rajagopalan and Handler, 1964b).

D. Kinetic Studies

1. Results from Conventional Studies

So far no kinetic studies have been reported for the rabbit liver enzyme in which the concentrations of both the electron donor and electron acceptor were varied. Such studies have been done with the enzyme from pig liver (Palmer, 1962b) and indicate a ternary complex mechanism, i.e., that the product is not dissociated from the reduced enzyme before the latter reacts with the electron acceptor. This behavior contrasts with the kinetics observed with xanthine oxidase, where a binary complex mechanism appears to operate (Fridovich, 1964; Massey *et al.*, 1969a).

2. Results from Rapid Reaction Studies and Anaerobic Titrations

An extensive series of experiments employing the rapid-freezing technique coupled with EPR measurements has been reported (Rajagopalan *et al.*, 1968a,b). Aldehyde oxidase is unusual in that the oxidized enzyme shows an EPR signal attributable to Mo(V) even before the addition of reducing substrate. The significance of this signal is uncertain, since it accounts for only a small percentage of the molybdenum present (Rajagopalan *et al.*, 1968a). On reduction with substrate, signals which can be ascribed to $FADH^{\bullet}$ and the reduced iron–sulfur chromophore(s) appear rapidly, as well as a molybdenum signal that is rather similar to the "rapid" ($\alpha\beta$) signal observed with xanthine oxidase. A second type of Mo signal is also apparent, particularly from power saturation studies, which appears to be formed on prolonged exposure of the enzyme with substrate. The rate of appearance of the various signals (except for the "slow" molybdenum signal) on anerobic reaction with *N*-methyl nicotinamide is sufficiently rapid for involvement of the reduced forms in catalysis. Turnover experiments, where both O_2 and substrate were present, showed a low steady-state level of the $FADH^{\bullet}$ and reduced iron–sulfur signals, and a much higher level of the "rapid" molybdenum signals. These results suggest that in electron transfer, the molybdenum is the site of electron input, and either the flavin or iron–sulfur chromophore is the site of reaction with O_2.

As discussed previously in the case of xanthine oxidase (see Section II,G,7), the interpretation of such results relies heavily on knowing how many electrons may be accepted by the enzyme, and whether nonfunctional enzyme is present. A number of anaerobic titration experiments have been reported which bear on these questions. In one experiment, where dithionite was used and reduction was monitored by EPR, full reduction appeared to be achieved with the uptake of approximately four electron equivalents per mole of E. FAD (Rajagopalan *et al.*, 1968a). In other experiments, where *N*-methylnicotinamide was used, and reduction was followed spectrophotometrically as well as by EPR, full reduction required the addition of 4–5 moles of substrate per active center, i.e., 8–10 electrons per mole E. FAD (Rajagopalan and Handler, 1964; Rajagopalan *et al.*, 1968a). Although the discrepancy between the results with the two reductants might be ascribed to problems of redox potential, the concept that only four electrons are donated to the enzyme

$$\mathrm{Mo(VI)} \rightarrow \mathrm{Mo(V)}, \qquad \mathrm{FAD} \rightarrow \mathrm{FADH_2}, \qquad \mathrm{(Fe/S)_{ox}} \rightarrow \mathrm{(Fe/S)_{red}}$$

(Rajagopalan *et al.*, 1968a,b) is difficult to accept in the light of more recent knowledge with other iron–sulfur proteins. In the above experiments there was evidence for more than one type of iron–sulfur chromophore, as judged by the different behavior of the appearance of $g = 1.94$ signal and the bleaching of the absorbance at 550 nm. Thus, in analogy to xanthine oxidase, it is likely that there are two iron–sulfur chromophores, each of which could accept an electron. Similarly, the so-called "fading" of the molybdenum signals (Rajagopalan *et al.*, 1968a,b) might be due to reduction of this metal to lower valence states than Mo(V).

With respect to the question of nonfunctional sites, the rapid-freezing results are consistent with this possibility. At reaction times where fully formation of FADH$^{\bullet}$ and "rapid" Mo signals were observed, the intensity of the $g = 1.94$ signal induced by substrate was only 40% that observed on addition of dithionite (Rajagopalan *et al.*, 1968b). This situation is reminiscent of that found with xanthine oxidase containing nonfunctional active sites (Massey *et al.*, 1970b).

E. Studies with Inhibitors and Deflavo Enzyme

The different effects of various inhibitors, depending on the electron acceptor used, has provided much information on the sites of interaction of various electron acceptors with the redox constituents of the enzyme. Inactivation of aldehyde oxidase by incubation with cyanide was first noted by Gordon *et al.* (1940), and inhibition with arsenite by Mahler *et al.* (1954). Menadione was also found to be a potent inhibitor, and significant inhibitions with some acceptors were found with the respiratory chain inhibitors: amytal, antimycin, and oligomycin (Rajagopalan *et al.,* 1962). These findings prompted Rajagopalan and Handler (1964b) to a thorough study of inhibitor–acceptor relationships, which lead to the three categories below:

1. Cyanide and *p*-chloromercuribenzoate: inhibition with all acceptors tested (dichlorophenol indophenol, ferricyanide, methylene blue, phenazine methosulfate, silicomolybdate, nitroblue tetrazolium, trinitrobenzene sulfonate, O_2, and cytochrome c).
2. Amytal and oligomycin: inhibition with all acceptors except dichlorophenol indophenol.
3. Antimycin A and menadione: inhibition only with trinitrobenzene sulfonate, nitroblue tetrazolium, O_2, and cytochrome c as electron acceptors.

These studies were interpreted in terms of a linear sequence of electron transfer from substrate through the various redox groups of the enzyme with O_2 reacting with the last constituent of this sequence. In order to explain the inhibition pattern, four redox groups were postulated, thus implying the participation of CoQ_{10} in the electron transfer process. While CoQ_{10} may indeed be involved, newer information gathered over the last few years permit a reinterpretation of these data involving only the Mo, FAD, and Fe/S chromophores as follows (scheme III, below).

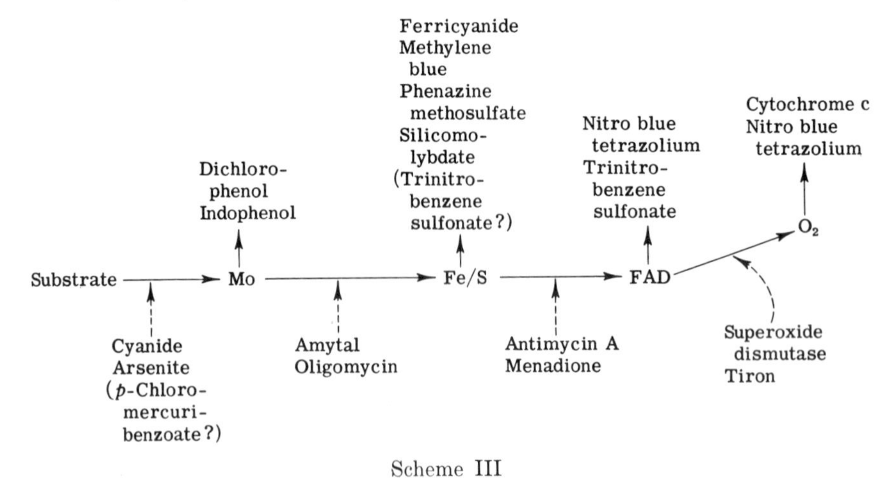

Scheme III

In scheme III, horizontal arrows represent electron transfer from substrate through the enzyme; vertical solid arrows, sites of electron flow to acceptors; and dashed arrows, sites of interference of electron flow by inhibitors. The rationale for postulating the flavin as the site of reaction with O_2 has been summarized in Section II,G,7; O_2 is reduced to O_2^- which then reacts to reduce cytochrome c (McCord and Fridovich, 1969, 1970; Ballou *et al.*, 1969) and presumably also nitro blue tetrazolium. In addition to such O_2-dependent reduction, nitro blue tetrazolium is also reduced anaerobically (Rajagopalan and Handler, 1964b). In view of the inhibitor behavior, it seems logical to assume that trinitrobenzene sulfonate and nitro blue tetrazolium may also react with reduced flavin directly. While no direct information is currently available, it would seem likely that cyanide reacts to destroy a persulfide linkage required for catalytic activity, as it does with xanthine oxidase (Massey and Edmondson, 1970). Possibly, in the case of aldehyde oxidase, this grouping also reacts with *p*-chloromercuribenzoate and arsenite. The validity of scheme III could be tested in part by acceptor-specificity studies with deflavo enzyme. While a deflavo:aldehyde oxidase has been prepared (Brady *et al.*, 1971; Rajagopalan *et al.*, 1970), it was too unstable to permit such studies.

Aldehyde oxidase, like xanthine oxidase, undergoes distinctive spectral changes on incubation with cyanide, arsenite, and methanol, all of which lead to inactivation or inhibition of activity (Coughlan *et al.*, 1969). These data have been interpreted as being due to reaction with the enzyme-bound molybdenum. However, in view of the recent finding of a persulfide group in xanthine oxidase (Massey and Edmondson, 1970), and the striking similarity of behavior of xanthine oxidase and aldehyde oxidase toward these inhibitors (Coughlan *et al.*, 1969), it would seem more likely that their action is associated with an active center persulfide in all cases.

Note Added in Proof. Recent work has indeed demonstrated the presence in rabbit liver aldehyde oxidase of an active site persulfide group (U. Branzoli and V. Massey, unpublished results).

REFERENCES

Ackerman, E., and Brill, A. S. (1962). *Biochim. Biophys. Acta* **56**, 397.
Alberty, R. (1956). *Advan. Enzymol.* **17**, 1.
Anderson, A. D., and Patton, R. L. (1954). *Science* **120**, 956.
Andrews, P., Bray, R. C., Edwards, P., and Shooter, K. V. (1964). *Biochem. J.* **93**, 627.
Avis, P. G., Bergel, F., and Bray, R. C. (1955). *J. Chem. Soc.* 1100.
Avis, P. G., Bergel, F., and Bray, R. C. (1956). *J. Chem. Soc.* 1219.
Avis, P. G., Bergel, F., Bray, R. C., James, D. W. F., and Shooter, K. V. (1956). *J. Chem. Soc.* 1212.

Ball, E. G. (1939). *J. Biol. Chem.* **128**, 51.
Ballou, D., Palmer, G., and Massey, V. (1969). *Biochem. Biophys. Res. Commun.* **36**, 898.
Beiler, J. M., and Martin, G. J. (1951). *J. Biol. Chem.* **192**, 831.
Beinert, H. (1960). *In* "The Enzymes" (P. D. Boyer, H. Lardy and K. Myrbäck, eds.), p. 399. Academic Press, New York.
Beinert, H. (1971). *In* "Flavins and Flavoproteins" (H. Kamin, ed.), p. 416. Univ. Park Press, Baltimore, Maryland.
Beinert, H., and Sands, R. H. (1960). *Biochem. Biophys. Res. Commun.* **3**, 41.
Bergel, F., and Bray, R. C. (1959). *Biochem. J.* **73** ,182.
Bergmann, F., and Dikstein, S. (1956). *J. Biol. Chem.* **223**, 765.
Bergmann, F., and Kwietny, H. (1958a). *Biochim. Biophys. Acta* **28**, 100.
Bergmann, F., and Kwietny, H. (1958b). *Biochim. Biophys. Acta* **28**, 613.
Bergmann, F., and Kwietny, H. (1959). *Biochim. Biophys. Acta* **33**, 29.
Bergmann, F., and Ungar, H. (1960). *J. Amer. Chem. Soc.* **82**, 3957.
Bergmann, F., Levin, G., and Kwietny, H. (1958). *Biochim. Biophys. Acta* **30**, 509.
Bergmann, F., Levin, G., and Kwietny, H. (1959). *Arch. Biochem. Biophys.* **80**, 318.
Bergmann, F., Kwietny, H., Levin, G., and Brown, D. G. (1960a). *J. Amer. Chem. Soc.* **82**, 598.
Bergmann, F. Kwietny, H., Levin, G., and Engleberg, H. (1960b). *Biochim. Biophys. Acta* **37**, 433.
Bergmann, F., Levin, G., Kwietny, H., and Ungar, H. (1961). *Biochim. Biophys. Acta* **47**, 1.
Booth, V. H. (1935). *Biochem. J.* **29**, 1732.
Booth, V. H. (1938). *Biochem. J.* **32**, 494.
Bradshaw, W. H., and Barker, H. A. (1960). *J. Biol. Chem.* **235**, 3620.
Brady, F. O., Rajagopalan, K. V., and Handler, P. (1971). *In* "Flavins and Flavoproteins" (H. Kamin, ed.), p. 425. Univ. Park Press, Baltimore, Maryland.
Bray, R. C. (1961a). *Biochem. J.* **81**, 189.
Bray, R. C. (1961b). *Biochem. J.* **81**, 196.
Bray, R. C. (1963). *In* "The Enzymes" (P. D. Boyer, H. A. Lardy and K. Myrbäck, eds.), p. 533. Academic Press, New York.
Bray, R. C. (1971). *In* "Flavins and Flavoproteins, (H. Kamai, ed.), p. 385. Univ. Park Press, Baltimore, Maryland.
Bray, R. C., and Knowles, P. F. (1968). *Proc. Roy. Soc. A* **302**, 351.
Bray, R. C., and Meriwether, L. S. (1966). *Nature (London)* **212**, 467.
Bray, R. C. and Swann, J. C. (1972). *In* "Structure and Bonding," Vol. 11, p. 107. Springer Verlag, Berlin.
Bray, R. C., and Vänngård, T. (1969). *Biochem. J.* **114**, 725.
Bray, R. C., and Watts, D. C. (1966). *Biochem. J.* **98**, 142.
Bray, R. C., Malmström, B. G., and Vänngård, T. (1959). *Biochem. J.* **73**, 193.
Bray, R. C., Pettersson, R., and Ehrenberg, A. (1961). *Biochem. J.* **81**, 178.
Bray, R. C., Palmer, G., and Beinert, H., *J. Biol. Chem.*, **239** (1964), 2667.
Bray, R. C., Chisholm, J., Hart, L. I., Meriwether, L. S., and Watts, D. C. (1966). *In* "Flavins and Flavoproteins" (E. C. Slater, ed.), p. 117. Elsevier, Amsterdam.
Bray, R. C., Knowles, P. F., and Meriwether, L. S. (1967). *In* "Magnetic Resonance in Biological Systems" (A. Ehrenberg, B. G. Malmström, and T. Vänngård, eds.), p. 249. Pergamon Press, Oxford.

Bray, R. C., Knowles, P. F., Pick, F. M., and Vänngård, T. (1968). *Biochem. J.* **107,** 601.
Bray, R. C., Knowles, P. F., and Pick, F. M. (1969). *FEBS Symp.* **16,** 267.
Bray, R. C., Pick, F. M., and Samuel, D. (1970). *Eur. J. Biochem.* **15,** 352.
Brown, G. B., Stevens, M. A., and Smith, H. W. (1958). *J. Biol. Chem.* **233,** 1513.
Brumby, P. E. (1963). Ph.D. Dissertation, Univ. of Sheffield.
Brumby, P. E., and Massey, V. (1963). *Biochem. J.* **89,** 46P.
Brumby, P. E., Miller, R. W., and Massey, V. (1965). *J. Biol. Chem.* **240,** 2222.
Bueding, E., and Jolliffe, N. (1946). *J. Pharmacol. Exp. Therap.* **88,** 300.
Carey, F. G., Fridovich, I., and Handler, P. (1961). *Biochim. Biophys. Acta* **53,** 440.
Cleland, W. W. (1963). *Biochim. Biophys. Acta* **67,** 173.
Coombs, H. I. (1927). *Biochem. J.* **21,** 1259.
Corran, H. S., Dewan, J. G., Gordon, A. H., and Green, D. E. (1939). *Biochem. J.* **33,** 1694.
Coughlan, M. P., Rajagopalan, K. V., and Handler, P. (1969). *J. Biol. Chem.* **244,** 2658.
Dalziel, K. (1957). *Acta Chem. Scand.* **11,** 1706.
Dalziel, K. (1962). *Biochem. J.* **84,** 244.
DeRenzo, E. C. (1956). *Advan. Enzymol.* **17,** 293.
Dikstein, S., Bergmann, F., and Henis, Y. (1957). *J. Biol. Chem.* **224,** 67.
Dixon, M. (1926). *Biochem. J.* **20,** 703.
Dixon, M. (1938–39). *Enzymologia* **5,** 198.
Dixon, M., and Keilin, D. (1936). *Proc. Roy. Soc. B* **119,** 159.
Dixon, M., and Thurlow, S. (1924). *Biochem. J.* **18,** 976.
Duggan, D. E., and Titus, E. (1959). *J. Biol. Chem.* **234,** 2100.
Edmondson, D. E., Massey, V., Palmer, G., and Elion, G. B. (1972). *J. Biol. Chem.* **247,** 1597.
Edmondson. D. E., Ballou, D., Van Heuvelen, A., Palmer, G., and Massey, V. (1973). *J. Biol. Chem.* In press.
Elion, G. B. (1966). *Ann. Rheum. Dis.* **25,** 608.
Elion, G. B., Kovensky, A., Hitchings, G. H., Metz, E., and Rundles, R. W. (1966). *Biochem. Pharmacol.* **15,** 863.
Feigelson, P., Davidson, J. D., and Robins, R. K. (1957). *J. Biol. Chem.* **226,** 993.
Felsted, R. L., and Chaykin, S. (1967). *J. Biol. Chem.* **242,** 1274.
Felsted, R. L., Chu, A. E. Y., and Chaykin, S. (1973). *J. Biol. Chem.* **248,** 2580.
Fridovich, I. (1962). *J. Biol. Chem.* **237,** 584.
Fridovich, I. (1964). *J. Biol. Chem.* **239,** 3519.
Fridovich, I. (1965a). *Arch. Biochem. Biophys.* **109,** 511.
Fridovich, I. (1965b). *Biochemistry* **4,** 1098.
Fridovich, I. (1966a). *J. Biol. Chem.* **241,** 3126.
Fridovich, I. (1966b). *J. Biol. Chem.* **241,** 3624.
Fridovich, I. (1967). *J. Biol. Chem.* **242,** 1445.
Fridovich, I., and Handler, P. (1957). *J. Biol. Chem.* **228,** 67.
Fridovich, I., and Handler, P. (1958a). *J. Biol. Chem.* **233,** 1581.
Fridovich, I., and Handler, P. (1958b). *J. Biol. Chem.* **231,** 899.
Fridovich, I., and Handler, P. (1958c). *J. Biol. Chem.* **233,** 1578.
Fridovich, I., and Handler, P. (1961). *J. Biol. Chem.* **236,** 1836.
Fridovich, I., and Handler, P. (1962). *J. Biol. Chem.* **237,** 916.

Garbett, K., Gillard, R. D., Knowles, P. F., and Stangroom, J. E., (1969). *Nature (London)* **215,** 824.
Gilbert, D. A., and Bergel, F. (1964). *Biochem. J.* **90,** 350.
Green, D. E. (1934). *Biochem. J.* **28,** 1550.
Green, D. E., and Beinert, H. (1953). *Biochim. Biophys. Acta* **11,** 599.
Green, R. C., and O'Brien, P. J. (1967). *Biochem. J.* **105,** 585.
Greenlee, L., and Handler, P. (1964). *J. Biol. Chem.* **239,** 1090.
Greenlee, L., Fridovich, I., and Handler, P. (1962). *Biochemistry* **1,** 779.
Gordon, A. H., Green, D. E., and Subrahmanyan, V. (1940). *Biochem. J.* **34,** 764.
Gurtoo, H. L., and Johns, D. G. (1971). *J. Biol. Chem.* **246,** 286.
Gutfreund, H., and Sturtevant, J. M. (1959). *Biochim. J.* **73,** 1.
Harris, J., and Hellerman, L. (1956). *In* "Inorganic Nitrogen Metabolism" (W. D. McElroy and B. Glass, eds.), p. 565. Johns Hopkins Press, Baltimore, Maryland.
Hart, L. I., and Bray, R. C. (1967). *Biochim. Biophys. Acta* **146,** 611.
Hart, L. I., McGartoll, M. A., Chapman, H. R., and Bray, R. C. (1970). *Biochem. J.* **116,** 851.
Hassall, H., and Greenberg, D. M. (1963). *Biochim. Biophys. Acta* **67,** 507.
Hemmerich, P., Massey, V., and Weber, G. (1967). *Nature* **213,** 728.
Hofstee, B. H. J. (1949). *J. Biol. Chem.* **179,** 633.
Hofstee, B. H. J. (1955). *J. Biol. Chem.* **216,** 235.
Horecker, B. L., and Heppel, L. A. (1949). *J. Biol. Chem.* **178,** 683.
Igo, R. P., and Mackler, B. (1960). *Biochim. Biophys. Acta* **44,** 310.
Igo, R. P., Mackler, B., and Duncan, H. (1961). *Arch. Biochem. Biophys.* **93,** 435.
Irzykiewicz, H. (1955). *Aust. J. Biol. Sci.* **8,** 369.
Kielley, R. K. (1955). *J. Biol. Chem.* **216,** 405.
Kimura, T., and Suzuki, K. (1967). *J. Biol. Chem.* **242,** 485.
Klenow, H. (1952). *Biochem. J.* **50,** 404.
Knowles, P. F., Gibson, J. F., Pick, F. M., and Bray, R. C. (1969). *Biochem. J.* **111,** 53.
Knox, W. E. (1946). *J. Biol. Chem.* **163,** 699.
Knox, W. E., and Grossman, W. I. (1946). *J. Biol. Chem.* **166,** 391.
Knox, W. E., and Grossman, W. I. (1947). *J. Biol. Chem.* **168,** 363.
Komai, H., and Massey, V. (1971). *In* "Flavins and Flavoproteins" (H. Kamin, ed.), p. 399. Univ. Park Press, Baltimore, Maryland.
Komai, H., Massey, V., and Palmer, G. (1969). *J. Biol. Chem.* **244,** 1692.
Krebs, E. G., and Norris, E. R. (1949). *Arch. Biochem. Biophys.* **24,** 49.
Krenitsky, T. A., Neil, S. M., Elion, G. B., and Hitchings, G. H. (1972). *Arch. Biochem. Biophys.* **150,** 585.
Landon, E. J., and Carter, C. E. (1960). *J. Biol. Chem.* **235,** 819.
Lichtenberg, D., Bergmann, F., and Neiman, Z. (1971). *J. Chem. Soc. C* 1676.
Lorz, D. C., and Hitchings, G. H. (1950). *Fed. Proc.* **9,** 197.
Lovenberg, W., Buchanan, B. B., and Rabinowitz, J. C. (1963). *J. Biol. Chem.* **238,** 3899.
Lowry, O. H., Bessey, O. A., and Crawford, E. J. (1949). *J. Biol. Chem.* **180,** 399.
Mackler, B., Mahler, H. R., and Green, D. E. (1954). *J. Biol. Chem.* **210,** 149.
Mahler, H. R., and Elowe, D. G. (1954). *J. Biol. Chem.* **210,** 165.
Mahler, H. R., Mackler, B., and Green, D. E. (1954). *J. Biol. Chem.* **210,** 465.
Mason, H. (1957). *Advan. Enzymol.* **19,** 1676.

Massey, V. (1958). *Biochim. Biophys. Acta,* **30,** 500.
Massey, V., and Edmondson, D. E. (1970). *J. Biol. Chem.* **245,** 6595.
Massey, V., Brumby, P. E., Komai, H., and Palmer, G. (1969a). *J. Biol. Chem.* **244,** 1682.
Massey, V., Strickland, S., Mayhew, S. G., Howell, L. G., Engel, P. C., Matthews, R. G., Schuman, M., and Sullivan, P. A. (1969b). *Biochem. Biophys. Res. Commun.* **36,** 891.
Massey, V., Komai, H., Palmer, G., and Elion, G. B. (1970a). *J. Biol. Chem.* **245,** 2837.
Massey, V., Komai, H., Palmer, G., and Elion, G. B. (1970b). *Vitamins Hormones,* **28,** 505.
McCord, J. M., and Fridovich, I. (1968). *J. Biol. Chem.* **243,** 5753.
McCord, J. M., and Fridovich, I. (1969). *J. Biol. Chem.* **244,** 6049.
McCord, J. M., and Fridovich, I. (1970). *J. Biol. Chem.* **245,** 1374.
McGartoll, M. A., and Bray, R. C. (1969). *Biochem. J.* **114,** 443.
McGartoll, M. A., Pick, F. M., Swann, J. C., and Bray, R. C. (1970). *Biochim. Biophys. Acta* **212,** 523.
Miller, R. W. (1970). *Can. J. Biochem.* **48,** 935.
Miller, R. W., and Massey, V. (1965). *J. Biol. Chem.* **240,** 1466.
Morell, D. B. (1952). *Biochem. J.* **51,** 657.
Morell, D. B. (1955). *Biochim. Biophys. Acta* **18,** 221.
Muraoka, S. (1963a). *Biochim. Biophys. Acta* **73,** 17.
Muraoka, S. (1963b). *Biochim. Biophys. Acta* **73,** 27.
Murray, K. N., Watson, J. G., and Chaykin, S. (1966). *J. Biol. Chem.* **241,** 4798.
Nakamura, S., and Yamazaki, I. (1969). *Biochim. Biophys. Acta* **189,** 29.
Nelson, C. A., and Handler, P. (1968). *J. Biol. Chem.* **243,** 5368.
Nilsson, R., Pick, F. M., and Bray, R. C. (1969). *Biochim. Biophys. Acta* **192,** 145.
Orme-Johnson, W. H., and Beinert, H. (1969). *Biochem. Biophys. Res. Commun.* **36,** 905.
Palmer, G. (1962a). *Biochim. Biophys. Acta* **56,** 444.
Palmer, G. (1962b). *Biochim. Biophys. Acta* **64,** 135.
Palmer, G., and Massey, V. (1969). *J. Biol. Chem.* **244,** 2614.
Palmer, G., and Massey, V. (1968). *In* "Biological Oxidations" (T. P. Singer, ed.), p. 263. Wiley (Interscience), New York.
Palmer, G., and Sands, R. H. (1966). *J. Biol. Chem.* **241,** 253.
Palmer, G., Bray, R. C., and Beinert, H. (1964). *J. Biol. Chem.* **239,** 2657.
Palmer, G., Brintzinger, H., and Estabrook, R. W. (1967) *Biochemistry* **6,** 1658.
Palmer, G., Müller, F., and Massey, V. (1971). *In* "Flavins and Flavoproteins" (H. Kamin, ed.), p. 123. Univ. Park Press, Baltimore, Maryland.
Peters, J. M., and Sanadi, D. R. (1961). *Arch. Biochem. Biophys.* **93,** 312.
Pick, F. M., McGartoll, M. A., and Bray, R. C. (1971). *Eur. J. Biochem.* **18,** 65.
Pick, R. M., and Bray, R. C. (1969). *Biochem. J.* **114,** 735.
Pullman, M. E., and Colowick, S. P. (1954). *J. Biol. Chem.* **206,** 121.
Rajagopalan, K. V., and Handler, P. (1962). *Biochem. Biophys. Res. Commun.* **8,** 43.
Rajagopalan, K. V., and Handler, P. (1964a). *J. Biol. Chem.* **239,** 1509.
Rajagopalan, K. V., and Handler, P. (1964b). *J. Biol. Chem.* **239,** 2022.
Rajagopalan, K. V., and Handler, P. (1964c). *J. Biol. Chem.* **239,** 2027.
Rajagopalan, K. V., and Handler, P. (1967). *J. Biol. Chem.* **242,** 4097.

Rajagopalan, K. V., and Handler, P. (1968). *In* "Biological Oxidations" (T. P. Singer, ed.), p. 301. Wiley (Interscience), New York.

Rajagopalan, K. V., Fridovich, I., and Handler, P. (1962). *J. Biol. Chem.* **237,** 922.

Rajagopalan, K. V., Handler, P., Palmer, G., and Beinert, H. (1968a). *J. Biol. Chem.* **243,** 3784.

Rajagopalan, K. V., Handler, P., Palmer, G., and Beinert, H. (1968b). *J. Biol. Chem.* **243,** 3797.

Rajagopalan, K. V., Brady, F. O., and Kanda, M. (1970). *Vitamins Hormones* **28,** 303.

Remy, C. N., Richert, D. A., Doisy, R. J., Wells, I. C., and Westerfeld, W. W. (1955). *J. Biol. Chem.* **217,** 293.

Richert, D. A., and Westerfeld, W. W. (1951). *Proc. Soc. Exp. Biol. Med.* **76,** 252.

Richert, D. A., and Westerfield, W. W. (1953). *J. Biol. Chem.* **203,** 915.

Richert, D. A., and Westerfeld, W. W. (1954). *J. Biol. Chem.* **209,** 179.

Sands, R. H., and Beinert, H. (1960). *Biochem. Biophys. Res. Commun.* **3,** 47.

Scott, R. B., and Brown, G. B. (1962). *J. Biol. Chem.* **237,** 3215.

Shaw, E., and Woolley, D. W. (1952). *J. Biol. Chem.* **194,** 641.

Shethna, Y. I., DerVartarnian, D. V., and Beinert, H. (1968). *Biochem. Biophys. Res. Commun.* **31,** 862.

Silberman, H. R., and Wyngaarden, J. B. (1961). *Biochim. Biophys. Acta* **47,** 178.

Singer, T. P., Kearney, E. B., and Bernath, P. (1956). *J. Biol. Chem.* **223,** 599.

Smith, S. T., Rajagopalan, K. V., and Handler, P. (1967). *J. Biol. Chem.* **242,** 4108.

Spector, T., and Johns, D. G. (1968). *Biochem. Biophys. Res. Commun.* **32,** 1039.

Spector, T., and Johns, D. G. (1970). *J. Biol. Chem.* **245,** 5079.

Stanulović, M., and Chaykin, S. (1971). *Arch. Biochem. Biophys.* **145,** 27.

Tagawa, K., and Arnon, D. I. (1968). *Biochim. Biophys. Acta* **153,** 602.

Taylor, J. D., Paul, H. E., and Paul, M. F. (1951). *J. Biol. Chem.* **191,** 223.

Totter, J. R., Burnett, W. T., Monroe, R. A., Whitney, I. B., and Comar, C. L. (1953). *Science* **118,** 555.

Totter, J. R., deDugros, E. C., and Riveiro, C. (1960b). *J. Biol. Chem.* **235,** 1839.

Totter, J. R., Medina, V. J., and Scoseriz, J. L. (1960a). *J. Biol. Chem.* **235,** 238.

Tsibris, J. C. M., Tsai, R. L., Gunsalus, I. C., Orme-Johnson, W. H., Hansen, R. E., and Beinert, H. (1968). *Proc. Nat. Acad. Sci. U.S.* **59,** 959.

Valerino, D. M., and McCormack, J. J. (1969). *Biochim. Biophys. Acta* **184,** 154.

Walker, W. H., Hemmerich, P., and Massey, V. (1967). *Helv. Chim. Acta* **50,** 2269.

Walker, W. H., Hemmerich, P., and Massey, V. (1970). *Eur. J. Biochem.* **13,** 258.

Wyngaarden, J. B. (1957). *J. Biol. Chem.* **224,** 453.

Wyngaarden, J. B., and Dunn, J. T. (1957). *Arch. Biochem. Biophys.* **70,** 150.

Author Index

Numbers in italics refer to the pages on which the complete references are listed.

C

D

E

F

L

M

Subject Index

C

Q

R

S

Molecular Biology

An International Series of Monographs and Textbooks

Editors

HAROLD A. SCHERAGA. Protein Structure. 1961

STUART A. RICE AND MITSURU NAGASAWA. Polyelectrolyte Solutions: A Theoretical Introduction, *with a contribution by Herbert Morawetz*. 1961

SIDNEY UDENFRIEND. Fluorescence Assay in Biology and Medicine. Volume I–1962. Volume II–1969

J. HERBERT TAYLOR (Editor). Molecular Genetics. Part I–1963. Part II–1967

ARTHUR VEIS. The Macromolecular Chemistry of Gelatin. 1964

M. JOLY. A Physico-chemical Approach to the Denaturation of Proteins. 1965

SYDNEY J. LEACH (Editor). Physical Principles and Techniques of Protein Chemistry. Part A–1969. Part B–1970. Part C–1973

KENDRIC C. SMITH AND PHILIP C. HANAWALT. Molecular Photobiology: Inactivation and Recovery. 1969

RONALD BENTLEY. Molecular Asymmetry in Biology. Volume I–1969. Volume II–1970

JACINTO STEINHARDT AND JACQUELINE A. REYNOLDS. Multiple Equilibria in Protein. 1969

DOUGLAS POLAND AND HAROLD A. SCHERAGA. Theory of Helix-Coil Transitions in Biopolymers. 1970

JOHN R. CANN. Interacting Macromolecules: The Theory and Practice of Their Electrophoresis, Ultracentrifugation, and Chromatography. 1970

WALTER W. WAINIO. The Mammalian Mitochondrial Respiratory Chain. 1970

LAWRENCE I. ROTHFIELD (Editor). Structure and Function of Biological Membranes. 1971

ALAN G. WALTON AND JOHN BLACKWELL. Biopolymers. 1973

WALTER LOVENBERG (Editor). Iron-Sulfur Proteins. Volume I, Biological Properties – 1973. Volume II, Molecular Properties – 1973

A. J. HOPFINGER. Conformational Properties of Macromolecules. 1973

R. D. B. FRASER AND T. P. MACRAE. Conformation in Fibrous Proteins. 1973

In preparation

OSAMU HAYAISHI (Editor). Molecular Mechanisms of Oxygen Activation